MATERIALS FOR OCEAN ENGINEERING

First Edition

Koichi Masubuchi

The M.I.T. Press
Massachusetts Institute of Technology
Cambridge, Massachusetts and London, England

ISBN 0 262 63034 6 (paperback)

Library of Congress catalog card number: 71-120729

Printed in the United States of America

MASSACHUSETTS INSTITUTE OF TECHNOLOGY

DEPARTMENT OF NAVAL ARCHITECTURE

AND MARINE ENGINEERING

CAMBRIDGE, MASS. 02139

ACKNOWLEDGEMENT

This text was prepared under the auspices of the
Department of Naval Architecture and Marine Engineering
at the Massachusetts Institute of Technology. Its preparation was supported in part by a grant made to M.I.T.
by the National Science Foundation under the terms of
the Sea Grant Project No. GH-1 for Curriculum Development
for Ocean Engineering Education, and in part by M.I.T.
funds.

The material contained in this volume was developed
and written by Professor Koichi Masubuchi.

Alfred H. Keil, Chairman

February, 1970

Author's Acknowledgement

Preparation of this text was initiated in December, 1968, three months after the author joined the faculty of the Department of Naval Architecture and Marine Engineering of M.I.T. The author is deeply indebted to Dr. A. H. Keil for encouraging him throughout the period of text development and draft preparation.

The author acknowledges the kindness of Professor J. H. Evans of the Department of Naval Architecture and Marine Engineering of M.I.T. for allowing the author to use a part of his textbook, entitled Ocean Engineering Structures, in Chapter 2 of this work; of Mr. W. S. Pellini, Superintendent of the Metallurgy Division of the U. S. Naval Research Laboratory and Senior Lecturer of the Department of N.A.M.E. of M.I.T. for allowing the author to use parts of a number of NRL reports written and edited by Mr. Pellini in various portions of the text, especially Chapters 3 and 12; and of Mr. A. H. Tuthill of the International Nickel Company for allowing the author to use some of the contents of an INCO publication in Chapters 6 and 14.

The author also wishes to thank Miss Judy Mahar for assisting him in preparing the draft, Miss Alice Habberton for typing the manuscript, and Mr. Keatinge Keays, Administrative Officer of the Department of N.A.M.E. for handling the details connected with the publication of the text.

Koichi Masubuchi
Cambridge, Massachusetts

February 17, 1970

TABLE OF CONTENTS

MATERIALS FOR OCEAN ENGINEERING

Chapter 6   Other Metals

Chapter 7   Fiberglass Reinforced Plastics and Other Filamentous Composites

Chapter 11   Sealing Materials

PART II   SELECTED MATERIAL PROBLEMS

Chapter 12   Fracture Toughness

Chapter 15   Welding

CHAPTER 1   INTRODUCTION

1.1   Scope of this Textbook

Ocean Engineering comprises the development and application
of technology for the advancement of ocean science and for the
exploration and utilization of the oceans and their resources,
including the resources on and in the ocean bottom. Ocean
engineering, therefore, is not an engineering discipline of the
type which, for instance, electrical, mechanical, chemical, and
nuclear engineering represent. It is an engineering effort
integrating many engineering disciplines and is directed at and
controlled by the ocean environment as aeronautical and astro-
nautical engineering are directed at and controlled by the
environment of air and outer space.

The term "ocean engineering" probably could be replaced
by a more descriptive term - "hydrospace engineering" - to
expand on the parallel to "aerospace engineering." In both
cases, the engineering of ground installations necessary to
support these systems is considered to be part of civil
engineering. Consequently, this report emphasizes materials
used in vehicles and floating structures rather than installations
fixed to the ocean bottom.

A number of components which are used in ocean engineering
have basically the same properties as in land-based engineering.
Materials in this category are excluded from this report. The
scope of this report revolves around those materials which are
uniquely fitted for ocean use. That is, materials utilized or

developed for their particular ability to withstand water
pressure and corrosion or any other special properties of ocean
engineering structures. A number of particular materials, their
unique properties and problems will occupy the majority of this
report.

In a broad sense, materials for ocean engineering are
comparable to materials for land-based engineering; thus an
enormous number of materials are used for ocean engineering.
They could include materials used for strength members of
ocean engineering structures, cables, materials for machinaries
and various pipes, etc. However, this book will focus primarily
on those materials used for strength members of ocean engineering
structures.

This book is prepared as a part of a series of textbooks
prepared at the Department of Naval Architecture and Marine
Engineering of the Massachusetts Institute of Technology which
covers various aspects of ocean engineering:

(1) "Ocean Engineering Structures," by J. H. Evans
    and J. C. Adamchak.

(2) "Water, Air and Interface Vehicles," by P. Mandell.

(3) "Stability and Motion Control of Ocean Vehicles," by
    M. A. Abkowitz.

(4) "Public Policy and the Use of the Seas," by N. J.
    Padelford.

Consequently, this book does not discuss various aspects of
ocean engineering which are covered in depth in the above textbooks.

1.2 Prerequisites for this Course

Since the study of ocean engineering materials is a complex technical field, it is assumed that students using this book are in a graduate program or an advanced undergraduate course. It is assumed, therefore, that students have some knowledge of ocean engineering, metallurgy, and structural mechanics as well as a working knowledge of mathematics, physics, and chemistry.

1.3 Construction of this Course

Chapter 2 presents an overview of structures and materials for ocean engineering.

Part I, which is composed of Chapters 3 through 11, covers various materials used for ocean engineering structures.

Part II, Chapters 12 through 15, covers selected material problems which are important but about which students may not have had enough background. These subjects include fracture toughness, fatigue fractures, corrosion, stress corrosion cracking and welding. In Part I, it is assumed that students have some knowledge of their particular areas. If this is not the case, however, it is recommended that students cover Part II first and then proceed to Part I.

In preparing this book, the following NRL Report has served as a guide:

> NRL Report 6167, "Status and Projections of Developments in Hull Structural Materials for Deep Ocean Vehicles and Fixed Bottom Installations"
> Edited by W. S. Pellini, November 4, 1964
> U. S. Naval Research Laboratory, Washington, D. C.

This NRL report should be very useful as a reference.

CHAPTER 2   AN OVERVIEW OF STRUCTURES AND MATERIALS FOR OCEAN ENGINEERING

By their very nature, materials are related intimately to the structures in which they are utilized. Service behavior of any ocean engineering structure is restricted greatly by the characteristics of the materials used. Development of materials with improved characteristics enables one to design and fabricate structures with better performance. Conversely, the demand for a new structure or a structure with better performance often plays a key role in the development of new materials or materials with improved characteristics. Consequently, it is necessary to consider materials not only in themselves, but in relation to the particular structures which they form.

This chapter covers the interface between structural design and materials engineering. Since a textbook on "Ocean Engineering Structures" has already been prepared by Evans and Adamchak, this chapter refers to their book as much as possible so that students will have a smooth transition from structural problems covered in the Evans-Adamchak book to material problems covered in the proceeding chapters of this book.

However, there are a number of articles which cover general aspects of materials for engineering. Some of the important articles are listed in the References.

2.1 Ocean Engineering Structures

Ocean engineering involves a wide range of structures
including:

(1) Subsurface habitations, observatories and production units

(2) Surface observatories and instrument platforms

(3) Offshore drilling and production platforms

(4) Submarine vehicles

A detailed discussion of the structures is given in a textbook
prepared by Evans and Adamchak.[5]

Subsurface Habitations, Observatories, and Production Units

With advancing technology, both the feasibility and
desirability of subsurface habitations, laboratories, and production
units steadily increase. Some of the uses of these underwater
structures are as follows:

> Oil production
>
> Fish farming and ocean agriculture
>
> Geological surveys
>
> Prospecting
>
> Scientific research.

For example, Sealab I was a four-man dwelling which was
submerged in 195 feet of water off Argus Island for 21 days in
July 1964 at the initiation of the U.S. Navy. It was cylindrical
in shape, 40 feet long and 10 feet in diameter.

Sealab II was larger than its predecessor, being 57 1/2 feet
long and 12 feet in diameter. It was designed for 10 men and was

submerged for 30 days at a depth of 205 feet in August 1965 off
La Jolla, California.

   Structures and Materials. Underwater structures for moderate
depth usually employ a pressure hull of stiffened cylindrical or
spherical configuration, or some combination of these two.

   Steels are most commonly used for such structures. However,
various other materials are used or are being considered, including
titanium, aluminum, glass, cement, and fibre reinforced plastics.

Surface Observatories and Instrument Platforms

   This category of surface structures covers all types of floating
structures from simple instrumented surface buoys to research
vessels and the new spar type floating platforms.

   Buoys. Instrumented buoys may be suitable for situations
where observations and measurements are to be performed at a
fixed location over a substantial period of time, and where the
quantity of equipment and personnel needed are not great. Buoys
are relatively inexpensive, and, unless they are too large, easily
mobile. Disadvantages include mooring problems, because buoys
may be unattended for long periods of time.

   Research Vessels. Research vessels may perform a great number
of functions; they may be used as on-the-site workshops, laboratories,
facilities for man-in-the-sea operations and storage of specimens.
Such veseels have almost unlimited mobility and the ability to
respond quickly to changes in prevailing conditions. Such
advantages are paid for in higher initial and operating costs,
however.

Spar-Type Platforms. A compromise between buoys and ships, spar-type platforms are elongated devices usually towed to the site in a horizontal position. On the site they are set in a vertical position by ballasting, so that about 80% of their length is beneath the water's surface. The platform's dynamic response is almost totally decoupled from wave action at the surface due to the combination of a large draft and a very small waterplane area. Consequently, the platforms are extremely stable with respect to magnitude of pitch, heave and roll motions. These are very attractive features for surface observatories. These advantages, however, are somewhat offset by disadvantages. In all attitudes of operation--horizontal, vertical and in transition--hydrostatic stability must be provided for the vessel. Furthermore, during the progression from horizontal to vertical, stability in regard to rotation about the longitudinal axis is likely to be required.

The federal government and private industry have built several spar type platforms to provide a stable platform with a low background noise level for making underwater acoustic measurements at sea. Two such platforms are the FLIP (Floating Instrument Platform), SPAR (Seagoing Platform for Acoustic Research), and POP (Perpendicular Oceanographic Platform).

Offshore Drilling and Production Platforms

Offshore work platforms provide, for people living above the ocean surface, the capability of working on and below the ocean surface and the ocean bottom. The functions such platforms may be called upon to perform are many. They may serve as platforms for prospecting and production of oil and other minerals, platforms for programs of scientific research, weather stations, and as manned navigation aids. Offshore platforms may be grouped into (1) bottom supported platforms and (2) floating platforms.

Figure 2-1 shows offshore structures in the past, present, and future presented in a paper by Lee.[14]

Bottom Supported Platforms. This category of platforms includes stationary platforms, submersible platforms and self-elevating or jack-up platforms. The first type of platform is used mainly where a more permanent structure is needed. Presently, they are limited to operation in depths of about 350 feet.

Submersible platforms are mobile with a buoyant hull used to float the rig on site with a fixed platform above. On location, the hull is sunk to the bottom.

Self-elevating platforms float on their platform-hull structure and are towed to location where pneumatic jacks, hydraulic jacks, or electric rack-and pinion drives lower the legs to the sea floor and then lift the hull above the sea surface to achieve the needed clearance. Once rigged, these platforms are highly stable in position and can continue operations in high seas.

Floating Platforms. This category of platforms includes semi-submersible platforms and ship-type platforms.

Semi-submersible platforms can sometimes be used in relating shallow water as bottom (submersible) units or in deep water as floating units. Semi-submersible rigs are so designed that when used as floating platforms, the primary buoyancy is well below the water surface, so it is relatively unaffected by the most

9

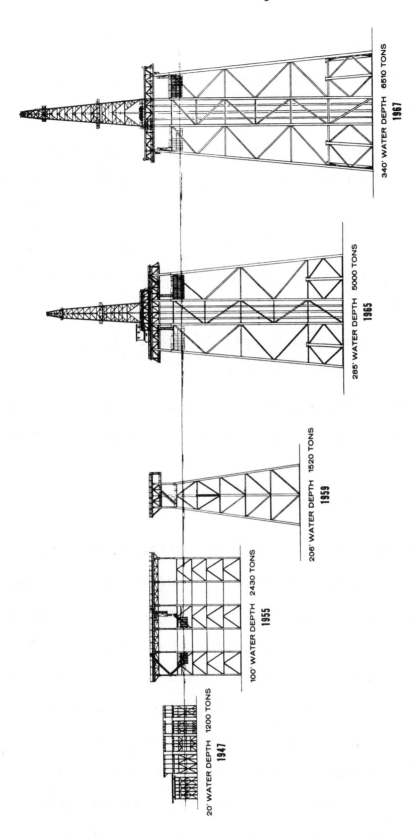

TWENTY YEARS OF PLATFORM DEVELOPMENT

20' WATER DEPTH 1200 TONS
1947

100' WATER DEPTH 2430 TONS
1955

206' WATER DEPTH 1520 TONS
1959

285' WATER DEPTH 5000 TONS
1965

340' WATER DEPTH 6510 TONS
1967

FIGURE 2-1 "Offshore Structures Past, Present and Future Design Considerations" by G. C. Lee(5, 14)

violent action of the surface waves.

Ship-type platforms, either catamarans or single hulls, are designed mainly for operation in deep water.

<u>Structures and Materials</u>. Offshore drilling and production platforms are frequently fabricated as tubular trusses. Steels are most commonly used.

A number of structural failures have occurred which indicate that tubular joints generally represent the most likely source of trouble. Figure 2-2 shows how tubular joint designs have developed recently.[14] In some cases, brittle fractures have occurred. Fatigue fractures also have been reported.

Weld failures have often occurred as a result of poor workmanship (possibly due to field connections), poor shop practices, or poor weld design.

Submarine Vehicles

Non-military submersible vehicles operating between the ocean surface and the ocean bottom have demonstrated rapidly increasing capabilities and sophistication in recent years. As a result, submarine vehicles are now actively performing a large number of work functions and are being seriously considered for others. Functions of submarine vehicles lie in the general areas of scientific research, military weaponry (both offensive and defensive), as well as in specific areas such as research and rescue, recreation, salvage operations, cable and pipe laying, etc.

For convenience, submersibles may be categorized as one of three types, manned untethered submersibles, manned tethered submersibles, and unmanned submersibles.

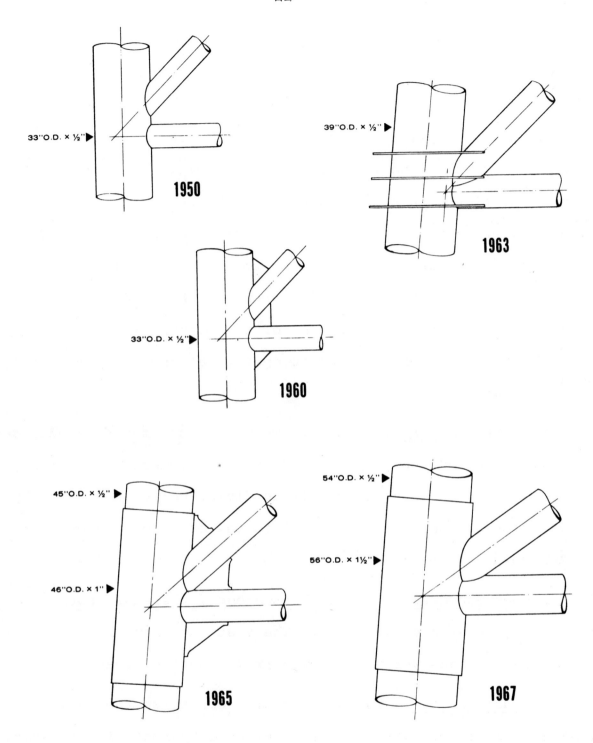

FIGURE 2-2   DEVELOPMENT OF A TUBULAR-JOINT DESIGN[5, 14]

Manned Untethered Submersibles. This category consists of
those vehicles which are most commonly called "submarines." Such
manned vehicles are entirely self-contained, carrying with them
their own power supplies for mobility, for surfacing and sub-
merging, and usually the means for life support. They employ no
direct connection with the surface.

Submarines vary in size from very small research or recreational
vessels carrying a crew of one or two to the large military vessels,
hundreds of feet in length and with crews in the hundreds.

Figure 2-3 is a general scheme of the Deep Submergence Search
Vehicle (DSSV) which is designed to operate at the maximum depth
of 20,000 feet.[4]

Manned Tethered Submersibles. This category may be subdivided
into self-propelled submersibles and passive submersibles. The
passive class includes such vessels as diving bells and the bathy-
spheres which are chambers lowered into the ocean on a support line
or drawn down from below.

Unmanned Submersibles. In this category there are both
tethered and untethered vehicles. The vast majority, however,
are tethered designs which are towed from a surface vessel.

Structural Design. The most crucial condition that any
submersible must cope with is the inevitable hydrostatic pressure.
This pressure varies from zero at the surface to 16,000 psi at the
maximum known depth of 36,000 feet.

The structural problem common to all submersibles is the
provision of adequate pressure bearing strength at or near the
the minimum cost in weight. For pressure hulls with a collapse

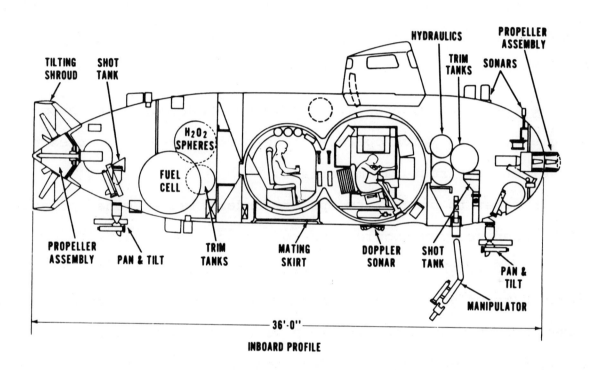

FIGURE 2-3   GENERAL SCHEMATIC OF 20,000 FT. DSSV[4]

depth greater than a few hundred feet, the provision of sufficient hydrostatic strength will more than satisfy longitudinal strength requirements, even when the vehicle is surfaced.

Structural discontinuities, access and window openings and hull penetrations for cable leads or the like require special compensation, careful detail design and the best workmanship. Cylindrical and spherical pressure hull configurations are almost universally used and great pains are necessary to strictly limit all out-of-roundness and residual stresses.

Selection of materials. Design and fabrication of submersibles, especially those which operate at deep sea, involve complex problems for selecting structural materials.

In order to be able to operate under high external pressures of the deep sea, one would like to choose a material with high strength-to-weight ratio. However, those materials with high strength-to-weight ratios also tend to exhibit undesirable features. These include high cost, problems with fabrication and/or joining, low ductility, low fatigue life, poor corrosion resistance, and a variety of undesirable behavioral characteristics. This is true of metals, both common and exotic, as well as of other structural materials such as glass, ceramics, composite materials, and concrete.

Figure 2-4 relates the weight/buoyancy ratios for different materials to the corresponding collapse depths for spherical shells with a thickness/radius ratio of 0.03.

Figure 2-5 presents similar data in a different manner. Shown here is a summary of operating depth potential of submersibles in different weight-to-displacement ratios and various materials.

15

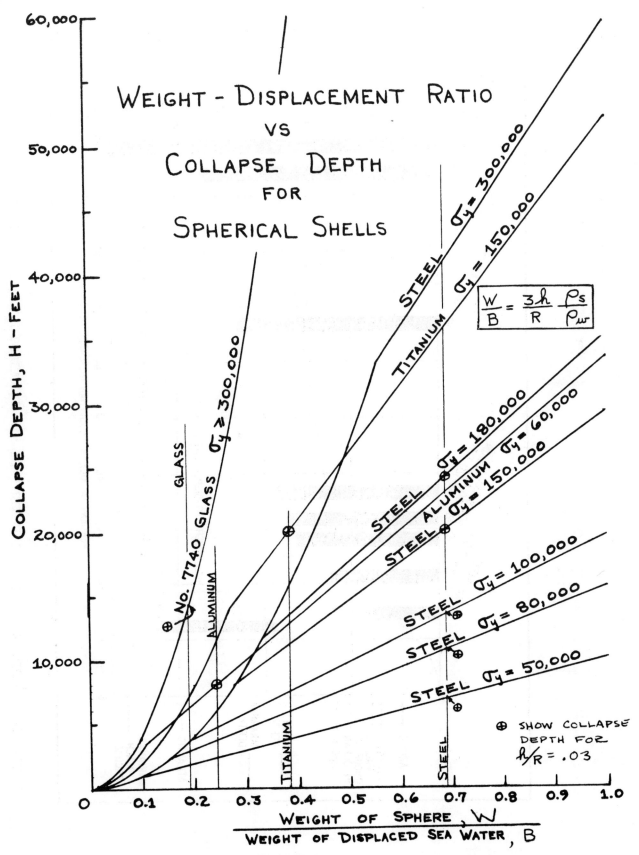

FIGURE 2-4   WEIGHT-DISPLACEMENT RATIO VS. COLLAPSE DEPTH FOR
SPHERICAL SHELLS [5]

16

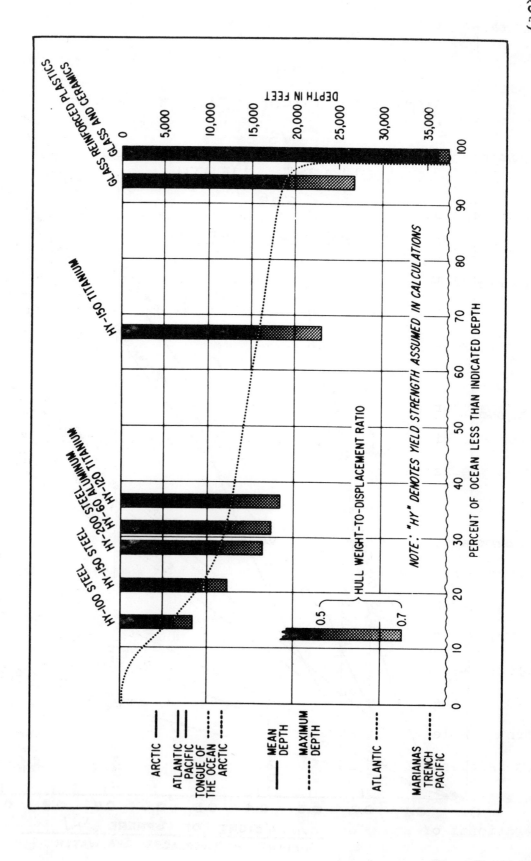

FIGURE 2-5 OPERATING DEPTH POTENTIAL OF SUBMERSIBLES MADE IN DIFFERENT MATERIALS. (20)

It is important to mention, however, that the material trade-off studies presented in Figures 2-4 and 2-5 are greatly simplified. These three areas of shell configuration, shell construction, and shell material are not completely independent variables as the data shown in these figures might lead one to believe. In fact, they are all somewhat independent in such a manner that one area cannot be considered without simultaneously considering the other two, and perhaps additional problems.

This simplified treatment is intended to illustrate the type of problems faced by engineers who are associated with the design and fabrication of not only deep submersibles but also any type of ocean engineering structure. As Evans and Adamchak[5] point out, computer-aided parametric studies and mathematical optimization techniques can be developed to help a designer with making various decisions.

Design Case Histories.* The characteristics of some 35 non-military, manned submarine vehicles are available and provide some insight into the structural design and material selection for submersibles.

Figure 2-6 shows operating depths of non-military submersibles plotted on a curve representing the relationship between the depth and the percentage of ocean area less than the indicated depth. Table 2-1 shows general characteristics of 35 non-military submersibles, and Table 2-2 shows their structural and material characteristics. Thirty-three of these vehicles are, at present, operational or are under construction.

---

* This section is a copy of the Evans-Adamchak book, pages 211 through 221.

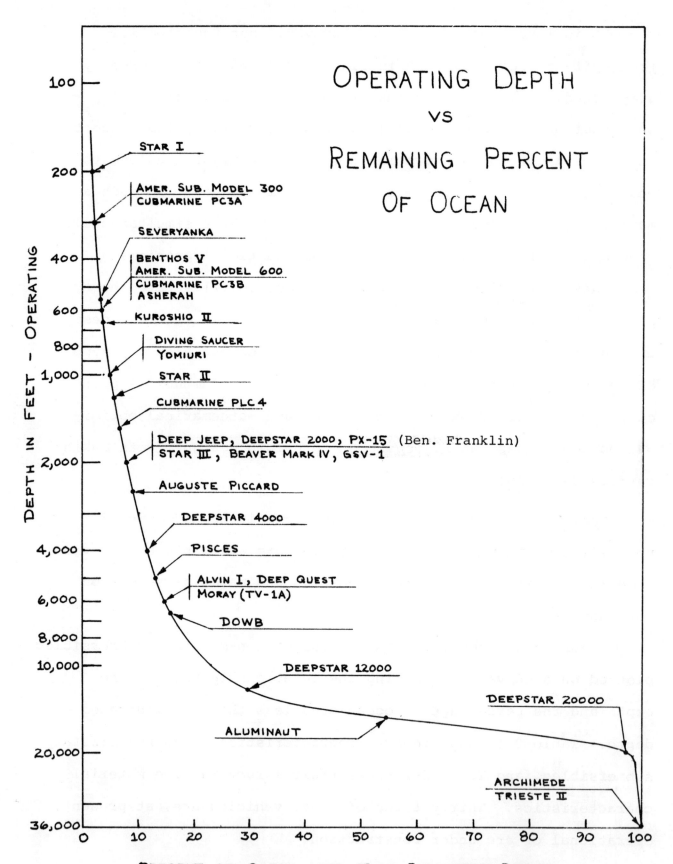

FIGURE 2-6   OPERATING DEPTHS OF NON-MILITARY SUBMERSIBLES[5]

TABLE 2-1 General Characteristics of Non-Military Submersibles [5]

Vehicle No.	Name	Owner/Operator	Dimensions (feet)	Operating Depth (feet)	Dry Weight-* Diving Condition (pounds)
1	Aluminaut	Reynolds International, Inc./Reynolds Submarine Service Corp.	51/8/14.25	15,000	162,000
2	Alvin I	Office of Naval Research/ Woods Hole Oceanographic Institution	22/8/13	6,000	29,100
3	American Submarine Model 300	American Submarine Co.	13/4.2/4.75	300	2,200
4	American Submarine Model 600	American Submarine Co.	13/5.5/5.2	600	3,500
5	Archimede	French Navy	69/13/26.5	36,000	438,800
6	Auguste Piccard	Swiss Nat. Exposition Corp.	93.5/19.7/24	2,500	368,200
7	Benthos V	Lear Siegler, Inc.	11.3/6.1/6	600	4,200
8	Cubmarine PC3A	Perry Submarine Builders, Inc./Ocean Systems, Inc.	19/3.5/6	300	4,790
9	Cubmarine PC3B	Perry Submarine Builders, Inc./Ocean Systems, Inc.	22/3.5/6	600	6,350
10	Cubmarine PLC4	Perry Submarine Builders, Inc./Ocean Systems, Inc.	24/4.5/8.7	1,500	17,900
11	Deep Jeep	Naval Ordnance Test Station, China Lake, California	10/8.5/8	2,000	8,000
12	Deep Quest	Lockheed Aircraft Co.	39.1/19/13.3	6,000	112,000

TABLE 2-1 (Continued)

#	Name	Organization	Dimensions		
13	Deepstar 2000	Westinghouse Electric Corp.	20/7/8.5	2,000	14,000
14	Deepstar 4000	Westinghouse Electric Corp.	18/11.5/7	4,000	18,000
15	Diving Saucer	OFRS - J. Y. Cousteau Westinghouse Electric Corp.	10/10/5.3	1,000	7,000
16	Dolphin AG (SS) 555	U. S. Navy	152/19/20	-----	1,568,000
17	DOWB	General Motors Defense Research Laboratory	16/8.5/6	6,500	14,720
18	Kuroshio II	Hokkaido Univ., Japan	36.7/7.15/10.4	650	25,800
19	Moray (TV-1A)	U. S. Naval Ordnance Test Station, China Lake, Calif.	33/5.3/5.3	6,000	22,400
20	NR-1	U. S. Navy/Special Projects Office	--------	------	-------
21	Pisces	International Hydrodynamics Co., Ltd., Vancouver, B. C.	16/11.5/9	5,000	14,560
22	Severyanka	USSR/All-Union Institute of Marine Fishery and Oceanography	240/22/15	550	-------
23	Star I	Electric Boat Co. General Dynamics Corp.	10.1/6/5.8	200	2,750
24	Asherah	General Dynamics Corp./Univ. of Pennsylvania Museum	17/7.7/7.6	600	8,600
25	Star III	Electric Boat Co. General Dynamics Corp.	24.5/6.5/9	2,000	18,300
26	Trieste II	U. S. Navy/COMSUBPAC	67/15/18	36,000	492,800
27	Yomiuri	Mitsubishi/Yomiuri Shimbun Newspaper, Tokyo, Japan	48/8.2/9.2	1,000	-------

TABLE 2-1 (Continued)

28	Star II	Electric Boat Co. General Dynamics Corp.	17.76/5.33/7.62	1,200	10,000
29	Beaver Mark IV	North American Aviation	25/9.5/8.5	2,000	27,000
30	Benjamin Franklin (PX-15)	Grumman Aircraft Eng. Corp.	48.73/18.5/20	2,000	291,000
31	Deepstar 12,000	Westinghouse Electric Corp.	18/11.5/7	12,000	18,000
32	Deepstar 20,000	Westinghouse Electric Corp.	28/8/8	20,000	36,000
33	Deep Diver	Perry Submarine Builders, Inc., Ocean Systems, Inc	22/-/-	1,350	18,500
34	Nai'a (Perry PC5C)	Perry Pacific Submersibles, Inc.	22/4.6/-	1,200	11,470
35	SURV	British Navy	10.85/6.3/9.5	1,000	13,440

TABLE 2-2 Structural and Material Characteristics of Non-Military Submersibles

Vehicle No.	Name	Payload (Pounds)	Crew	Pressure Hull - Type and Diameter (feet)	RN$_w$	RN$_D$
1	Aluminaut	6,000	4-6	Aluminum alloy 7079-T6 integrally stiffened cylinder with hemispherical heads, 8.0 o.d., 7.0 i.d.	0.46	0.77
2	Alvin I	1,200	2	HY-100 steel sphere, 7.0 o.d., (1.33" thick).	0.68	0.88
3	American Submarine Model 300	450	2	A-36 steel, (0.375" thick).	----	----
4	American Submarine Model 600	750	2	A-36 steel, (0.5" thick).	----	----
5	Archimede	4,000	3	Ni-Cr-Mo forged steel sphere, 7.87 o.d., 6.9 i.d.	0.38	0.73
6	Auguste Piccard	20,000	40	Steel cylinder with hemispherical heads	----	----
7	Benthos V	400	2	Mild steel sphere, 5.0 o.d., (0.625" thick)	0.17	0.65
8	Cubmarine PC3A	750	2	A285 Steel	----	----
9	Cubmarine PC3B	950	2	A212 steel, (0.5" thick)	----	----
10	Cubmarine PLC4	1,500	4	T-1 Steel	----	----
11	Deep Jeep	200	2	HY-50 steel sphere, 5.0 o.d.	0.30	0.67
12	Deep Quest	3,400	4	Two intersecting maraging steel sphere	----	----
13	Deepstar 2000	1,000	3	HY-80 steel cylinder with hemispherical heads, 5.0 o.d., (0.75" thick)	0.17	0.56
14	Deepstar 4000	600	3	HY-80 steel sphere, 6.42 o.d., (1.2" thick)	0.56	0.83

TABLE 2-2 (Continued)

No.	Name			Description		
15	Diving Saucer	75	2	Mild steel ellipsoidal, 6.5 major d., 4.8 minor d., (0.75" thick)	0.32	0.73
16	Dolphin AG(SS) 555	----	22	HY-80 cylinder with hemispherical heads	----	----
17	DOWB	1,021	2	HY-100 steel sphere, 6.68 i.d., (0.915" thick)	1.30	1.09
18	Kuroshio II	----	4-6	Mild steel plate	----	----
19	Moray (TV-1A)	200	2	Two aluminum A-356-T6 spheres, 5.0 o.d.	0.32	0.69
20	NR-1	----	---	HY-80 steel	----	----
21	Pisces	1,500	2	Two Algoma 44 steel spheres	----	----
22	Severyanka	----	6-8	------------	----	----
23	Star I	200	1	A212 steel sphere, (0.375" thick)	----	----
24	Asherah	250	2	A212 steel sphere, 5.0 i.d., (0.625" thick)	----	----
25	Star III	1,000	2	HY-100 steel sphere, 5.5 i.d., (0.5" thick)	0.17	0.56
26	Trieste II	20,000	3	Ni-Cr-Mo forged steel sphere, 7.0 o.d., (3.5" thick) or forged steel sphere, 7.2 o.d., (4.72" thick)	0.24	0.64
27	Tomiuri	----	6	High tensile strength steel cylinder	----	----
28	Star II	250	2	HY-80 steel sphere, 5.0 i.d., (0.625" thick)	0.14	0.54
29	Beaver Mark IV	2,000	4-5	Two HY-100 steel sphere connected by 2.08 d. tunnel, 7.0 d., (0.481" thick), 5.5 d., (0.387" thick)	0.24	0.64
30	Benjamin Franklin	11,200	6	HY-80 steel ring stiffened cylinder with hemispherical heads, (1.375" thick).	----	----

TABLE 2-2 (Continued)

#	Name			Description		
31	Deepstar 12,000	-----	3	----------------------------	----	----
32	Deepstar 20,000	1,000	3	Steel (180,000 psi) sphere, 7.0 o.d., (1.85"thick)	1.83	1.22
33	Deep Diver	-----	4	T-1 steel ring stiffened cylinder, 4.5 d., (0.5" thick)	0.06	0.40
34	Nai'a (Perry PC5C)	1,000	3	A212 Grade B steel, (0.5" thick)	----	----
35	SURV	-----	2	Steel cylinder, 5.0 d.	0.09	0.48

Operations on the Continental Shelf, as usually defined, are limited to depths of less than about 600 feet or 200 meters (656 feet). The continental slopes fall off to about 1200 feet before meeting the ocean's basins. Only the very small percentage of bottom area included in the trenches lies below 20,000 feet and the greatest known depth is about 36,000 feet.

Of the 31 vehicles in operation and whose operating depths are known, the pressure hulls of those 12 with operating depths up to 1000 feet are constructed of ordinary carbon steel of one type or another. The configuration of the pressure hulls is hardly uniform, consisting of spheres, stiffened cylinders, various unconventional shapes, and in one case an elliptical shell. At these relatively shallow depths, the hydrostatic pressure is not so severe, and designers can sacrifice some of the structural efficiency to take advantage of operational and/or economical features offered by other structural configurations.

As depth capability increases, however, the effect of increasing hydrostatic pressure can be seen, both on configuration and material. The 10 vehicles in the operating range of 1200 to 2500 feet use high-strength steels such as HY-80, HY-100, and "T-1." Their hull forms are the more conventional but more structurally efficient types, including stiffened cylindrical shells with hemispherical end caps, spherical shells, and two spheres connected by a cylindrical shell.

At operating depths in the range of 4000 to 6500 feet, the trend toward the more structurally efficient hull forms continues. Out of 6 vehicles in this group, 4 have spherical shells, 1 has two separated spherical shells, and 1 has two intersecting spheres.

High-strength steels maintain their monopoly as construction material, but with one exception: the separated sphere hull uses a high strength aluminum alloy, A-356-T6.

For operating depths greater than 6500 feet, where lies more than 84% of the ocean bottom, few vehicles are to be found and few generalities can be drawn. Only three of the vehicles now in existence have a depth capability of greater than 6500 feet, one having a limit of 15,000 feet and the other two of 36,000 feet. These must be considered individually. The Aluminaut is a relatively large vehicle. Its hull is close to 50 feet long and is made up of 11 forged cylindrical sections 8 feet in diameter with hemipherical heads. The sections are of 7079-T6 aluminum alloy 6.5 inches thick and are bolted together. The two vehicles capable of 36000 feet, Trieste II and Archimede, are fairly similar. Both use forged steel spheres for pressure hulls, relatively small in diameter (about 7 feet) but very thick (approximately 5-6 inches). Because of the negative buoyancy of the pressure hulls, resulting from the severe strength requirements, both require large floats for lift, making them also relatively larger than most other research submarine vehicles. "Fail-Safe" operation consists of jettisoning whatever solid ballast remains aboard at the time. In case of an emergency, the pressure capsule, if cast loose, could not by itself return personnel to the surface. They are in fact, a unique class of vessel and have limited performance capabilities, except for their operating depth.

A few general conclusions can be drawn from these data. The first is that except for very shallow depths, the more structurally

efficient spheres, stiffened cylinders, and combinations of these
two types have been used exclusively, for submarine pressure hulls.
The second is, that in spite of all the interest and development
work on special materials, steel of one type or another, with very
few exceptions, is the material of which pressure hulls are presently
constructed. And third, there is a definite lack of practical
vehicles with operating depths in the vicinity of 20,000 feet. A
vehicle with such depth capability would be able to explore 99%
of the ocean floor as well as the ocean depths above it.

2.2 Requirements for and Selection of Materials
in Ocean Engineering

The selection of materials for any given ocean engineering
application is partly the problem of determining which of the
many characteristics are, in sum, most suitable to the specific
problem. This process is a primary design cpnsideration for marine
structures, particularly in view of the wide variety of applications
presently found in this field, the diversity of structural characteristics demanded of each, as well as the hostile environment in
which they are required to exist and operate.

Required Properties

The following pages discuss some of the important properties
materials must possess to be used successfully for strength members
of ocean engineering structures.

Strength-to-weight Ratio. The weight density of a material
is frequently a critical characteristic, since structural weight
is so often a major design consideration. In many cases it is not

the absolute density itself which is important but a strength-to-weight ratio, usually represented by the ratio of either yield stress or ultimate stress to the weight density. Such a parameter is usually employed in cases where maintaining a certain level of strength of the minimum structural weight is desirable.

Among various ocean engineering structures, submarine hulls present the most crucial problems. Figure 2-5 shows curves representing the calculated performance of near-perfect spherical pressure hulls in various materials.[20] Shown here are relationships between the collapse depth and the ratio of collapse or buckling stress (which is dependent on geometry) to density. The advantage of materials with high strength-to-weight ratio is obvious especially at a greater depth.

It is important to mention that techniques for fabricating submarine hulls with various materials included in Figure 2-6 are not necessarily available at the present. The Navy classifies materials according to background and experience.

Category 1 materials include those alloys such as HY-80 and HY-100 for which there is an abundance of technical data and operational experience. Category 2 materials include those alloys such as HY-130, Maraging (190) steel, HP 9-4-25, and annealed Ti-Al-4V. There is also an abundance of data, but experience in the operations environment is limited. Category 3 contains those materials for which there is little technical data and experience. Several Category 3 materials have high collapse-stress-to-density ratios. They include heat treated titanium alloys, ultrahigh-strength steels, glass, ceramics such as aluminum oxide, advanced metal-matrix and resin-matrix composites, and dual-alloy, diffusion-bonded plates.

Figure 2-7 shows a projection of maximum strength-to-weight characteristics of pressure hulls for Navy small submersibles.[4]

Fracture Toughness. Fracture toughness is a measure of a material's ability to absorb energy through plastic deformation before fracturing. Several technical terms including "ductility," "notch toughness," and "fracture toughness" are used to describe the resistance of a material to fracture. Notch toughness, for example, refers to the ability of a material to resist brittle fracture in the presence of a metallurgical or mechanical crack or notch. In general, the more energy ebsorbed, the more ductile or tough the material is said to be. Chapter 12 discusses fracture toughness in detail.

Fracture toughness often becomes a critical problem when a material with high strength is considered, because there is a general tendency for fracture toughness to decrease with increasing strength. Notch toughness also becomes a critical problem when a structure is subjected to a low temperature.

Fatigue Strength. Loads or deformations which will not cause fracture in a single application can result in fracture when applied repeatedly. The mechanism of fatigue failure is complex, but it basically involves the initiation of small cracks, usually from the surface, and the subsequent growth under repeated loading.

Chapter 13 covers subjects related to fatigue failures including (1) high-cycle, low-stress fatigue, (2) low-cycle, high-stress fatigue, and (3) corrosion fatigue.

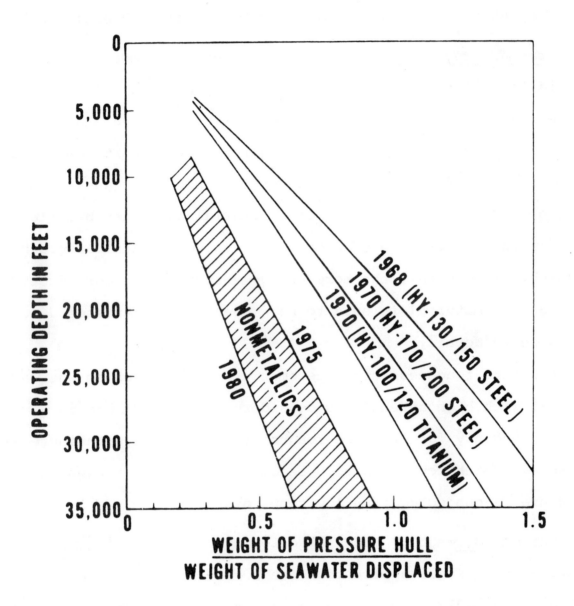

FIGURE 2-7   PROJECTION OF MAXIMUM STRENGTH-TO-WEIGHT
CHARACTERISTICS OF PRESSURE HULLS FOR SMALL
SUBMERSIBLES [4]

Resistance Against Corrosion and Stress Corrosion Cracking.
Materials used for structural components exposed to seawater must
have adequate resistance against corrosion and stress corrosion
cracking.

Corrosion is the destructive attack of a metal by chemical
or electrochemical reaction with the environment. Stress corrosion
cracking, on the other hand, is the fracture of a material under
the existence of both stress and certain environments. These
subjects are covered in detail in Chapter 14.

Other Properties. Other material characteristics which merit
consideration include ease of fabrication, weldability, durability,
maintenance, general availability, and finally (but not least
important), cost. With several possible modes of failure to be
anticipated in each element of a structure and weight and/or cost
to be minimized (or perhaps other performance characteristics to be
optimized) trade-off studies must be resorted to before a final
optimum choice of material can be made for any specific application.

Chapter 15 covers welding, which is an important fabrication
process.

Commonly Used or Promising Structural Materials

Structural materials which are commonly used at present
and which are promising in the future fall into four main categories:
ferrous metals, nonferrous metals, nonmetals, and composites.

Steels. Steels show promise mainly because of the extremely
high strengths which new heat treatment techniques are making
possible. These new steels include such types as HY-80, HY-100,

HY-150, and the maraging steels. Yield stresses range from 80,000 psi for HY-80 to approximately 400,000 psi for some maraging steels. A tendency toward brittle behavior and low notch toughness, in addition to only moderate enhancement of fatigue life are the major drawbacks of these high-strength steels.

As the strength level increases, steels tend to become more difficult to weld without cracks and other defects. Some high-strength steels also are sensitive to stress corrosion cracking.

Chapter 3 discusses various steels for marine structures.

Aluminum. Aluminum is of interest mainly because of the low density. Some of the new aluminum alloys are competitive with some steels in yield and ultimate strength, but have better corrosion resistance. As with steel, as strengths increase, aluminum alloys show the tendency toward increased brittleness, lower notch toughness, and questionable fatigue life.

Some high-strength aluminum alloys, especially those which are heat treated, also are difficult to weld, and are sensitive to stress corrosion cracking.

Chapter 4 discusses aluminum and its alloys for marine structures.

Titanium. Titanium combines a relatively low density with very high strength, excellent fatigue properties and corrosion resistance, and anti-magnetic properties. A severe problem with titanium is stress-corrosion cracking. Especially for deep applications, titanium alloys are generally considered to be the most promising materials in the hydrospace field despite present high cost.

Chapter 5 covers titanium and its alloys for marine structures.

Other Metals. Various metals other than steels, aluminum alloys, and titanium alloys are used or can be used for various components of ocean engineering structures. They are covered in detail in Chapter 6.

Composite Materials. Composite materials are made of filaments of some material specifically oriented in a matrix material. The filaments may be of either a metallic or non-metallic material. Glass and boron are commonly considered. Such fiber composites are being developed with very high strength to weight ratios. As with glass and ceramics, the problems of fastening and joining and of delamination under pressure in long term use is currently a major problem with composites.

Chapter 7 covers glass reinforced plastics (GRP) and other filamentous composites.

Glass and Ceramics. Glass and ceramics are of interest because of their extremely high strengths in compression. They also demonstrate excellent corrosion resistance qualities. Glass, in addition, offers the advantage of transparency. The chief drawbacks of glass and ceramics are their brittle behavior and low fracture toughness.

Chapter 8 covers glass and ceramics.

Other Materials. Plywood and concrete have been suggested for use in underwater structures. The main advantage of both is their relatively low cost. In addition, concrete possesses good compressive strength, good availability, resistance to corrosion, and excellent formability. Its chief disadvantage is its limited tensile strength. Chapter 9 covers concrete.

Materials for Non-Structural Uses

Ocean engineering platforms, habitations, and vehicles contain many non-structural members and parts. They include pipes for various uses, pumps, valves, cables, etc. Obviously a wide range of materials is used for these applications.

Chapter 6 covers various metals used for marine applications. Chapters 10 and 11 cover buoyancy materials and sealing materials, respectively, which are unique for marine applications.

REFERENCES

(1) Bannerman, D. B., "Development of ABS Rules for Construction of Offshore Drilling Rigs," Marine Technology, pp. 58-60, January 1969.

(2) Cousteau, J., "Exploring the Sea," Industrial Research, pp. 42-52, March 1966.

(3) Craven, J. P., "The Design of Deep Submersibles," Paper presented to Society of Naval Architects and Marine Engineers, June 18-21, 1968.

(4) Evans, J. Harvey, and Adamchak, John C., "Ocean Engineering Structures," M.I.T. Press, 1969.

(5) Fabian, Robert J., "Deep Sea Hull Materials - the Quest Goes On", Materials Engineering, April 1968, pp. 28-36.

(6) Fisher, F. H., and Spiess, F. N., "FLIP - Floating Instrument Platform," Journal of the Acoustical Society of America, Vol. 25, p. 1633, 1963.

(7) Glosten, L. R., "Ocean Platforms - With Reference to Scripps Institution FLIP," Transactions of the Institute of Marine Engineering, Canadian Division, No. 19, March 1965.

(8) Groves, Don, "Materials in the Sea," Naval Engineers Journal, February 1968, pp. 35-45.

(9) Groves, Don, "Ocean Materials," Naval Engineers Journal, April 1968, pp. 185-203.

(10) International Nickel Company, Inc., "Guidelines for Selection of Marine Materials," Copyright 1966, pp. 34-37.

(11) Krenzke, M. A., and Reynolds, T. E., "Structural Research on Submarine Pressure Hulls at the David Taylor Model Basin," Journal of Hydronautics, Vol. 1, No. 1, July 1967.

(12)

Lederman, Peter B., "Materials: Key to Exploiting the Oceans," *Chemical Engineering*, June 3, 1968, pp. 105-113.

(13)

Lee, G. C., "Offshore Structures, Past, Present, and Future Design Considerations," Proceedings of OECON Offshore Exploration Conference-1968, published by Offshore Exploration Conference, Palos Verdes Estates, California.

(14)

Liebowitz, H., "Comprehensive Appraisal of Structural Mechanics Requirements for Deep Submergence Vehicles," American Institute of Aeronautics and AStronautics, 1966.

(15)

Momsen, C. B., Jr., "POP - A Low Cost Perpendicular Ocean Platform," presented at the 10th Annual Joint Meeting of SNAME California Sections, Santa Barbara, October 1966.

(16)

Pellini, W. S., "Summation and Interpretation of the Materials Sub-Panel Topical Input Reports," NRL Report 6167,"Status and Projections of Developments in Hull Structural Materials for Deep Ocean Vehicles and Fixed Bottom Installations," U. S. Naval Research Laboratory, November 1964, pp. 1-15.

(17)

Roseman, D. P., "Vehicles for Ocean Engineering," Paper presented to the Society of Naval Architects and Marine Engineers, June 18-21, 1968.

(18)

Scott, M. I., and Parsey, "Moored Ocean Platforms," *Marine Systems*, May/June 1967; from conference on Technology of the Sea and Sea-bed, U. S. Atomic Energy Authority, April 5-7, 1967.

(19)

Shankman, A. D., "Materials Technology for Hydrospace Pressure Hulls," Metal Progress, Vol. 93, No. 3, March 1968, pp. 69-73.

(20)

Shen-D'Ge, N. J., "Structures/Materials Synthesis for Safety of Oceanic Deep-Submergence Bottom-Fixed Manned Habitat," *Journal of Hydronautics,* Vol. 2, No. 3, pp. 120-130, July 1968.

PART I. MATERIALS USED FOR
OCEAN ENGINEERING

CHAPTER 3 <u>STEELS</u>

This chapter covers various steels which have been or may be used for marine structures. These steels include: carbon steels, low-alloy high-strength steels, quenched-and-tempered steels, and maraging steels.

3.1 Historical Perspective and Current Trends

<u>Past and Present</u>

The age of wooden ships drew to a close in the middle of the 19th century with the introduction of iron as a construction material. By the early 1900's, iron also became obsolete as a shipbuilding material, and since then, steel has dominated the field. Although other construction materials have been developed since that time, steel remains one of the most widely used ocean engineering materials.

Low carbon steels are the most widely used for ocean engineering structures. However, high-strength steels have been used increasingly. High-strength heat-treated steels with excellent notch toughness are important for structural components which are subjected to low temperatures.

As an example of how steels have been used for ships and ocean engineering structures, the following describes briefly how steels used for submarines have been developed.

Prior to the early 1940's, combat submarines in the United States were fabricated largely from low carbon steel, a material with a tensile yield strength of about 32,000 psi. With the advent of World War II, however, steels were developed in response to demands for stronger materials.

Between 1940 and 1958, high-tensile-strength (HTS) steel with a 50,000 psi yield strength was used in most submarine structures, including one famous example, the "Nautilus." By 1958, and continuing into the 1960's, HY-80 was predominantly in use. This steel, which has a specified minimum tensile-yield strength of 80,000 psi, will be discussed in detail in a separate section of this chapter. Today, HY-80 is still the basic fabrication steel for submarine hull structures.

HY-100, a steel with 100,000 psi minimum yield strength and very similar to HY-80, was the next high-strength steel developed, followed by HY-130(T). The latter was at first called HY-140, however it was discovered that only 130,000 psi yield strength can be guaranteed in welds, so the name was changed. HY-130 and its filler wire are being developed under a U.S. Navy contract. The Navy plans to use the alloy for the pressure hull of the DSRV, (Deep Submergence Rescue Vehicle) being fabricated by Lockheed Missile and Space Company.

Current Trends

Table 3-1 shows approximate mechanical and physical properties of steels with various strength levels which are currently used and which may be used in the future for ships and ocean engineering structures.[14] These steels include:

Quenched and tempered steels: HY-80, HY-100, and HY-150.

Maraging steels: MA-180, MA-250, and MA-300.

TABLE 3-1   APPROXIMATE MECHANICAL PROPERTIES OF VARIOUS HIGH-STRENGTH STEELS[14]

TYPE	ULTIMATE TENSILE STRENGTH (ksi)	TENSILE YIELD STRENGTH (ksi)	ELONGATION, % (2 in.)	REDUCTION OF AREA	COMPRESSIVE YIELD STRENGTH (ksi)
HY-80	103	88	22	65	98
HY-100	118	105	22	65	115
HY-150	165	150	18	60	165
MA-180	190	180	17	58	200
MA-250	255	250	12	50	280
MA-300	305	300	10	50	340

Density - .283 to .285 lbs./in.^{3} (.287 - .292 lbs./in.^{3} for maraging)

Modulus - 29 to 30 x 10^6 psi (28 to 29 x 10^6 psi for maraging)

Poisson Ratio - .33 (.32 for maraging)

Discussions of these steels are given in the following sections.

3.2 Low Carbon Steels and High-Strength Steels With Less Than 80,000 Psi Yield Strength

Low carbon steels and high-strength steels with less than 80,000 psi yield strength are among the most widely used ocean engineering materials.

Base Metal

In this section, discussions are made largely in reference to ship-hull steels, because most ship hull steels are applicable to many ocean engineering structures.

United States merchant vessels are constructed in accordance with requirements established by the U. S. Coast Guard and the American Bureau of Shipping. Naval combatant vessels and many merchant-type naval vessels are constructed in accordance with U. S. Navy specifications.

The American Bureau of Shipping requirements can be found in its Rules for Building and Classing Steel Vessels, which is revised annually. Tables 3-2 and 3-3 show ABS requirements in 1968 for ordinary and high-strength, respectively, hull structural steel. The present rules include high-strength steels of about 47,000 to 50,000 psi yield strength.

The ABS specifications for hull steel since 1948 recognize variations in notch toughness due to thickness of plates by specifying grades. Requirements for notch toughness are specified for steels under Grades D, E, and EH. Typical applications for ABS steels are given in Table 3-4.

42

TABLE 3-2 ABS REQUIREMENTS FOR ORDINARY STRENGTH HULL STRUCTURAL STEEL

GRADES	A	B	C	CS	D	E	R
PROCESS OF MANUFACTURE	For All Grades: Open Hearth, Basic Oxygen or Electric Furnace						
DEOXIDATION	Any method.	Semi-killed or killed.	Fully killed, fine grain practice.	Fully killed, fine grain practice.	Semi-killed or killed.	Fully killed, fine grain practice.	Semi-killed or killed.
AUSTENITE GRAIN SIZE	...	...	5 or finer[1]	5 or finer[1]	...	5 or finer[1]	...
CHEMICAL COMPOSITION (Ladle analysis)							
Carbon, %	...	.21 max.	.23 max.[4]	.18 max.	.21 max.[5]	.18 max.[5]	...[5]
Manganese, %	...	.80–1.10[2] [6]	.60–.90[2]	1.00–1.35	.60–1.40[5]	.70–1.40[5]	2.5xCmin.[5]
Phosphorus, %	.05 max.	.05 max.	.05 max.	.05 max.	.05 max.	.05 max.	.05 max.
Sulphur, %	.05 max.	.05 max.	.05 max.	.05 max.	.05 max.	.05 max.	.05 max.
Silicon, %	...	...[3]	.10–.35	.10–.35	.35 max.	.10–.35	...
HEAT TREATMENT	...	...	Normalized over 34.9 mm (1.375 in.) thick.	Normalized	...	Normalized	...
TENSILE TEST Tensile strength Elongation	For all Grades: 41–50 kgs. per sq. mm.[7] or 58000–71000 lbs. per sq. in. or 26–32 tons per sq. in. For all Grades: 21% in 200 mm (8 in.)[7] (See 39, 3, 4 (c) and 39, 3, 4 (d)) or 24% in 50 mm (2 in.) or 22% in 5.65 $\sqrt{A}$ (A equals area of test specimen)						
BEND TEST	For all Grades: 180° around a diameter equal to 3 times thickness of the specimen.[8]						
IMPACT TEST CHARPY STANDARD V-NOTCH							
Temperature	...	...	...	...	0°C (32°F)	−10°C(14°F)	...
Energy, avg., min.	...	...	...	...	4.8 kg. m (35 ft. lbs.)	6.2 kg. m. (45 ft. lbs.)	...
No. of specimens	...	...	...	...	3 from each 40 tons	3 from each plate	...
STAMPING	AB/A	AB/B	AB/C (as rolled) AB/CN (normalized)	AB/CS	AB/D	AB/E	AB/R

NOTES:

(1) To be determined by the McQuaid-Ehn Method (ASTM E112-63) on material representing each ladle of each heat.
(2) Upper limit of manganese may be exceeded provided carbon content plus ⅙ manganese content does not exceed .40%.
(3) When the silicon content is .10% or more (killed steel) the minimum manganese content may be .60%.
(4) For normalized Grade C plates the maximum carbon content may be .24%.
(5) Carbon content plus ⅙ manganese content shall not exceed .40%.

(6) Where the use of cold flanging quality has been specially approved (3.1) the manganese content may be reduced to .60–.90%.
(7) The tensile strength of cold flanging steel shall be 39–46 kg/mm² (55000–65000 psi) and the elongation 23% min. 200 mm (8 in.).
(8) The bend test requirements for cold flanging steel are:
   19.1 mm (0.750 in.) t and under..180° flat on itself.
   Over 19.1 mm (0.750 in.) t to 31.8 mm (1.250 in.) t, incl. 180° around diam. of 1 x specimen thickness.
   Over 31.8 mm (1.250 in.) t..180° around diam. of 2 x specimen thickness.

TABLE 3-3   ABS REQUIREMENTS FOR HIGHER STRENGTH HULL STRUCTURAL STEEL

GRADES	AH	BH	CH	EH	RH
PROCESS OF MANUFACTURE	For all grades: Open Hearth, Basic Oxygen or Electric Furnace.				
DEOXIDATION	Semi-killed or killed.	Killed	Fully killed, fine grain practice.	Fully killed, fine grain practice.	Semi-killed or killed.
AUSTENITE GRAIN SIZE	......	......	5 or finer[1]	5 or finer[1]	......
CHEMICAL COMPOSITION[2] (Ladle Analysis)					
Carbon, %	.20 max.	.20 max.	.20 max.	.20 max.	.20 max.
Manganese, %	......	.90–1.50	1.00–1.50	1.10–1.50	3.5×C. min.
Phosphorus, %	.05 max.	.05 max.	.05 max.	.05 max.	.05 max.
Sulphur, %	.05 max.	.05 max.	.05 max.	.05 max.	.05 max.
Silicon, %	.50 max.	.15–.50	.15–.50	.15–.50	......
HEAT TREATMENT	......	Normalized over 19.1mm (0.750 in.)	Normalized	Normalized	......
TENSILE TEST Tensile strength Yield point, min. Elongation	For all grades: 50–60 kg. per sq. mm or 71000–85000 lbs. per sq. in. or 32–38 tons per sq. in. 33 kg. per sq. mm or 47000 lbs. per sq. in. or 21 tons per sq. in. 19% in 200mm (8 in.) or 22% in 50mm (2 in.) or 20% in 5.65 $\sqrt{A}$ (A equals area of test specimen)				
BEND TEST	For all grades: 180° around diameter equal to 3 times thickness of the specimen				
IMPACT TEST CHARPY STANDARD V-NOTCH					
Temperature	......	......	......	−10°C(14°F)	......
Energy; avg., min.	......	......	......	6.9 kg. m. (50 ft. lbs.)	......
No. of specimens	......	......	......	3 from each plate	......
STAMPING	$\frac{AB}{AH}$	$\frac{AB}{BH}$	$\frac{AB}{CH}$	$\frac{AB}{EH}$	$\frac{AB}{RH}$

NOTES:

(1) To be determined by the McQuaid-Ehn Method (ASTM E112-63) on material representing each ladle of each heat.
(2) Small amounts of certain alloying elements will be present but shall not exceed the following amounts:

Copper	.35%	Chromium	.25%
Nickel	.25%	Molybdenum	.08%

44

TABLE 3-4 TYPICAL APPLICATIONS FOR ABS STEELS[20]

Ordinary-Strength Structural Steel, Plates

Grade A - Acceptable up to 1/2 inch inclusive in thickness.

Grade B - Acceptable up to 1 inch inclusive in thickness.

Grade C - Acceptable up to 2 inch inclusive in thickness, except where Grades CS or E are required; plates over 1-3/8 inch thickness shall be normalized when used in important structural parts.

Grade D - Acceptable up to 1-3/8 inch inclusive in thickness.

Grades CS and E - Acceptable up to 2 inch inclusive in thickness; intended primarily for application where superior notch tpughness is desired in specifically designated strakes above 1.09 inch thickness in certain designs.

Grade R - Acceptable up to 2 inch inclusive in thickness for specially approved locations in individual ships; not acceptable in thickness over 1/2 inch in vessels of primarily welded construction for the bottom or bilge plating, the sheer strake or strength deck plating within the midship portion or other members which may be subject to comparatively high stresses.

Higher Strength Structural Steel, Plates

Grade AH - Acceptable up to 1/2 inch inclusive in thickness.

Grade BH - Acceptable up to 1 inch inclusive in thickness.

Grade CH - Acceptable up to 2 inch inclusive in thickness except where Grade EH is required.

Grade EH - Acceptable up to 2 inch inclusive in thickness; intended primarily for applications where superior notch toughness is desired in specifically designated strakes above 1.09 inch thickness in certain designs.

The U. S. Navy Specification MIL-S-22698A, Steel Plate, Carbon, Structural for Ships, is in substantial agreement with the American Bureau of Shipping specifications for ordinary-strength hull steels as follows:

1. Grade HT, a carbon steel with minimum yield strength 42,000 to 50,000 psi depending upon thickness.

2. QT 50, a carbon manganese steel heat-treated by quenching with minimum yield strength 50,000 to 70,000 psi.

In regard to the last steel mentioned, QT 50, quenching is required to prevent a transformation of the high temperature austenite phase to undesirable microstructure constituents which are a normal result of transformations on "slow" cooling in temperatures 400°-1250° F. for one to several hours. The highest tempering temperature results in the lowest strength and maximum fracture toughness, and the converse is true.

In addition to the above steels which have been developed for ship-hull applications, there are many other steels which can be used or have been used for some marine applications. Tables 3-5 and 3-6 provide lists of various steels which can be used for ship structures. These tables are prepared from Technical and Research Bulletin No. 2-11a, "Guide for the Selection of High-Strength and Alloy Steels," published by the Society of Naval Architects and Marine Engineers.[20]

Steel plates and shapes listed in these tables are for ship structures requiring an increase of strength not originally available with standard ship steels. The values tabulated are for steels of 1-inch thickness and may vary with mil practice. Further

TABLE 3-5   STEELS, 30,000 TO 49,000 PSI, YIELD STRENGTH MINIMUM(20)

TYPE	FOR REFERENCE ABS CLASS "B"	ASTM A-255 GRADE A FLANGE QUALITY	ASTM A-242	ASTM A-441	MIL-S-16113C GRADE HT
MECHANICAL PROPERTIES:					
Yield Strength, psi-min.	(not specified) (about 32,000)	(Pressure Vessels) 40,000	46,000	46,000	47,000
Tensile Strength, psi	58,000 min.	70,000 min.	67,000 min.	67,000 min.	88,000 max.
Elongation in 8-in.-Percent	28	17	19	19	20
Approx. NDT range Deg F.	-20 to +40	+20 to +90	-20 to +40	0 to +70	-60 to +20
Available Thickness Range-in.	1/2 to 1	3/16 to 4	3/16 to 4	3/16 to 4	No limit
Relative Cost, Dec. 1963[a] Ref. ABS Class "B"	1	1.77	1.20	1.20	1.67
Heat Treatment	As rolled	As rolled[b]	As rolled[b]	As rolled[b]	Normalized
Weldability	Good	Good	Good	Good	Good
CHEMISTRY: Ladle[c]					
C max.	0.21	0.18	0.22	0.22	0.18
Mn	0.80-1.10	1.45 max.	1.25 max.	1.25 max.	1.30 max.
P max.	0.05	0.035	----	0.04	0.04
S max.	0.05	0.04	0.05	0.05	0.05
Si	----	0.15-0.30	----	0.030 max.	0.15-0.35
Cr	----	----	----	----	0.15 max.
Ni	----	----	----	----	0.25 max.
Mo	----	----	---- (d)	----	0.06 max.
V	----	0.09-0.14	----	0.02 min.	0.02 min.
Cu	----	----	----	0.02 min.	0.35 max.
Ti	----	----	----	----	0.005 min.

a. Includes chemistry and mechanical property extras only, except Grade HT includes heat treatment extra.
b. Plates 3/4 in. and above should be normalized for ship structural applications.
c. Where no figure is given, that particular element will occur only in residual amounts-no planned addition is made.
d. Can vary based on material source.

TABLE 3-6   STEELS, 50,000 TO 79,000 PSI, YIELD STRENGTH MINIMUM(20)

47

TYPE	CA C-Mn	ASTM A-302 Grade B	C-Mn-V (or Cb)	C-Mn-V (or Cb)	CB C-Mn	(MILS 13326) 70,000 Class
MECHANICAL PROPERTIES:						
Yield Strength, psi-min.	50,000	50,000	50,000	55,000	60,000	70,000
Tensile Strength, psi-min.	70,000	80,000	65,000	70,000	80,000	90,000
Elongation in 8-in.-Percent	19	15	18	15	19	17
Approx. NDT Range Deg F.	-10 to +40	-20 to +50	-60 to +30	-40 to +40	-75 to -40	-65 to -30
Available Thickness Range-in.	3/16-1 1/4	1/4 min.[a]	3/16-1 1/2	3/16-1 1/4	3/16-2	3/16-2
Relative Cost, Dec. 1963 Ref. ABS Class "B"	1.12	1.77	1.12	1.15	1.43	1.80
Heat Treatment	As rolled[b]	As rolled	As rolled	As rolled	Q & T	Q & T
Weldability	Good	Special	Special	Special	Good	Special
CHEMISTRY: Ladle						
C max.	0.22[d]	0.23	0.24	0.24[d]	0.20[c]	----
Mn	1.35	1.15-1.50	1.40[d]	1.40[d]	1.40[d]	----
P max.	0.040	0.035	0.040	0.040	0.040	0.040
S max.	0.050	0.040	0.0505	0.050	0.050	0.040
Si	0.15-0.30	0.15-0.30	0.15-0.30	0.15-0.30	0.15-0.30	Compositions vary with Manufacturer
Cr	----	----	----	----	----	
Ni	----	----	----	----	----	
Mo	----	0.45-0.60	----	----	----	----
V	----	----	0.02 min. (or Cb)[e]	0.02 min. (or Cb)[e]	----	----
Cu	----	----	----	----	----	----
Ti	----	----	----	----	----	----

a. May be heat treated by normalizing subject to purchase agreement. Plates over 2-in. in thickness
b. Plates 3/4-in. and above should be normalized for ship structure applications. require heat treatment.
c. Aluminum deoxidized fine grain practice.   d. Indicates **maximum.**
e. Plates containing columbium above 1/2-in. should be normalized for ship structural applications.

TABLE 3-7 STEELS, 80,000 TO 120,000 PSI YIELD STRENGTH MINIMUM(20)

TYPE	Proprietary Grades[a]	MIL-S-16216 Gr HY-80	ASTM A543 Cl. 1	ASTM A543 Cl. 2
MECHANICAL PROPERTIES:				
Yield Strength, psi-min.	80,000	80,000	85,000	100,000
Tensile Strength, psi-min.	b	b	105,000	115,000
Elongation in 2-in.-percent	18	20	16	16
Approx. NDT Range Deg F	-40 or lower	-130 or lower	-120 or lower	-90 or lower
Available Thickness range-in	3/16 to 2-1/2c	3/16 to 8	3/16 to 4	over 4
Relative Cost, Dec. 1963, Ref. ABS Class "B"	2.0	3.5	3.0	3.0
Heat Treatment	Q & T	Q & T	Q & T	Q & T
Weldability	Special	Special	Special below 0.21C	Special below 0.21C
CHEMISTRY: Ladle				
C	0.21 max.	0.18	0.23 max.	0.23 max.
Mn	---	0.10-0.40	0.40 max.	0.40 max.
P Max1	0.040	0.025	0.035	0.035
S Max1	0.040	0.025	0.040	0.040
Si	---	0.15-0.35	0.20-0.35	0.20-0.35
Cr	---	1.00-1.80	1.50-2.00	1.50-2.00
Ni	---	2.00-3.25	2.60 3.25	3.00-4.00
Mo	---	0.20-0.60	0.45-0.60	0.45-0.60
V	---	---	0.03 max.	0.03 max.
Cu	---	---	---	---
Ti	---	---	---	---
Zr	---	---	---	---

a. This column includes proprietary grades individually indentified by chemistry, mechanical properties, and thickness range.

b. Approximately 10,000 to 15,000 psi above yield strength.

c. Normal range available. Plates to 8-in. thick are available in some proprietary grades.

TABLE 3-7 (Continued)

TYPE	ASTM A517-67 a	MIL-S-13326 Class 90	MIL-S-16216 Gr. HY-100	MIL-S-13326 Class 120
MECHANICAL PROPERTIES:				
Yield Strength, psi-min.	100,000	90,000	100,000	120,000
Tensile Strength, psi-min.	115,000	b	b	b
Elongation in 2-in.-percent	16	No limit spec.	18	No limit spec.
Approx. NDT Range Deg F	-50 or lower	-40 or lower	-100 or lower	-20 or lower
Available Thickness Range-in.	3/16 to 2-1/2	No limit spec.	3/16 to 3	No limit spec.
Relative Cost, Dec. 1963, Ref. ABS Class "B"	2.0	2.0	3.5	3.5
Heat Treatment	Q & T	Q & T	Q & T	Q & T
Weldability	Special	Special	Special	Special below 0.21C
CHEMISTRY" Ladle				
C	0.21 max.	No Chemistry Specified	0.20	No Chemistry Specified
Mn	---		0.10-0.40	
P Max1	0.035		0.025	
S Max1	0.040		0.025	
Si	---		0.15-0.35	
Cr	---		1.00-1.80	
Ni	---		2.25-3.50	
Mo	---		0.20-0.60	
V	---		---	
Cu	---		---	
Ti	---		---	
Zr	---		---	

details of these steels are available in ASTM or manufacturers' specifications.

Steels with less than 60,000 psi minimum yield strength are usually available in the as-rolled or normalized condition, while steels with minimum yield strength over 60,000 psi are usually available in the quenched-and-tempered condition.*

Notch-Toughness Requirements. Ship-hull steels and similar ocean engineering steels must have a suitable degree of notch toughness and weldability in addition to conventional mechanical properties such as ultimate tensile strength, yield strength, and elongation.

Notch toughness is important to avoid brittle fracture of welded structures. Numerous research programs have been conducted during the past 25 years on various aspects of brittle fracture of welded structures and notch toughness of steel. Notch toughness of steel is discussed in detail in Chapter 12.

Tables 3-5 through 3-7 provide notch toughness data given in NDT (nil ductility transition) temperatures of various steels.

Welding

Detailed discussions on welding processes and problems related to welding fabrication are given in Chapter 15. This section discusses briefly welding processes commonly used for fabricating steel structures.

---

*This trend is true for steels with minimum yield strength over 80,000 psi (see Table 3-7).

Welding of Low-Carbon Steels. Shielded metal-arc welding
continues to be the major welding process for fabricating ships and
ocean engineering structures in low carbon steels. Electrodes for
use in welding ship-hulls should meet the requirements of the AWS-ASTM
Tentative Specifications for Mild-Steel-Covered-Arc Welding Electrodes
or the American Bureau of Shipping rules for the approval of Electrodes for Manual Arc Welding. The E60xx series is used for all
hull construction, except for E6012 and E6013, which are not approved
for any joints in shell plating, strength decks, tank tops, bulkheads
and longitudinal members of large vessels or in galvanized material
because of their slightly lower ductility.

Electrodes of the E7015, E7016, and E7018 classes are often
used where improved mechanical properties in the weld are desirable.
The trend is toward the use of electrodes having iron-powder
coatings, because they offer high deposition rates.

Various automatic and semi-automatic processes are often used
in the fabrication of ships and ocean engineering structures.
Such processes include submerged arc, gas metal-arc, electroslag,
and electrogas processes. Welding processes with high deposition
rate, such as submerged arc, electroslag and electrogas processes,
often provide weld metals with rather low notch toughness. This
problem is discussed in detail in Chapter 15.

Welding of High-Strength Steels. The U. S. Navy provides an
electrode specification, MIL-E-2220011 with the following classi-
fications and intended uses:

Type MIL-7018: for welding of medium carbon steels such as Classes A, B, C and grade HT (under 5/8-inch thickness)

Type MIL-8018: for welding grade HT (5/8-inch and thicker).

In welding high-strength steels there are two factors to be emphasized.[11] First, select the proper electrode to meet the strength requirements for the welded joint. Second, select adequate preheat to assure that a sound weld will be produced. In cases of high restraint it is recommended that higher preheats be used. Better results are obtained when low-hydrogen electrodes are used with the higher strength grades. Satisfactory welds can be obtained using these electrodes at much lower preheats than are necessary for conventional electrodes.

The term "low-hydrogen" developed because these electrodes are produced to a maximum limit of moisture in the electrode coating. The specified maximum moisture content in the E70xx and E80xx class low-hydrogen electrodes are 0.6 and 0.4 per cent, respectively. For electrodes with higher strength levels the maximum moisture allowed is lowered even further.[11]

Hydrogen has been associated with both underbead cracks and toe cracks. The source of hydrogen can be either in the electrode coating or on the surface of the plates to be welded. Therefore, it is important to keep the plate surface dry, as well as to use an electrode with a low hydrogen content.

For effective use the low-hydrogen electrodes must be properly stored and kept dry. If there is any doubt about the moisture level of the electrodes, they can be rebaked at approximately 800 F before use. Porosity at the start of a weld is one indicator that the electrode coating has picked up moisture. After baking the electrodes, make sure that they are placed in holding ovens until they are ready to be used. The welder is usually supplied with just the number of electrodes that he expects to use within a 4-hour period. Electrodes held out of the oven for a longer period of time are returned to be dried out.

High-strength steels can also be welded with various other processes. However, further discussions of welding high-strength steels are not included here, because the subject is considered to be outside the scope of this book. For those who are interested in this subject, a recent book from the American Society for Metals (ASM) entitled "Welding High-Strength Steels" is recommended.[22]

3.3 Quenched and Tempered Steels

Table 3-7 lists several steels with 80,000 to 120,000 psi minimum specified yield strength. To obtain high strength, good notch toughness and good weldability, all steels listed here are quenched and tempered.

Quenching is conducted to prevent the transformation of the high-temperature austenite phase into undesirable microstructure constituents which are a normal result of "slow" cooling in the 1100-800° F range. Tempering involves reheating for one to several hours to temperatures between 400 to 1250 F.

Steel plates and shapes at the strength level given in Table 3-7 for industrial applications have evolved in the United States since World War II. This has made possible new and inproved design concepts. Many of the steels listed here are grades

proprietary to the major plate manufacturers. The exceptions are
HY-80 and HY-100 made to MIL-S-16216 which are used principally for
naval ships. Specification MIL-S-13326 covers four strength levels,
including 90,000 and 120,000 psi. Steel plates at higher strengths
are available from several steel manufacturers. Some of these
steels are not covered yet by ASTM designations.

Among the steels listed, HY-80 is the most commonly used
for submarines. ASTM A517-67 is also commonly used for commercial
applications including pressure vessels, storage tanks and merchant
ships.* Compared with HY-80, ASTM A517-67 contains less nickel and
is less costly; about twice as expensive as ABS B steel while HY-80
is about 3.5 times as expensive as ABS B steel.

Since HY-80 is the most well known among the steels listed
in Table 3-7, the following discussions primarily concern HY-80
steel.

Development of HY-80 Steel

Prior to the development of HY-80, high tensile steel (HTS)
was primarily used for construction of naval vessels. To justify
the introduction of the new steel, it had to be proved that the new
material would have substantial advantages and not introduce significant new defects.[7] To understand HY-80 in relation to the material
it replaced, HTS, it is desirable to review the chemical composition
of each.

HTS requires the following maximum limits in chemical composition:**

---

*"T-1" steel is an ASTM A514/A517 steel produced by the U. S. Steel
Corporation.
**Chapter 12 covers effects of chemical composition on notch tough-
ness of steel.

1. 0.18% for carbon to assure notch toughness and weldability

2. 0.04% for phosphorous because it has a detrimental effect
   on the steel's notch toughness and weldability

3. 0.05% for sulphur, which is undesirable because when combined
   with iron as FeS, it becomes a liquid at normal rolling and
   forging temperatures. This "hot-shortness" can be prevented
   by either less sulphur, or more manganese. When MnS rather
   than FeS forms, the "hot-shortness" is avoided since MnS has
   a higher melting point.

4. 1.30% manganese to increase strength and notch toughness

5. 0.15% - 0.35% for silicon which is used for deoxidation.[7]

Attempts were made to develop new forms of HTS through the
addition of small alloys. However, this resulted in only small
increases in strength and toughness, so a new approach was required.[7]

HY-80 was created as a low carbon steel which derives its
strength and toughness from quenching and tempering, unlike HTS
which has its strength in as-rolled condition. In other words, HY-80
is a "hardenable" steel which obtains its properties from heat
treatments.[7]

Chemical Composition and Mechanical Properties of HY-80 Steel

Chemical Composition. Table 3-7 shows specification limits
of HY-80 chemical composition. The maximum limits of chemicals
for successful production of HY-80 are:

1. 0.18% for carbon, which is the same requirement for HTS

2. 0.15 - 0.35% for silicon, which is used for deoxidation

3. 0.025% each for sulphur and phosphorous, not to exceed
   0.045 altogether. This strict control of sulphur and

phosphorous requires that more care than usual be taken
during the steel-making process.

4. 0.10 - 0.40% for manganese which is again used for sulphur
control, rather than strength. An amount over 1.0% of Mn
would cause embrittled steel during heat treatments.

5. Molybdenum, slight quantities for lowering the temper
embrittlement

6. Nickel, slight quantities for toughness.[7]

Table 3-8 shows the nominal chemical composition of HY-80
steel.

Mechanical Properties. Table 3-9 shows specification limits
for mechanical properties of HY-80 steel. Values of specified yield
strengths are 80,000 to 100,000 psi for plates less than 5/8 inch
thick and 80,000 to 95,000 psi for plate 5/8 inch and over. Minimum
specified Charpy V-notch impact energy values at -120° F are 50 ft-lb
plates 1/2 inch to 2 inches inclusive and 30 ft-lb for plates over
2 inches.

Figure 3-1 shows typical Charpy V-notch energy bands for
production of HTS and HY-80.[7] The figure shows that notch toughness
of HY-80 is much superior to that of HTS.

Production of HY-80 Steel.

HY-80 steel is produced through an open hearth or electric
furnace method. The result is that the steel is fully killed
and finely grained. The final heat treatment of quenching and
tempering is needed for strength and toughness. This final process
sets HY-80 apart from previously used steels.[7]

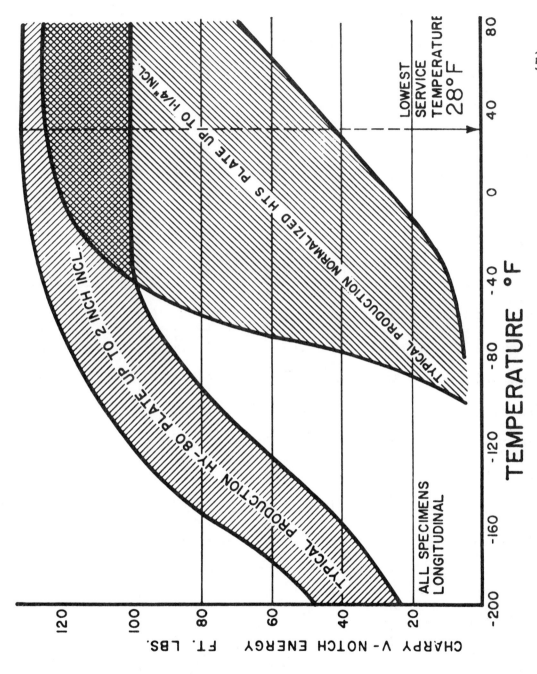

FIGURE 3-1 TYPICAL CHARPY V-NOTCH ENERGY BANDS FOR PRODUCTION HTS AND HY-80 (7)

TABLE 3-8   NOMINAL CHEMICAL COMPOSITION OF HY-80 STEEL[7]

Thickness	C	Mn	P	S	Si	Ni	Cr	Mo
Thin Plate (Up to 1-1/4 in. Incl.)	.14	.30	.015	.020	.20	2.35	1.15	.30
Thick Plate (Over 1-1/4 in. to 3 in. Incl.)	.16	.30	.015	.020	.20	2.85	1.55	.45
Insert Plate (Over 3 in. to 6 in. Incl.)	.17	.30	.015	.020	.20	3.10	1.65	.55

The heat requirements during the steel-making process are a temperature of not less than 1100° F and a microstructure at a mid-thickness of the plate containing no less than 80% martensite.[7]

To be used widely and successfully, the material must be available in a plate form at first. When the steel was in its early days of development, it could be made into a flat plate form, but the problem of heat treatment for other than flat shapes had to be solved. By 1960, extruded structural tees were developed. In 1961, rolled shapes were available as well as forgings and castings. Today, we have a full family of HY-80 shapes and forms.[7]

Welding HY-80 Steel

In welding any material, the goal is to produce weld metals with properties the same as those of the base metal. It is not easy with oven common materials, but with a high-strength notch-tough steel such as HY-80, it is extremely difficult. So far it has not been possible to develop electrodes which produce weld metals as notch tough as HY-80 base plate. This problem is discussed in detail in Chapter 15. This section decribes briefly covered electrodes used for welding HY-80 steel.

When HY-80 was first developed during 1944-46, a low-hydrogen electrode was used for welding this steel. The electrode, however, did not produce a weld metal with enough strength; and also its best Charpy V-notch impact energy was only about 20 ft-lb at 0° F. Between 1955 and 57, covered electrodes for manual welding were developed. Later, electrodes for gas-shielded metal-arc welding and an electrode-flux combination for submerged arc welding were developed. Recently, a heat treatable electrode identified as

TABLE 3-9   SPECIFICATION LIMITS OF HY-80 MECHANICAL PROPERTIES[7]

PROPERTY	PLATE THICKNESS	
	Less than 5/8 in.	5/8 in. and over
Ultimate Strength (psi)	For information	For information
Yield Strength at 0.2% Offset (psi)	80,000 to 100,000	80,000 to 95,000
Min. Elongation in 2 in. (per cent)	19	20
Reduction in area (per cent) Longitudinal Transverse	-- --	55 50

CHARPY V-NOTCH ENERGY REQUIREMENTS

PLATE THICKNESS	SPECIMEN SIZE (mm.)	FOOT POUNDS (Min.)	TEST TEMPERATURE (Degrees F.)
1/4 in. to 1/2 in Excl.	10 x 5	For information	-120
1/2 in. to 2 in. Incl.	10 x 10	50	-120
Over 2 in.	10 x 10	30	-120

MIL 8218Y QT has been developed that produces strength and toughness through the same heating and quenching method as HY-80.[7]

Table 3-10 lists electrodes currently used for welding HY-80 and their application in submarine construction.[7] In addition to the electrode type designation, the appropriate Military Specifications, welding processes, welding positions, and applications are included in the table. Regarding electrodes for shielded metal arc process, low-hydrogen iron powder type xx18 are used most widely. Table 3-10 also includes electrodes for inert-gas metal arc (semi-automatic and automatic) and submerged arc processes.

Table 3-11 shows the specification limits of chemical compositions for deposited weld metals. These limits include:

1. Carbon is limited to a maximum of 0.10% to provide an adequate resistance to weld cracking and to improve notch toughness

2. Phosphorous and sulphur are limited to a maximum of 0.03% each, also to provide a resistance to weld cracking and to improve notch toughness

3. More than five times as much manganese as sulphur is added to prevent hot shortness. Manganese also contributes to strength and toughness.

4. Silicon is permitted to a maximum of 0.6%. Silicon, which is a deoxidizer, promotes the fluxing action of the slag and provides immunity from porosity. Silicon, however, has a tendency toward decreasing toughness and resistance to hot cracking in amounts greater than that required as a deoxidizer.

TABLE 3-10   HY-80 WELDING ELECTRODES AND APPLICATION(7)

ELEC. TYPE	SPEC.	PROCESS	POSITION	APPLICATION
MIL-11018	MIL-E-22200/1	Shielded metal arc	All	All
MIL-10018	MIL-E-22200/1	Shielded metal arc	All	Fillet, Fillet Groove or Groove Joints
MIL-9018	MIL-E-22200/1	Shielded metal arc	All	Limited Use
MIL-B88	MIL-E-19822	Semi-Automatic or Automatic Metal Inert-Gas Arc	Flat or Horizontal	All
MIL-EB82*	MIL-E-22749	Submerged Arc	Flat	All
MIL-MI88*	MIL-E-22749n	Submerged Arc	Flat	All
MIL-8218Y QT	MIL-E-22200/5	Shielded metal arc	All	Limited to Procedure Approval

*Granular Flux Particle Size 10 x 50.

**Granular Flux Particle Size 12 x 150.

TABLE 3-11 SPECIFICATION LIMITS OF DEPOSITED WELD METAL CHEMICAL COMPOSITIONS

ELEMENT	FOR HTS				FOR HY-80				
	MIL-7018	MIL-8018	MIL-9019	MIL-10018	MIL-11018	MIL-B88	MIL-EB82	MIL-MI88	MIL-8218YQT
Carbon	.12	.12	.10	.10	.10	.08	.12	.06	.10-.15
Manganese	.40-1.25	.40-1.10	.60-1.25	.75-1.70	1.30-1.80	1.15-1.55	.80-1.25	1.00-1.50	.80-1.15
Phosphorous	.030	.030	.030	.030	.030	.025	.020	.010	.030
Sulphur	.030	.030	.030	.030	.030	.025	.020	.010	.030
Silicon	.80	.80	.80	.60	.60	.35-.65	.80	.50	.30-.60
Nickel	.25	.80-1.10	1.40-1.80	1.40-2.10	1.25-2.50	1.15-1.55	.80-1.25	1.40-1.90	1.50-2.00
Chromium	1.5	.15	.15	.35	.40	--	.30	.10-.30	.90-1.20
Molybdenum	.35	.35	.35	.25-.50	.30-.55	.30-.60	.15-.60	.20-.40	.45-.75
Vanadium	.05	.05	.05	.05	.05	.10-.20	.05	.05	.02
Copper							.40-1.10	.10-.30	
Titanium								.10	
Zirconium								.10	
Aluminum								.10	

NOTE: Per cent-single values are maximum

5. Relatively large amount of nickel is added, because it provides toughness without effecting strength

6. Vanadium is limited to a maximum of 0.05%. Although there is no unanimity of opinion concerning the effect of vanadium, it is believed to contribute to a decrease in toughness. The exception is MIL-B88 wire for which 0.1 to 0.2% vanadium is permitted for strengthening and for which thermal stress relief is prohibited.

7. Chromium is limited to less than about 1%. Chromium is a less effective strengthener in as-deposited weld metal than manganese, molybdenum, and nickel, and it has an adverse effect on toughness in amounts in excess of about 1%.

8. Molybdenum contributes to the stability of weld metal strength under the variety of heating and cooling conditions typical of welding. Experience has shown that these beneficial effects can be obtained without adverse toughness if molybdenum is kept within the range of 0.15 to 0.75%.

The foregoing ranges and limits apply to the as-deposited weld metal composition - not to electrode material composition. As such, they can only be checked after deposition of weld metal in a groove of base material.

Table 3-12 shows specification limits of mechanical properties of as-deposited weld metals. A family of xx18-type electrodes has been developed which can provide a range of yield strength from 60,000 to 100,000 psi. It is possible, therefore, to select from Table 3-12 an electrode which will undermatch, match, or overmatch the strength of HY-80 base metal as desired or required by the design.

TABLE 3-12 SPECIFICATION LIMITS OF DEPOSITED WELD METAL MECHANICAL PROPERTIES

PROPERTY	FOR HTS					FOR HY-80			
	MIL-7018	MIL-8018	MIL-9018	MIL-10018	MIL-11018	MIL-B88	MIL-EB82	MIL-MI88	MIL-8218YQT
Ultimate Strength (psi)	70,000	80,000	90,000	100,000	110,000	-------	-------	110,000	-------
Yield Strength at 0.2% Offset (psi)	60,000-75,000	70,000-82,000	78,000-90,000	90,000-102,000	95,000-107,000	88,000	82,000	88,000	82,000
Elongation in 2 in. (per cent)	24	24	24	20	20	14	16	20	18
Charpy V-Notch Impact Energy (Ft/Lb) At-10°F	20	20							
At-60°F			20	20	20	20		30	20
At-80°F							20		

NOTE: Single values are minimum.

As has been indicated earlier, welds that have properties the same as those of the base metal are desired. Since welds may contain some flaws of defects which escape detection, it has been considered desirable as a practice to have a weld metal slightly stronger than the base metal, or the weld metal overmatch the base metal. This will offset the possible loss in strength in the weld metal and provide superior performance of a weldment as a whole. For this reason, a trend is to use an electrode which slightly overmatches the base metal.

As pointed out earlier, it has not been possible to develop an electrode which provides a weld metal as notch tough as HY-80 base metal, see Tables 3-9 and 3-12.

Quenched-and-Tempered Steels With Over 100,000 Psi Minimum Yield Strength

Several steels have been developed which have minimum yield strengths of over 100,000 psi. Table 3-7 lists mechanical properties and chemical compositions of some quenched-and-tempered steels with minimum yield strength of up to 120,000 psi.

Rigorous efforts have been and are being made to develop steels with higher and higher strength. Table 3-13 presents a highly simplified summary of compositional aspects of weldable, high-strength steels. The first four are quenched and tempered steels, while the last two are maraging steels which will be covered in the next section.(3.4). As the strength level increases it becomes increasingly difficult to maintain sufficient fracture toughness. The discussion given in the following pages comes from a recent NRL report by Pellini.[13]

TABLE 3-13 <u>WELDABLE HIGH-STRENGTH STEELS</u>

Type	Yield Strength Range (Ksi)	Section Size Limit* (in.)	Heat Treatment & Melting Practice	Primary Alloying Elements (%)								Impurity Level (%)
				C	Mn	Ni	Cr	Mo	Co	V	B	P&S
Commercial Low Alloy	90-125	2	Q&T (Air)	0.15	1.0	1.0	0.50	0.50	--	0.05	0.005	0.03
				0.15	1.5	--	--	0.50	--	--	0.005	0.03
Ni-Cr-Mo	80-100	6	Q&T (Air)	0.15	0.30	3.0	1.5	0.50	--	--	--	0.02
	130-140	6	Q&T (Air)	0.10	0.80	5.0	0.50	0.50	--	0.10	--	0.008
High Ni+Co	160-200	6	Q&T (VAR to VIM+VAR)	0.20	0.30	9.0	0.80	1.0	4.0	0.10	--	0.005
12-5-3 Marage	160-200	6+	Q&A (VAR to VIM+VAR)	0.02	--	12.0	5.0	3.0	--	--	Ti+Al 0.10 to 0.30	0.005
18-8-4 Marage	200-240	6+	Q&A (VAR to VIM+VAR)	0.02	--	18.0	--	4.0	8.0	--	Ti+Al 0.10 to 0.30	0.005

Codes: Q&T Quench and Temper
   Q&A Quench and Age

Air Air melting under slags
VAR Vacuum Arc Remelt
VIM Vacuum Induction Melting
+   For best properties

*For optimized properties.

Pellini's Analysis of Trends in Fracture Toughness. Figure 3-2 illustrates the temperature displacements of drop-weight tear (DT) test energy transition curves for quenched-and-tempered alloy steels of 110 Ksi yield strength, as compared to a carbon-manganese steel of 48 Ksi yield strength. This displacement in the transition temperature range is obtained with retention of high levels of fracture toughness. With increasing yield strength to levels of 200 Ksi, there occurs a progressive decrease of the shelf level (strength transition), an increase in the transition temperature range, and finally the elimination of the temperature transition features. The low-level dashed curve labeled 110 Ksi represents a commercial steel which has been optimized with respect to cost rather than maximum fracture toughness properties.

The general effects of increasing strength level on the temperature and strength transitions are presented schematically in the three-dimensional plot of Figure 3-3. The vertical scale references the DT test energy. One of the horizontal axes defines the transition temperature range features, and the other the strength transition features. Two surfaces are indicated in this plot. The outer surface relates to the best (premium) quality steels that have been produced to date for the respective yield strength levels. The inner surface, which lies at generally lower DT test energy values, pertains to commercial products of lowest practical cost. The letter steels are produced with minimum (Marginal) alloy content for the section size involved and are melted by conventional practices which result in relatively large amounts of nonmetallic inclusions. Both aspects evolve for reasons dictated by minimization

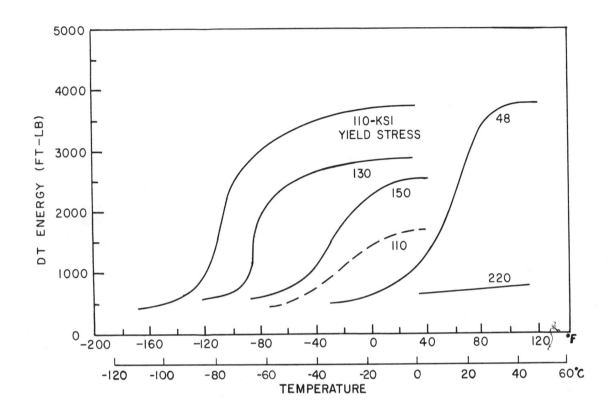

FIGURE 3-2   ILLUSTRATING THE GENERAL TRENDS OF INCREASING YIELD
STRENGTH (SOLID CURVES) ON THE 1-IN. DT TEST TEMPERATURE
TRANSITION RANGE AND ON SHELF LEVEL FEATURES.   THE
EFFECTS OF INCREASING VOID SITE DENSITY ARE INDICATED BY
COMPARISON OF THE TWO CURVES FOR THE 110-KSI STEELS.
THE SOLID CURVE RELATES TO LOW VOID-SITE-DENSITY METAL
AND THE DASHED CURVE TO HIGH VOID-SITE-DENSITY METAL.

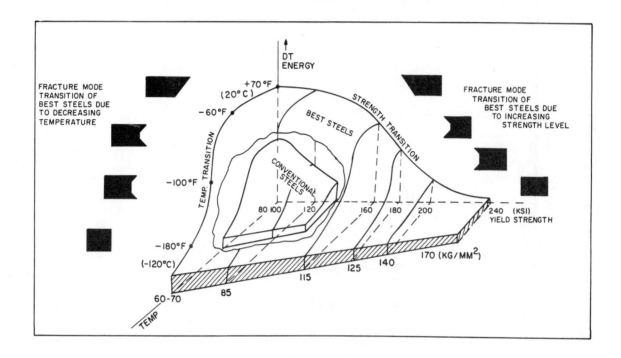

FIGURE 3-3   SCHEMATIC ILLUSTRATIONS OF THREE-DIMENSIONAL DT TEST
              ENERGY SURFACE EVOLVED BY COMBINED EFFECTS OF TEMPERATURE-
              AND STRENGTH-INDUCED TRANSITIONS.   THE NATURE OF CHANGES
              IN FRACTURE APPEARANCE FOR THE 1-IN. DT SPECIMENS ARE
              ILLUSTRATED BY THE DRAWINGS.   TRANSITIONS FROM DUCTILE
              TO BRITTLE FRACTURE ARE DEVELOPED AS A CONSEQUENCE OF
              DECREASING TEMPERATURE OR INCREASING STRENGTH LEVEL.

of costs.

Low-alloy contents restrict the nature of metallurgical transformations to less than optimum microstructural states and, therefore, result in higher transition temperature ranges as compared with
the high-alloy grades of higher cost. The relatively low cleanliness
of the low-cost commercial steels provides sites for void incubation
which decreases the level of shelf fracture toughness. Accordingly,
the strength transition to the brittleness levels of plane strain
fracture toughness is attained at lower yield strengths as compared
to premium products of high cleanliness. These combined effects
are evident by comparison of the two steels of 110 Ksi yield strength
level, represented by the solid and dashed curves of Figure 3-2.
The low position of the inner surface (Figure 3-3) for the commercial
steels is dictated by these factors. It is important to note that
the metallurgical factors which control the temperature and strength
transitions are different and independent in major degree. The
metallurgist must therefore consider a wide range of options in
evolving steels which are optimized with respect to temperature and
shelf transitions, as well as to strength level, section size,
cost, weldability, etc.

A simple, yet highly significant, zoning of metallurgical type
is developed in Figure 3-4 by tracing the effects of increasing
strength level on the shelf transition features of various generic
classes of steels. A series of metallurgical quality corridor zones,
which are related to the melting and processing practices used to
produce the steels, become evident. The lowest corridor zone involves
relatively low-alloy commercial Q&T steels produced by conventional

72

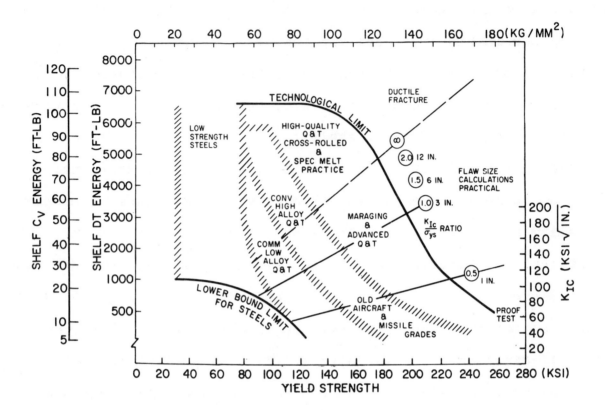

FIGURE 3-4   METALLURGICAL ZONING OF THE RATIO ANALYSIS DIAGRAM
WHICH DEFINES THE GENERAL EFFECTS OF MELTING AND
PROCESSING FACTORS ON THE STRENGTH TRANSITION.   THE
THREE CORRIDORS OF STRENGTH TRANSITION RELATE TO
METALLURGICAL QUALITY (VOID SITE DENSITY) WHICH CONTROLS
MICRO-FRACTURE PROCESSES AND, THEREBY, THE MACROSCOPIC
FRACTURE TOUGHNESS OF THE METAL.   THE LOCATION OF
GENERIC ALLOY STEEL TYPES ARE INDICATED BY THE
NOTATIONS.

low-cost melting practices. The corridor is defined by the strength
transition of these steels to the 0.5 ratio level, as the result of
heat treatment to yield strength levels in the order of 130 to 150
ksi. Optimizations of the alloy content, coupled with improved
melting and processing practices, elevate the corridors to higher
levels. The strength transition to the 0.5 ratio is shifted accordingly to higher levels of yield strengths.

Recent metallurgical investigations of high-strength steels
have emphasized processing and metal purity aspects rather than
purely physical metallurgy considerations of transformations. New
scientific knowledge of void growth mechanisms and of the importance
of void-site-density factors clearly indicate that control of metal
bridge ductility can only be effective within limits. As these
limits are reached by optimization of microstructures, it then is
essential to obtain further increases in ductility by suppression
of void initiation aspects. Void sites are provided by the presence
of non-metallic particles of microscopic size featuring noncoherent
metal grain boundaries. The extent to which such void sites can be
eliminated is related to a number of factors including melting and
deoxidation practices as well as the P and S impurity contents.

Welding of High-Strength Steels. With increasing strength
levels, welding of higher strength steels presents various problems.
One of the most difficult problems is how to obtain weld metals
which have both strength and toughness comparable to those of the
base plate. For steels in the yield strength range of 130 to 180 ksi,
the research effort expended for the weld metal development ordinarily
exceeds that for the base material by several fold.

Pellini[13] has prepared Figure 3-5 which provides the weld metal zoning of the Ratio Analysis Diagram (RAD).* This presentation highlights the strong effects of metal quality factors as related to the welding procedure. For attainment of high corridor features it is essential to use weld wire of high metallurgical quality (equal to the low-void-site-density of high corridor base metals) and then to protect the weld metal pool by the use of inert gas shielding. The Gas Tungsten Arc (GTA) method is superior to the Gas Metal Arc (GMA) method in these respects. Because of poor protection to atmospheric gases, Shielded Metal Arc (SMA) welds are limited to the low corridor irrespective of improvments in wire quality. Adjusting weld alloy compositions to control microstructures is not sufficient--the metal quality aspects related to void coalescence processes must be controlled also if the highest corridor relationships are to be utilized in practice.

The heat affected zone (HAZ) of high strength metals also poses difficult metallurgical problems because of the off-standard nature of the heat treatments developed in the welding cycles. In this case there is no recourse to changing compositions as for the weld metal. For the high-alloy base materials which follow high RAD corridors, the HAZ problem is minimized, and excellent properties are obtained with latitude in the control of welding parameters. The high-alloy contents promote the development of optimum metallugrical structures over fairly wide ranges of weld heat treatment variables, as compared to the low-alloy steels.

---

*Refer to Chapter 12 for a further explanation of the RAD.

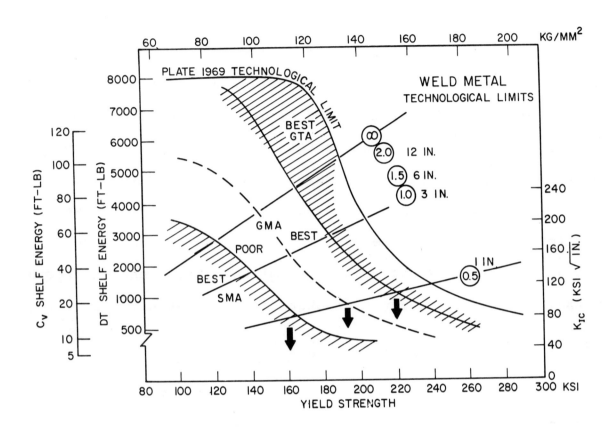

FIGURE 3-5   ZONING OF THE RATIO ANALYSIS DIAGRAM WHICH DEFINES
WELD METAL FACTORS OF FIRST ORDER IMPORTANCE TO
STRENGTH TRANSITION FEATURES. THE STRENGTH TRANS-
ITION CORRIDORS RELATE TO WELD METAL DEPOSITED BY
METAL ARC (SMA), INERT GAS SHIELDED METAL ARC (GMA),
AND INERT GAS SHIELDED TUNGSTEN ARC (GTA). THE BOLD
ARROWS INDICATE THE STRENGTH LEVEL OF TRANSITION TO
THE 0.5 RATIO VALUE.

For the low-alloy commercial steels which follow the low-corridor realtionships, the HAZ problem becomes more difficult to resolve. This results from the marginal alloy contents which are barely adequate for developing desirable microstructures under controlled mill heat treatment conditions. Thus, weld HAZ for such metals may be seriously degraded by off-optimum welding procedures. The shelf characteristics for the HAZ of steels featuring only 90 to 120 ksi yield strength may then drop to the very low levels described previously for the ultrahigh strength steels. The shelf level differential between the base metal and the HAZ for such cases is so large (plane stress to plane strain) that preferential fracture in the HAZ becomes the expected fracture mode in structures.

3.4 Maraging Steels

Maraging steels are a family of metals which have been devloped relatively recently. As a group, maraging steels are fairly different from other steels we have discussed in terms of chemical compositions, steel making processes and welding fabrication.

Although the use of maraging steels for marine applications has been very limited, they may be used for some future marine vehicles. For example, maraging steels are being considered for the pressure hull of the Deep Submergence Search Vehicle (DSSV). This section presents a rather brief discussion of maraging steels.

Development of Maraging Steels

In 1960, International Nickel Company introduced a line of maraging steels with yield strengths of 150-200 ksi. Recently, these steels have reached over 300 ksi.[4] The family of maraging steels includes five commercial steels classified into two groups according to their hardening elements (see Table 3-14):

1. 18% grades using cobalt-molybdenum additions
2. 20% Ni and 25% Ni grades using titanium-aluminum-columbium additions.[4]

Table 3-15 shows mechanical properties of annealed maraging steels.

Production of Maraging Steels

Maraging steels are characterized by a very low carbon content (0.03% is the maximum) and relative ductility and low yield strength when solution-annealed. Strength is attained during an aging process at 900° F by the precipitation of Ni_3Mo and inter-metallic compounds including aluminum and titanium and by cobalt reactions probably involving ordering. Improved toughness is also achieved. Figure 3-6 shows the heat treatment used for the 18% Ni maraging steel. Somewhat different heat treatments are used for 20% and 25% Ni maraging steels.

Cooling. Cooling occurs from temperatures of 1500-1900° F. The cooling rate from the austenitizing temperature is not critical, however, because undesirable high-temperature products such as pearlites do not form. 25% and 20% Ni alloys changes to martensite at temperatures below room temperature. 25% grade steels must be

TABLE 3-14   COMPOSITION OF MARAGING STEELS (%)[a] (4)

ALLOY ADDITION	18% Ni			20% Ni	25% Ni
	200,000 Psi	250,000 Psi	300,000 Psi		
Ni......	17-19	17-19	18-19	18-20	25-26
Ti......	0.15-0.25	0.3-0.5	0.5-0.7	1.3-1.6	1.3-1.6
Al......	0.05-0.15	0.05-0.15	0.05-0.15	0.15-0.35	0.15-0.35
Co......	8-9	7-8.5	8.5-9.5	-	-
Mo......	3.0-3.5	4.6-5.1	4.7-5.2	-	-
Cb......	-	-	-	0.3-0.5	0.3-0.5

[a]Other elements: C, 0.03 max; Mn, 0.10 max; Si, 0.10 max; S, 0.01 max; P, 0.01 max; B, 0.003 added; Zr, 0.02 added; Ca, 0.05 added.

TABLE 3-15   MECHANICAL PROPERTIES OF ANNEALED MARAGED STEELS (4)

ALLOY	18% Ni[a]	20% Ni[b]	25% Ni[a]
Yld Str (0.2% offset), 1000 psi..............	110	115	40
Ten Str, 1000 psi.....	140	152	132
Elong (in 1 in.) % ...	17	8	30
Red. of Area %........	75	-	72
Hardness, R_c..........	28-32	26-35	10-15

[a]Bar specimen 0.252 in. dia, gage length 1 in.

[b]Flat specimen 0.145 in. thick, gage length 2 in.

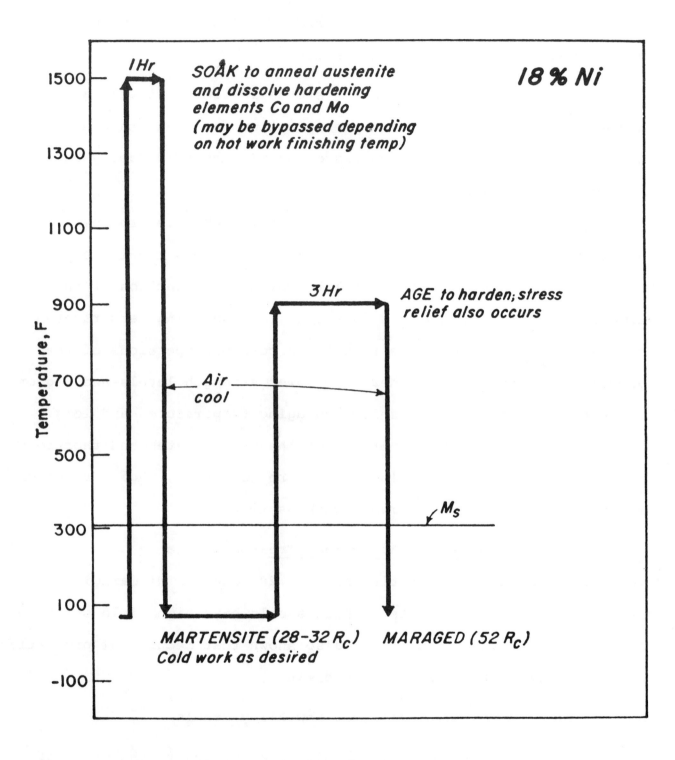

FIGURE 3-6   HEAT TREATMENT USED FOR 18% Ni MARAGING STEEL[4]

kept refrigerated at -100° F. 18% and 12% Ni grades, on the other hand, transform above room temperatures.

Aging. The aging of maraging steels is similar to the age hardenable aluminum alloys and precipitation hardened stainless steels. It is through the process of aging that this family of steels derives its strength.

Aging is carried out at 850-950° F for 15 minutes to several hours depending upon the particular steel. The end result is a combination of strength and ductility which is a function of the time/temperature process used in aging. A high aging temperature coupled with long aging time results in coarse dispersions of the precipitate particles and therefore lower strength levels and higher fracture toughness. Conversely, low aging temperatures and long aging time gives a fine dispersion of the precipitates and therefore higher strength levels and lower fracture toughness. Aging reactions may be induced by alloy elements such as titanium.

Mechanical Properties After Heat Treatment. Table 3-16 shows the mechanical properties of 18%, 20% and 25% Ni maraging steels after heat treatments. These steels have high yield strengths and good notch toughness. It is also shown that mechanical properties are not greatly affected by cold working.

Advantages of Maraging Steels Over Low-Alloy Steels

Although specific properties differ according to the grade of steel, the following are some advantages of maraging steels over conventional low-alloy steels:

1. Useful yield strengths to and above 300,000 psi

2. Low nil ductility temperature

TABLE 3-16   MECHANICAL PROPERTIES OF THE 18% NICKEL MARAGING STEEL AFTER HEAT TREATMENT

YIELD STRENGTH

BAR STOCK	200,000 Psi[a]	250,000 Psi		300,000 Psi[b]
		900 F, 3 Hr[c]		
	900 F, 3Hr[c]	Air Melt	Vac Melt	900 F, 3 Hr (direct maraged)
Yld Str, 1000 psi	190-210	240-268	–	295-303
Ten Str, 1000 psi	200-220	250-275	–	297-306
Elong, %	14-16	10-12	–	12
Red. in Area, %	65-70	48-58	–	60
Charpy V-Notch Impact Energy, ft-lb				
70 F	60-110	18-26	25-30	–
-320 F	30-60	12-15	–	–
Endur Limit, Ksi	95	95-100	–	–

SHEET[d]	900 F, 3 Hr[c]	50% CW + 900 F, 3 Hr[c]	900 F 3 Hr[c]	50% CW + 900 F, 3Hr[c]	900 F, 3 Hr[c]	50% CW + 900 F, 3 Hr[c]
Yld Str, 1000 psi						
70 F	203	244	252	286-290	280-290	300-309
-320 F	286	298	–	–	–	–
Ten Str, 1000 psi						
70 F	224	246	262	289-292	290-294	303-312
-320 F	291	303	–	–	–	–
G_c, in-lb/sq in.[h]						
70 F	1880	1680	2000	1240-1690	1510-1710	1100-1800
-320 F	2720	1800	–	–	–	–

[a]Air melted [b]Vacuum melted [c]Ann. 1 hr., 1500 F; cold worked, maraged as shown

[d]All vacuum melted. Data based on NASA 1-in. edge notch specimen, 0.06 in. thick. 0.0005 in root radius K_t>20.

3. In the case of the 18% nickel grade, exceptional resistance to stress corrosion by sea water

Advantages in fabrication and heat treatment are:

4. Thorough hardening without quenching

5. Simple heat treating procedure

6. Good formability without prolonged softening treatments

7. Good machinability--comparable to that of 4340 steel having equal hardness

8. Low distortion during the strengthening treatment (maraging) after forming or machining

9. Good weldability; no preheating is required even when welding fully heat treated material

10. Freedom from decarburization[4]

Section Size Effects

Maraging steels have proved insensitive to section size when strength properties are concerned, but very sensitive in regard to fracture toughness properties. One major problem involved in section size considerations is that large grains form at the high rolling temperatures necessary for producing thick plates. Moreover, maraging steels are stiffer at rolling temperatures than quenched and tempered steels and therefore it is harder to work the center regions of the thick plates.

Weldability

Maraging steels have better welding properties than many other types of steels because the weld and the heat affected zone (HAZ) are "soft" as welded and therefore should resist

83

cracking. The disadvantage for welding is that <u>the weld region must be heated to 850-950° F.</u> (the aging temperature) to obtain the necessary strength level. This is a problem in large structures; small structures can be placed physically into a furnace. Large structures must have a localized flame or electric strip heaters. Or, possibly, the whole structure may be heated by an internal heating apparatus.

Ni grades of 18% tend to develop subsurface cracks in areas contiguous to the heat affected zone because of the pressure of macroscopic, sheet-like inclusions which probably show aggregations of complex titanium composition through cracks which are parallel to the plate surface. This situation worsens with an increase in the plate thickness.

Other welding problems include an incomplete transformation of the austenite to martensite which results in lower strength in the heat affected zone.

Fracture Toughness

Fracture toughness usually declines with an increase in strength, as shown in Figure 3-4. Composition and heat treatment conditions must be optimized in order to obtain maximum fracture toughness.

REFERENCES

(1) Bannerman, D. B., Jr., "Development of ABS Rules for Construction of Offshore Drilling Rigs," Marine Technology, pp. 58-59, January 1969.

(2) Crimmins, P. P., and Tenner, W. S., "How to Weld Thick Plates of 18% Maraging Steel," Metal Progress, pp. 57-62, May 1965.

(3) Dawson, T. J., "Survey of Notch Steels Shows Fabrication Problems," Welding Engineer, pp. 75-78, April 1961.

(4) Decker, R. F., Yeo, R. B. G., Eash, J. T., and Bieber, C. G., "The Maraging Steels," publication of the International Nickel Company, Inc., New York, New York.

(5) Gross, J. H., and Stout, R. D., "Steels for Hydrospace," a publication of the United States Steel Corporation, Applied Research Laboratory.

(6) Hall, A. M., "The Status of Ultrahigh Strength Steels Today," Metal Progress, pp. 178-192, August 1965.

(7) Heller, Capt. S. R., Jr., Fioriti, and Vasta, J., "An Evaluation of HY-80 Steel as a Structural Material for Submarines, Part I," Naval Engineers Journal, pp. 29-44, February 1965.

(8) Hucek, H. J., Elsea, A. R., and Hall, A. M., "Evolution of Ultrahigh-Strength Hardenable Steels for Solid-Propellant Rocket-Motor Cases," DMIC Report 154, Defense Metals Information Center, Battelle Memorial Institute, May 25, 1961.

(9) Kihara, H., and Yamamoto, N., "Recent Developments in Managements and Production Methods in Japanese Shipyards," a paper presented before the 1968 Spring Meeting of the Society of Naval Architects and Marine Engineers.

(10) Linnert, G. E., "Welding High Strength Quenched and Tempered Alloy Steels," Metal Progress, pp. 95-102,August 1965.

(11)
    Martin, D. C. "Welding of High-Strength Steels," Lecture presented at a Special Summer Session on "Welding Fabrication in Shipbuilding and Ocean Engineering," M.I.T., August 19, 1969.

(12)
    Martini, E. K., "HY 130-150 Steel Weldment for Submarine Hulls," <u>BuShips Journal</u>, pp. 8-11, September 1965.

(13)
    Owen, Capt. T. B., and Sorkin, G., "Metallurgical Materials Problems," <u>BuShips Journal</u>, pp. 3-12, July 1962.

(14)
    Pellini, W. S., "Evolution of Engineering Principles for Fracture-Safe Design of Steel Structures," NRL Report 6957, U. S. Naval Research Laboratory, September 1969.

(15)
    Pellini, W. S., "High-Strength Steels," NRL Report 6167, "Status and Projections of Developments in Hull Structural Materials for Deep Ocean Vehicles and Fixed Bottom Installations," U. S. Naval Research Laboratory, pp. 16-30, November 1964.

(16)
    Rathbone, A. M., Gross, J. H., and Dorschu, K. E., "A Tough Weldable Steel With a 140,000 Psi Yield Strength," <u>Metal Progress</u>, pp. 63-67, May 1965.

(17)
    Rules for Building and Classing Steel Vessels, American Bureau of Shipping, 1968.

(18)
    Schapiro, L., "Steels for Deep Quest and DSRV-1," <u>Metal Progress</u>, pp. 74-77, March 1968.

(19)
    Shankman, A. D., "Materials Technology for Hydrospace Pressure Hulls," <u>Metal Progress</u>, pp. 69-73, March 1968.

(20)
    Stout, R. D., and Tense, A. W., "Selecting Higher Strength Steels for Pressure Vessels, <u>Metal Progress</u>, pp. 147-158, August 1965.

(21)
    Technical and Research Bulletin No. 2-11a, "Guide for the Selection of High Strength and Alloy Steels," The Society of Naval Architects and Marine Engineers, September 1968.

(22)
    <u>Welding Handbook</u>, Fifth Edition, Section Five, Chapter 88 (Ships), <u>The American Welding Society</u>, 1967.

(23)
<u>Welding High Strength Steels</u>, American Society for Metals, 1969.

CHAPTER 4   ALUMINUM AND ITS ALLOYS

4.1 History and Present Trends in the Use of Aluminum and Its Alloys

Although aluminum and its alloys have been known since the late 1800's, their successful use in marine engineering is rather recent. Since aluminum promises extensive applications for ocean engineering, this chapter will consider the background, properties and applications of the metal.

Early Applications of Aluminum

Table 4-1 presents a condensed chronology of aluminum in marine construction.[3] The first use of aluminum in ships occurred around 1890, very shortly after steel was introduced in shipbuilding. In 1899, aluminum was used in U. S. Navy torpedo boats. Two such craft, the "Dahlgren" and the "T.A.M. Craven", both built by Bath Iron Works, utilized aluminum for its light weight, regardless of the higher cost.[11] Yarrow & Co. built "La Foudre," a 60-foot aluminum torpedo boat in Britain in 1894. In this early ship, the aluminum alloy contained 6% copper. All scantlings were increased 25% over that allowed for steel.[11]

Aluminum applications were added to many ships, but successful alloy and fabrication processes known then were inadequate, and aluminum joined iron as an obsolete building material.[3]

The Washington Disarmament Conference revived interest in aluminum as a construction material when, in 1922, it limited the total naval displacements. This judgement encouraged naval architects to consider aluminum for its light weight. Hand-in-

TABLE 4-1:   A CONDENSED CHRONOLOGY OF ALUMINUM IN
MARINE CONSTRUCTION [3]

$$\frac{1891}{1963}$$

1890-Naphtha launch "Zepher"
built of aluminum by Escher
Wyss & Co., Zurich. Length:
17 ft. Speed: 16 to 19 mph.

1891-World's first seagoing
aluminum-hulled vessel, yacht
"Mignon", built in France.
40 ft. long, 6-ft. beam.

1891-Two aluminum torpedo
boats built for French Navy.

1893-"La Foudre", 60-ft., 11-
ton torpedo boat with aluminum
hull, built for the French
Navy by Yarrow & Co. Trial
speed: 20.5 kts. hp: 275-300.

1893-An aluminum hulled German
steamer plied the waters of
the Victoria Nyanza.

1894-"Jules Davoust", an
aluminum boat, was sent to the
Niger for hydrographic purposes.
40 ft. long, 6-ft. beam. 2 1/2
ft. draft. Weight: 4,400 lb;
capacity: 11 tons.

1895-"Defender", successful
United States entry in the
America's Cup races, was
aluminum from the waterline up.

1898-"The Doktor Karl Peters",
built by Escher Wyss & Co.,
was shipped to the Victoria
Nyanza. It had an aluminum
hull and topsides.   Length:
43 ft. Speed: 9.5 mph.
Cost: $10,000.

1899-U.S.N. torpedo boats
"Dahlgren" and "T.A.M.
Craven" built by Bath Iron
Works with aluminum galleys,
observation towers, hatch
covers and cowls. Aluminum
angles, plates and rivets
were used.

1900-Samuel Ayer & Son.
Nyack, N. Y., built the steam
yacht "Arrow" with aluminum
deck, deck beams, frames,
stack, hatches and fittings.
130 ft. long, 12 1/2-ft.
beam. "Arrow" won a race
from New York to Atlantic
Highlands at approximately
22 mph and was considered the
fastest yacht in the world.

1922-Washington Disarmament
Conference, limiting total naval
displacement, stimulated the
use of aluminum in warships
throughout the world.

1928-U.S.S. light cruiser
"Houston" built with aluminum
deckhouses, using plate, shapes
and rivets of alloy 2017.
Duralumin.

1933-Major U.S.N. building
program begun. New aluminum
structural alloys extensively
used in detroyers, cruisers and
carriers for bridges, deckhouses,
catapults, elevators, masts, yards.
Reasons for aluminum: higher
speeds, improved stability and
nonmagnetic properties.

1934-Structural aluminum applied
to three Mystic Steamship
Company colliers to solve weight
and corrosion problems.

1937-Three New York ferryboats
built with full-length aluminum
deckhouses, shade decks and pilot
houses to increase passenger
capacity, reduce weight and fuel
consumption. Additional ferries
with aluminum structures have been
built since then because of savings
in fuel and maintenance. Condition
of the aluminum on these ferries
in use today (1963) is excellent,
and three additional vessels have
been ordered.

1939-Norwegian vessel M. V.
"Fernplant" built with aluminum-
alloy superstructure. Fourteen
tons of aluminum replaced 40 tons of
steel. Since aluminum proved
eminently satisfactory, three
sister ships subsequently so
equipped. Age of aluminum in
merchant ships begins.

1940-100 U.S.N. warships with
extensive aluminum superstructure
applications in service.

1950-More than 200 merchant
ships in service with substantial
aluminum topside structures
and equipment.

1952-"S.S. United States" built
with 2,000 tons of aluminum,
saving 8,000 tons in weight.
Sets transatlantic speed records,
establishes pattern for major liners
built since then.

1956-First modern all-welded
aluminum personnel boat built
by Sewart Seacraft, Inc. High
speed and low operating cost
led to construction by 1963
of more than 200 aluminum
personnel boats.

1958-First modern fleet of
purse seine boats in service.
Maintenance savings and increased
earning capacity led to wide-
spread adoption of aluminum
for fishing vessels of all types.

1959-223-ft. German tanker,
"Aluminia" becomes largest
vessel with all-aluminum hull.

1960-More than 1,000 merchant
ships afloat with substantial
aluminum structural installations.

1962-Success of aluminum 90-
ton H. S. "Denison" spurs U.S.
hydrofoil development.

1963-Navy's newest aluminum
vessel, a 22 1/2-ton Hydroskimmer, cruises at 70 knots
while air-supported 1 1/2 ft.
above the ocean.

hand with this decision came new developments in aluminum to meet strength and corrosion resistance requirements for marine structures. [3]

In 1928, the USS "Houston", a light cruiser, used the alloy Duralumin in its deckhouses. This innovation ushered in a new era of aluminum uses including "Salt Lake City," a 10,000 ton cruiser; "Portland," and "Phoenix," both using aluminum applications for light weight. [11] One of the first vessels designed and built from a true aluminum alloy, the "Diana II," was constructed and used by the British during World War II. The vessel is still in use today. [9]

By 1940, aluminum was used on about 100 U.S. warships. Later, a guided missile destroyer, the USS "Dewey" was constructed with aluminum superstructures. [9] The English again introduced new uses of aluminum in 1953 with the "Moray Mhor," a motor yacht containing the first all welded aluminum hull. [9] Up until the 1950's, aluminum had been applied primarily to miscellaneous constructions and small boats. Now, designers turned their attentions to aluminum applications in large structures.

Present Trends and New Applications

Present applications of aluminum in marine structures are discussed briefly.

Military and Commerical Ships. Aluminum alloys have long been used in superstructures of both military and commercial ships.

Large tonnages have gone into ship superstructures in vessels ranging from destroyers to aircraft carriers. In many destroyers, the whole structure above the main deck consists of aluminum. For example, in the U.S. Navy's first guided missile destroyer leader, the "Dewey," 368,000 lbs. of aluminum are used, most of which is 5456-H321 plate and 5086-H32 sheet.[11] One of the Navy's largest aircraft carriers, the USS "Independence," contains 2,250,000 lbs. of aluminum in its construction. Had steel been the basic material used, the weight would have been 2,000,000 lbs. more than it is now.[11] The U.S. Navy now has all-aluminum gunboats up to 165 feet in length, and is considering all-aluminum destroyers.

Aluminum alloys also have been used in merchant ship structures, again primarily in superstructures. More than 4,000,000 lbs. of aluminum alloys were used in the USS "United States," completed in 1952. Table 4-2 shows the major applications of aluminum in this vessel.[11] More recently, the all-aluminum vessel, the "Sacal Borincano," is a commercial vessel designed to haul truck trailers from Miami to Puerto Rico.[6]

Submarines. Like many other submersibles, submarines must contain materials capable of combining strength and lightness in a deep ocean environment. From 40,000 to 50,000 lbs. of aluminum were used in each of the George Washington-class nuclear powered submarines to lighten the structures and increase their cruising range and speed.[11]

Deep Submergence Vehicles. The "Aluminaut," conceived and built by the Reynolds Metals Company, was one of the first deep submersibles using aluminum. Made from 7079-T6 forgings, this cylindrical vessel is 50 feet long with an internal diamter of 6-1/2 feet. It has the greatest internal volume of any vessel

TABLE 4-2   APPLICATION OF ALUMINUM USED IN THE S.S. "UNITED STATES"[11]

Superstructure (4 decks high, 660 ft. long)	1,100 tons
Life-saving equipment (24 boats, davits, winches)	200 tons
Stack enclosures and masts	200 tons
Ventilation ducts	125 tons
Railings	10 miles long
Furniture	20 tons

of its type in existence today. The "Aluminaut's" design depth/
capability is about 15,000 feet and it can cruise at 3-1/2 knots.[6]

At the time the "Aluminaut" was built, 7079-T6 was the only
aluminum material available which would meet the designer's
requirements for an aluminum alloy with a compressive yield strength
of 60,000 lbs. per square inch in a 6-1/2 inch section. This alloy
does not perform unprotected without the deterioration that is
characteristic of alloys of lesser strength. Like steel, it must
be protected and its performance is dependent on how good this
protection is. The boat hull has a multilayer coating of poly-
urathane. Anodes of 606 aluminum alloy also protect the structure.[6]

Examination of the "Aluminaut's" hull shows that no deterioration
has taken place from corrosive attack. The success in protecting
such a structure in a high-strength aluminum alloy has major
significance in determining where aluminum may be applied in deep
submergence structures.[6]

Off Shore Structures. Aluminum alloys in the form of anodes
are being used extensively in protecting large steel off-shore
structures below the water line. This provides an effective means
of stopping the submerged corrosion process, although the process
at and above the water line must be halted in other ways.[6]

New Applications. Two major developments in the use of
aluminum include the hydrofoil craft for antisubmarine warfare
and the hover craft. The SKMR-1 is a 22.5 ton hydroskimmer
consisting of 5456 aluminum alloy extrusions, four Solar Saturn
gas turbines for thrust up to 80 m.p.h. and it rides on an air
cushion of 18".[11]

The USS "High Point" (PCH-1) is a hydrofoil subchaser 115 feet long with all aluminum construction except for steel foils. Other new applications range from 10,000 lb. mine sweepers to 400,000 lb. amphibious transport docks and aluminum-hulled PT's.[11]

4.2 Metallurgy and Properties of Aluminum Alloys

Pure aluminum can be alloyed readily with many other metals to produce a wide range of physical and mechanical properties. Table 4-3 describes the major alloying elements in the wrought aluminum alloys.[15] The four digit system of classifying aluminum alloys is used by the Aluminum Association:

      1st digit = major constituents

      2nd digit = the modification

      3rd and   = alloy
      4th digits

In the 2xxx-8xxx alloy classification group, the last two digits have no special significance, but are used to identify different aluminum alloys in the group.[18]

1xxx series. This series is demonstrated by alloy 1100, which is 99% pure aluminum, non-heat treatable, soft and ductile, highly corrosion resistant, formable and weldable.[11]

2xxx series. The "aircraft alloys," as this group is called, has high strength, and is heat treatable, but lower ductility than 1xxx. Furthermore, it requires spot or seam welding as opposed to fusion welding. It is less corrosion resistant than 1xxx, but may be good if mechanical fastening for parts can be used.[11]

3xxx series. This class is non-heat treatable. 3003, used extensively for military structures, has Mn added and is 20%

TABLE 4-3   DESIGNATIONS FOR ALLOY GROUPS*

MAJOR ALLOYING ELEMENT	DESIGNATION
99.0% minimum aluminum and over	1xxx
Copper	2xxx
Manganese	3xxx
Silicon	4xxx
Magnesium	5xxx
Magnesium and Silicon	6xxx
Zinc	7xxx
Other elements	8xxx
Unused Series	9xxx

*Aluminum Association designations.

stronger than 1100.

4xxx series. The addition of silicon in amounts up to
12% will yield aluminum alloys particularly suitable as a filler
material for welding and brazing, because they have low melting
points. Aluminum alloys containing silicon also are used for
casting and forging; however, casting alloys have different numerical
designations.

5xxx series. This is the most important material. It is
a weldable Mg-Mn alloy, formable, highly corrosion resistant, has
a high weld zone ductility and high strength. Advances in 5052,
5083, 5086, and 5456 welding make realistic the welding of aluminum
in applications where welds are subject to heavy dynamic loading.[10]

6xxx series. A heat treatable material, this alloy has
sheets, plate and extrusions 2/3 to 3/4 less in strength than the
2xxx series. However, it is easily available, weldable and cheap,
which makes it practical for structures with low dynamic loading
factors.[11] Alloys 6063 and 6061 are two of the most versatile of
aluminum alloys of intermediate strength. They have outstanding
corrosion resistance and none of the problems of corrosion inherent
in many high strength materials.[6]

7xxx series. This group contains the highest strength. It
is used in aircraft although it involves some forming difficulties.
7079 in particular hardens to a great depth.[11]

Alloys and Heat Treatments. Alloys are classified further
into two broad categories: heat treatable and non-heat treatable.
The first is stronger and more expensive because of the heat
treatment involved. It is used for high structure strength when

heavy dynamic loading is a factor. Spot and seam welding rather than fusion welding may be used with these alloys.[11]

The non-heat treatable group is used when corrosion resistance is important and for fabrication of welded assemblies encountering dynamic stress. It is easily welded, the weld is less brittle and still strong. One such alloy is 5083, a high magnesium weldable alloy that is highly resistant to shock loading.[11]

Alloy Composition. Alloys usually involve some addition of copper, magnesium, silicon, manganese, and zinc, with aluminum in varying degrees. These additions provide hardness, toughness and high strength, ductility, formability, weldability, and machinability.[11]

Copper, magnesium, silicon, manganese and zinc all increase the material's strength and hardness, but each adds particular properties to the alloy:

"Silicon in a casting alloy, for instance, increases fluidity in pouring molten metal, eliminates hot shortness in casting (the tendency to crack or craze during cooling), increases casting strength after aging. Magnesium and silicon together in an alloy combine chemically to form magnesium-silicide, which makes an alloy capable of being heat treated. The 356-type alloys are of this kind. They are heat-treatable alloys that can be fusion welded satisfactorily. Copper, magnesium, silicon, and zinc, when used together in a heat-treatable alloy, serve to increase aging strength.[11]

Aluminum Alloys for Marine Applications

The metallurgy of marine aluminum alloys has developed

greatly since World War II. Formerly, Type 6061 served as the primary marine alloy. It proved to have good corrosion resistance, but its ductility after welding was low. Consequently, the alloy required riveting. Since the increase of welded ocean engineering structures, 6061 has had restricted use.[9]

Today, aluminum alloys in the 5xxx series and the 7xxx series are used extensively in marine applications; therefore, these groups are discussed in more detail.

The 5xxx group involves magnesium as a major additive. The 7xxx group, on the other hand, is composed of magnesium, zinc, magnesium and zinc, or copper.[18]

The mechanical properties of these alloys, beyond that obtained by solid solutions, include:

1. strain hardening

2. artificial aging

Table 4-4 shows mechanical properties of aluminum alloys for marine applications.

The 5xxx series has been used for more than ten years in marine structures. It is valuable for its high resistance to corrosion in the water. These materials are non-heat treatable and derive their mechanical properties from combinations of chemistry and work hardening. In the 5xxx series in general, annealing takes place between 650°-800° F. with cooling in still air. Usually is is then strained to get various strength levels and finally, it is thermally stabilized. Stability treatments vary, but usually involve hot forming at 425° F. Such treatment attains stability but lowers the strength.[18]

The 7xxx series, also used for ocean engineering structures, is solution heat treated at 850° F. With a soaking time depending

TABLE 4-4   MECHANICAL PROPERTIES OF ALUMINUM ALLOYS
FOR MARINE APPLICATIONS (1-INCH PLATE)[18]

Alloy Designation and Temper	Tensile Ultimate Strength (ksi)	Tensile Yield Strength (ksi)	Percent El. in 2-in.	Compressive Yield Strength (ksi)
5083-0	40	18	16	18
5083-H112	40	18	12	19
5083-H113	44	31	12	26
5086-H34	44	34	10	32
5086-H112	35	16	10	16
5454-H32	36	26	12	24
5454-H34	39	29	10	27
5454-H112	31	12	11	12
5456-H321	46	33	12	27
X7002-T6	61	50	9	53
X7005-T6	45	36	7	36
X7106-T6	55	50	--	50
X7106-T63	56	50	9	50
X7039-T6	65	55	13	58
X7039-T61	62	51	14	--
X7139-T63	63	55	13	--
X7005-T63	47	38	7	38

Note: "X" refers to experimental.

upon the thickness of the material. After soaking, it is
immediately cold-water quenched.[18] This treatment gives the
exceptional strength properties inherent in the 7xxx series alloys.

4.3 Design Considerations

Since aluminum has different mechanical, physical and chemical
characteristics than other construction materials, these differences
must be considered in designing ocean engineering structures.

Advantages and Problem Areas of Aluminum

First a brief discussion is made of advantages which aluminum
has over steel as well as some problem areas of aluminum for marine
application.

Some of the advatages of aluminum are:

"1. Cheaper cost over the life of the structure if all
     factors are taken into account

 2. A lighter structure for the same job

 3. Ease of fabrication

 4. Ability to perform without protective coatings

 5. Longer life

 6. Pleasing appearance."[6]

Some of the above points can be illustrated by considering an
aluminum and a steel plate one foot square and 1/4 inch thick. The
aluminum plate would weigh 3.8 lbs. and the steel about 10 lbs.
The aluminum plate would cost about $1.90 and the steel plate about
$1.00. If the cost of maintaining the steel were added to the
original cost, the cost for the steel plate over a 10 year period
would be $2.62. Under certain circumstances the aluminum plate
would require little maintenance resulting in lower overall cost

than a steel structure. However, under some other circumstances, (including improper handling) aluminum will respond poorly to marine applications. Since aluminum is anodic to most other metals, if electrolytic cells are set up in a structure, the aluminum will be preferentially attacked. Like many other metals, aluminum does not stand up well under poultice conditions. Crevices that allow the accumulation of foreign materials and form poultices should be avoided.

Mechanical Properties and Weight

Figure 4-1 shows a typical stress-strain diagram for various materials in marine applications.[11] They include:

Aluminum alloy 5083-H113 base metal

Weldment of 5083-H113 welded with 5183 filler wire

Ship steel plate, ASTM-131

Reinforced polyester laminates with grain.

Table 4-5 presents a comparison of weight and strength in these materials.

Strength. Maximum tensile strength and yield strength of aluminum are comparable to those of low-carbon steel.[9] Or, to cite another comparison, most 5000 series aluminum alloys have a strength of 31,000-63,000 psi. Most steels used in ocean structures have 58,000-71,000 psi. However, on the weight basis, aluminum is stronger.[3]

Weight. Aluminum weighs about half as much as steel with the same strength and deflection. Therefore, a smaller amount or weight of aluminum can be used in a structure in order to achieve the same strength as steel.[9] Table 4-6 shows weight savings

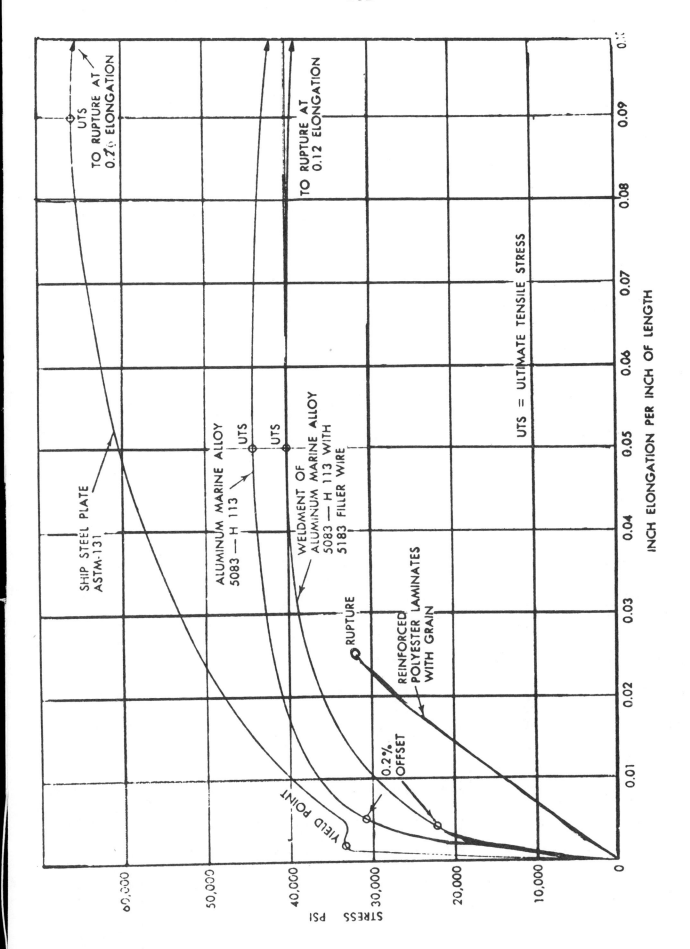

FIGURE 4-1 TYPICAL STRESS-STRAIN DIAGRAMS FOR VARIOUS MARINE APPLICATIONS (11)

TABLE 4-5   COMPARISON OF WEIGHT AND STRENGTH OF VARIOUS MATERIALS

FOR MARINE APPLICATIONS (11)

COLUMN 1	COLUMN 2	COLUMN 3	COLUMN 4	COLUMN 5	COLUMN 6	COLUMN 7
		Typical Strength (psi)				
Material	Weight lb/cu.in.	Ultimate Tensile	Yield Tensile	Ultimate Shear	Tensile Modulus of Elasticity (x10⁶)	Strength to-Weight Col.3/Col.2 (x1000)
(1) Aluminum alloy:						
5083-H113	0.096	46,000	33,000	27,000	10.3	480
5086-H34	0.096	47,000	37,000	27,000	10.3	490
6061-T6	0.098	45,000	40,000	30,000	10.0	460
Ship Steel ASTM-A131	0.29	66,000	33,000		29.0	230
Mild Steel ASTM-A100	0.28	56,000	30,000	42,000	29.0	200
Cooper, hard sheet	0.32	46,000	40,000		17.0	140
(2) Reinforced polyester laminates:						
With grain	0.062	32,000		13,000	1.4	520
Across grain	0.062	21,000		14,000	1.1	340

The Tensile Modulus of Elasticity column values use the LaTeX notation $(\times 10^6)$.

Table   4-6   Weight Saving in S. S. "Sunrip" (3)

	ALUMINUM WEIGHT	WEIGHT SAVED
	short tons	short tons
Superstructure		
Midshiphouse	64.0	104.0
Afthouse	9.0	19.0
Funnel	2.0	4.0
Hatch beams	15.0	11.0
Hatch boards	15.5	22.5
Lifeboats	2.5	3.0
Davits	1.5	2.5
Ventilation trunking	0.75	1.5
Windows	0.75	.15
Piping	12.25	27.0
Miscellaneous	12.0	21.0
TOTAL:	135.25	217.0

Over-all aluminum-to-steel weight ratio - 0.384

obtained by the use of aluminum in the 450-ft. bulk carrier SS "Sunrip."[3]   The 135 tons of aluminum resulted in a weight saving of 217 tons.

Modulus of Elasticity. The modulus of elasticity of aluminum is about one third that of steel. This can be an advantage or disadvantage depending upon the situation. If section deflection, as opposed to strength, is the most important consideration, a low modulus of elasticity is a limiting factor. On the other hand, a low modulus of elasticity is good for eliminating expansion joints on dock houses and similar circumstances. It also helps aluminum resist damage from collision impacts. Aluminum "gives" more than steel before permanent deflection is created.[3]

"Deflection is usually the restraining factor in aluminum design. Once the deflection requirements are met, the section modulus is normally ample. If equal deflection is required, and there are no corrosion allowances, aluminum will require a moment of inertia three times that of steel. This, of course, assumes the same span and loading.[5]

$$I_A = \frac{I_s E_s}{E_A} \ \text{in.}^4$$

where:

I_A = moment of inertia, aluminum, in.^{4}

I_s = moment of inertia, steel, in.^{4}

E_s = modulus of elasticity, steel, psi

E_A = modulus of elasticity, aluminum, psi."

Corrosion Resistance

Aluminum naturally forms a self-protecting corrosion-resistant oxide film on its surface.[11] This oxide film adhers tightly to the base metal and does not flake or chip easily as do the oxides formed by the reaction of other metals, such as steel, to oxygen. Moreover, the aluminum oxide film is self-renewing. If the film is pierced in any way, a new aluminum oxide film forms immediately to protect the base metal.

Tables 4-7, 4-8, and 4-9 show the corrosion resistance of 5083, 5086, and other aluminum alloys in marine environments. None of the materials shown here were protected in any manner. Resistance to corrosion was determined experimentally by the change in mechanical properties and by measuring the depth of individual pits on test panels after prolonged periods of exposure. Tests have shown that most aluminum alloys in sea-water will undergo localized pitting to an average depth of two or three mils in one or two years. With longer exposure, corrosion continues, but the rate of increase in depth diminishes with time. This has been referred to as the "self-stopping" nature of corrosion on aluminum and is considered to be due to the formation of protective corrosion products over the small pits.[11]

Galvanic Corrosion. Although aluminum is virtually indestructable in salt water or in the marine environment in general, galvanic corrosion may occur when aluminum alloys are combined with other metals in the presence of an electrolyte. Some attacks occur when aluminum is combined with most shipbuilding metals unless the joints are protected adequately.[4]

TABLE 4-7   CORROSION RESISTANCE OF UNPROTECTED ALUMINUM
ALLOYS IN SEAWATER

DAYTONA BEACH, FLORIDA[11]

ALLOY & TEMPER	EXPOSURE PERIOD, YRS.	MAXIMUM MEASURED PIT DEPTHS, MILS	PER CENT CHANGE IN TS
3003-H14	8	7.0	-1
Alclad 3004-H18	8	2.5*	0
5050-H34	8	12.0	-3
5052-H34	8	10.5	-2
5052-H36	6	23.0	-2
5086-H34	6	34.0	0
6061-T4	8	14.0	-8

*Depth of pitting confined to the cladding.

TABLE 4-8   RESISTANCE OF ALUMINUM ALLOY WELDMENTS TO CORROSION IN SEAWATER, FIVE YEARS EXPOSURE[11]

ALLOY & TEMPER	TEST CONDITION*	Per Cent Change in Strength	
		TENSILE STRENGTH	YIELD STRENGTH
5083-H113	Beads Intact	-6	0
	Beads Removed	0	-2
	Non-Welded	-2	-14
5086-H32	Beads Intact	-2	0
	Beads Removed	-8	0
	Non-Welded	-2	0
5154-H34	Beads Intact	0	0
	Beads Removed	-2	-3
	Non-Welded	-2	0
5356-H112	Beads Intact	-2	0
	Beads Removed	-1	0
	Non-Welded	-2	-1

*MIG welded with 5356 filler alloy.

TABLE 4-9   CORROSION RESISTANCE TO TIDE RANGE SEAWATER IMMERSION, SEVEN YEARS [11]

ALLOY & TEMPER	LOCATION TENSILE SAMPLES	Per Cent Change in Strength	
		TENSILE STRENGTH	YIELD STRENGTH
5083-0	Totally Immersed	-2	0
	Water Line	-3	-2
	Splash Zone	-6	-2
5083-H34	Totally Immersed	0	0
	Water Line	0	0
	Splash Zone	-3	0
5086-0	Totally Immersed	-2	0
	Water Line	-2	0
	Splash Zone	-3	0
5086-H34	Totally Immersed	0	0
	Water Line	0	0
	Splash Zone	-1	0

The most harmful reactions occur when aluminum is combined with copper, brass, or bronze. Possible harm is risked when it is combined with mild steels. The most common joint is between the steel boundary base and the aluminum deckhouses. Insulating systems are used in such cases.[9]

Protection against Corrosion. Painting or coatings do little to protect aluminum against the ocean environment since it is little affected by sea-water. However, painting surfaces of different metals with zinc chromate paints retards galvanic corrosion. Or, as is more usually the case, the surfaces of the metals are separated with gaskets, washers, sleeves and insulating materials like neoprene, alumalastic, fairprene, pressite and micarta.[11]

Cost Considerations

Although the strength of aluminum is not a limiting design factor, the cost of attaining this strength may be limiting. As mentioned earlier, a widely used aluminum, such as the 5xxx series, has approximately the same strength capacity (63,000 psi) as commonly used steels (71,000 psi). However, on a weight basis, aluminum is much stronger. Consequently, the judgement on whether to use aluminum rests on economic factors--whether the advantages of using aluminum outweight the higher costs.[3]

Fracture Toughness

Aluminum which has a face-centered cubic atomic arrangement is ductile even at low temperatures. Aluminum alloys generally are less notch sensitive than many steels. However, there has been concern recently about fracture toughness, i.e. resistance against fracture, of aluminum alloys in the presence of a sharp notch. At present, only limited data are available concerning the fracture toughness of aluminum alloys in marine applications.

Figure 4-2 shows a summary of V-notch Charpy impact energy of aluminum alloys as a function of strength level.[10] The figure shows the following:

(1) For aluminum alloys, the impact energy does not change drastically with temperature as observed for steel.

(2) As the strength level of an aluminum alloy increases, the impact energy generally decreases.

(3) Compared with steel and titanium, impact energy for an aluminum alloy is considerably low.

Figure 4-3 shows the optimum material trend line (OMTL) for aluminum alloys.[10] The figure shows the following values:

V-notch Charpy impact energy, Cv, ft-lb

Drop-weight tear test energy, DT, ft-lb

Plane strain fracture toughness, Kzc, Ksi $\sqrt{in.}$

Fracture toughness decreases as the yield strength increases. Generally speaking, 5000 series alloys have the highest fracture toughness followed by 6000 series, 2000 series, and 7000 series.

4.4 Fabrication Considerations

Aluminum was not accepted early for use in making structures because builders were uncertain of its workability and were afraid that its cost and fabrication difficulties would prove prohibiting. Experience and new fabrication methods proved this to be untrue, and today, many builders quote less for aluminum

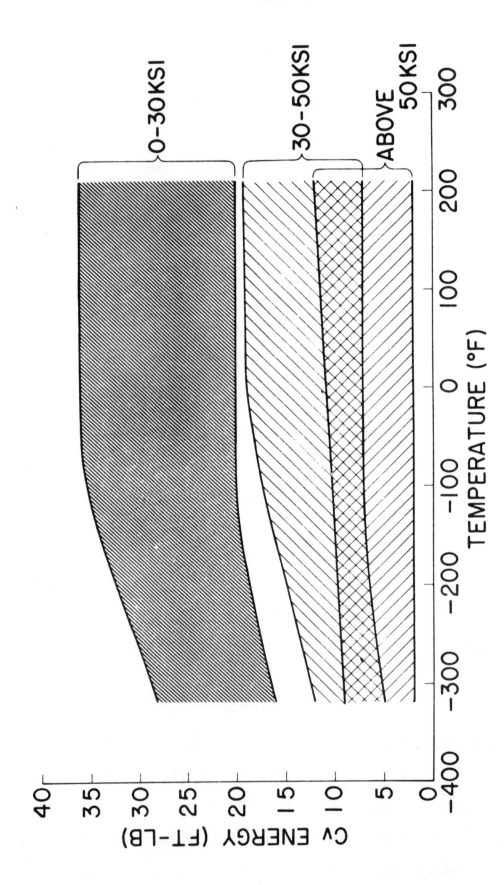

FIGURE 4-2   SUMMARY OF V-NOTCH CHARPY IMPACT ENERGIES OF ALUMINUM ALLOYS AS A
FUNCTION OF STRENGTH LEVEL[10]

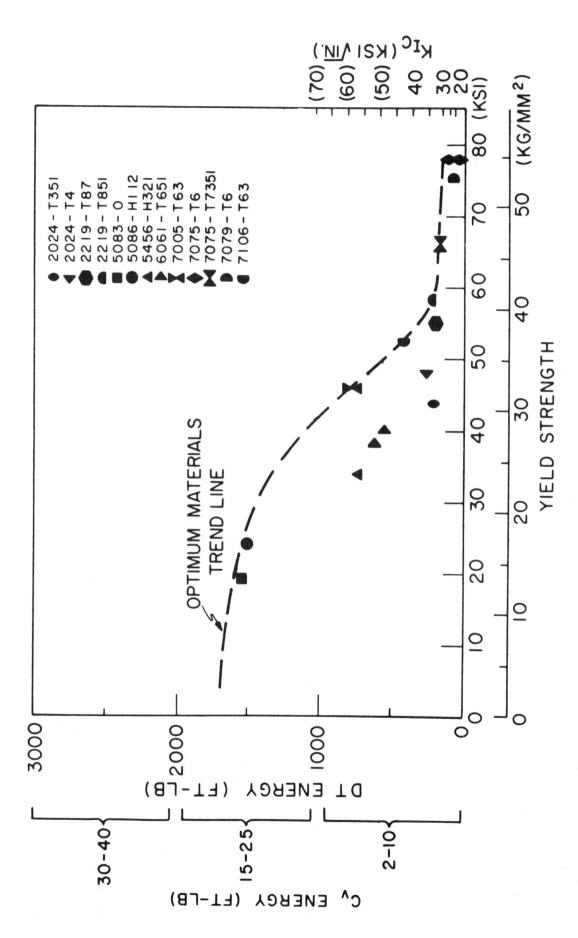

FIGURE 4-3 OMTL DIAGRAM FOR ALUMINUM ALLOYS. THE DIAGRAM PRESENTS RELATIONSHIPS
TO C_v, DT, AND K_{IC} SCALES. (10)

than the equivalent in steel construction (exclusive of material costs).[3]

Forming

Aluminum can be formed on the same equipment as steel, but some specific properties of aluminum need to be noted:

1. Aluminum has a spring-back characteristic, therefore parts should be overformed

2. The thermal expansion of aluminum should be considered. It is three times as conductive and two times as expansive as steel.[11]

Aluminum alloys can be flanged, bent, rolled, drawn, pressed, stretched, spun and expanded.[11]

Cutting

Aluminum can be cut relatively easily by saw. However, it cannot be cut with an oxyacetylene torch because it forms oxides which have melting points much higher than that of aluminum plate. Aluminum can be cut by a tungsten arc and a plasma arc.[17]

Welding

The advances in welding aluminum greatly affected its use as a construction material. Until suitable welding techniques were developed for aluminum, use of the material was greatly restricted. New processes allow high-speed welding with manual or automatic equipment which is standardized and readily available. Many times the power source for steel welding can be applied to aluminum welding.[3]

Two welding processes are most used with aluminum--gas tungsten arc welding (GTA) and gas metal arc welding (GMA).[11]

113

GTA. The process involves no flux, but is slow and not used extensively in marine structures. It is used on thickness down to .040 in. The usual commercial thickness is .062 in. to 1/4 in. If the material being welded is aluminum, a thickness greater than 1/4 in. must be preheated.[3]

GMA. The process is faster and cheaper and requires less manual dexterity that GTA. It involves minimum property reduction of adjacent materials and minimum distortion of the weld. It is used for thicknesses of .081 in. and higher. No preheating is required. This type of welding is most used for aluminum fabrication.[3] Table 4-10 presents recommended practices for welding of aluminum alloys.[11] Table 4-11 shows typical mechanical properties of MIG welded joints in aluminum alloys 5086 and 5083.[11]

Problems in Welding Aluminum. As described above, aluminum alloys are definitely weldable. However, this does not mean that no problems exist in welding aluminum. Major problems in welding aluminum include:

(1) Porosity in welds. Compared with steel, aluminum alloys are more active and thus are prone to weld porosity.

(2) Shrinkage and distortion. Aluminum alloys, compared with steel, have higher heat conductivity, larger thermal expansion coefficients, and a lower modulus of elasticity. Therefore, welds in aluminum alloys have more shrinkage and distortion.

(3) Loss of strength in the heat-affected zone. A reduction in the strength of the heat-affected zone has been experienced in weldments in aluminum alloys, especially

TABLE 4-10 RECOMMENDED PRACTICES FOR GMA WELDING OF ALUMINUM ALLOYS[11]

MATERIAL THICKNESS INCHES	WELDING POSITION	JOINT DESIGN	CURRENT AMPS-DC	ARC VOLTAGE	FILLER WIRE DIA. INCHES	ARGON* GAS FLOW CFH	NO. OF PASSES
3/32	Flat	None	70-110	18-22	0.030	30	1
	Flat	None	100-120	18-22	3/64	30	1
1/8	Flat	None	110-130	20	3/64	30	1
	Horiz.&vert.	None	100-120	20	3/64	30	1
	Overhead	None	100-120	20	3/64	40	1
1/4	Flat	None or single bevel	200-225	26-28	1/16	40	1
	Horiz.&vert.	Single bevel	170-190	26-28	1/16	45	2 or 3
	Over head	Single bevel	180-200	26-28	1/16	50	2 or 3
	Flat	None	220-250	28-30	1/16	80	2
3/8	Flat	Single or double bevel	230-320	26-28	1/16	50	1 or 2
	Horiz.&vert.	Single or double bevel	180-235	26-28	1/16	50	3
	Overhead	Single or double bevel	200-240	26-28	1/16	50	5
	Flat	None	260-280	28-30	1/16	80	2
1/2	Flat	Single or double bevel	280-340	26-30	3/32	50	2 or 3
	Horiz.&vert.	Single or double bevel	210-250	26-30	1/16	50	3 or 4
	Overhead	Single or double bevel	225-275	26-30	1/16	80	8 to 10
	Flat	None	280-320	30-32	1/16	80	2
1	Flat	Single or double bevel	320-420	26-30	3/32	60	4 to 5
	Horiz.&vert.	Single or double bevel	225-285	26-30	1/16	60	4 to 6
	Overhead	Single or double bevel	225-285	26-30	1/16	80	15 or more
	Flat	None	390-400	35-37	1/16	80	2
2	Flat	Single or double bevel	350-450	26-30	3/32	60	12 or more
3	Flat	Single or double bevel	350-450	26-30	3/32	60	20 or more

*Gas flows for helium are slightly higher than for argon.

TABLE 4-11   TYPICAL MECHANICAL PROPERTIES OF INERT-GAS METAL-ARC WELDED JOINTS IN ALUMINUM ALLOYS 5086 and 5083. (11)

ALLOYS, TEMPER AND GAGE	TENSILE STRENGTH psi	YIELD STRENGTH* psi	ELONGATION % in 2 in.	JOINT EFFICIENCY %**	LOCATION OF FRACTURE
Tested with beads in place					
5086,0, 1/4 in.	38,000	18,000	15.4	100	Parent plate, fusion line
5086,H112, 1/4 in.	38,000	19,000	14.0	100	Parent plate, fusion line
5086,H34, 1/4 in.	38,000	21,000	8.3	80	Parent plate, fusion line, weld metal
5086,H112, 1/2 in.	39,000	21,000	11.4	100	Fusion line
5086,H34, 1/2 in.	39,000	21,000	12.0	84	Fusion line
5086,H112, 3/4 in.	41,000	21,000	16.7	100	Fusion line, parent plate
Tested with beads machined off					
5086,0, 1/4 in.	35,000	17,000	12.5	100	Fusion line
5086,H112, 1/4 in.	37,000	17,000	14.3	94	Fusion line, weld metal
5085,H34, 1/4 in.	37,000	18,000	12.9	78	Fusion line, weld metal
5086,H112, 1/2 in.	39,000	20,000	16.5	100	Fusion line
5086,H112, 3/4 in.	39,000	20,000	16.8	100	Fusion line, weld metal
Tested with beads in place					
5083,0, 1/4 in.	43,000	20,000	16.2	100	Fusion line, parent plate
5086,H113, 1/4 in.	46,000	24,000	16.6	100	Fusion line
5083,H113, 1/2 in.	45,000	22,000	12.5	88	Fusion line
5083,H113, 3/4 in.	45,000	23,000	16.0	97	Fusion line
Tested with beads machined off					
5083,0, 1/4 in.	40,000	20,000	15.3	97	Weld metal
5083,H113, 1/4 in.	42,000	22,000	14.0	91	Weld metal
5083,H113, 1/2 in.	42,000	21,000	16.3	93	Weld metal
5083,H113, 3/4 in.	42,000	21,000	18.3	90	Weld metal

*At 0.2% offset. **Based on the typical tensile strength shown in the Kaiser Aluminum Sheet and Plate Book, Second Edition, 1958.

heat-treated alloys.

A recent RSIC report by Masubuchi[1] describes the results of extensive studies on welding aluminum supported by the National Aeronautics and Space Administration. The studies are primarily on alloys 2014 and 2219 which are used extensively in the fabrication of fuel and oxidizer tanks of the Saturn V space vehicle. Results of these studies should be useful for welding of aluminum alloys for ocean engineering.

Forging and Casting

Forging. Aluminum forgings are dependable, and also reduce costs and shorten machine time by eliminating dead weight, and by lowering sectional thickness. Forgings are a good aluminum for dynamic and heavy impact load conditions.[11]

Castings. Aluminum can be cast by various foundry processes which are less costly for aluminum than most materials. This is because aluminum has a low density, low melting temperature and handles easily. Aluminum cast components are formed by die casting, permanent mold casting, sand casting and specialized casting practices. Newly developed high-purity, high-strength casting alloys, 355, 356, and 357 give high physical properties in casting that are corrosion-resistant. Therefore, castings can compete with other fabrication processes, including forgings.[11]

REFERENCES

(1) Aluminum Association, "The Story of Aluminum," New York, New York.

(2) Aluminum Association, "Uses of Aluminum," New York, New York.

(3) Aluminum Company of America, Marine Sales Division, Aluminum Afloat, 1964.

(4) Aluminum Development Association, "Aluminum in Contact with Other Materials," ADA Information Bulletin 21, London, England, December 1955.

(5) Angermayer, K., "Structural Aluminum Design," Reynolds Metals Company.

(6) Brooks, C. L., "Aluminum Alloys and Their performance in Marine Environments," paper presented at the Symposium on Materials Applications in Oceanography, Pacific Symposium, Honolulu, Hawaii, July 2, 1968. Reprinted by Reynolds Metals Company.

(7) Brooks, C. L., "Aluminum Alloys for Pressure Hulls," American Society for Metals, Metals Engineering Quarterly, August, 1965, vol. 5, No. 3, p. 19.

(8) Holtyn, C. H., "Aluminum--The Age of Ships," a paper presented at the Annual Meeting of the Society of Naval Architects and Marine Engineers, New York, New York, pp. 356-391, November 10-11, 1966.

(9) Holtyn, C. H., "Big Barges Spearheaded Drive to Boost All-Aluminum Hulls in Heavy Marine Industry," Marine Engineering Log, February 1962.

(10) Judy, R. W. Jr., Goode, R. J., and Freed, C. N., "Fracture Toughness Characterization Procedures and Interpretations to Fracture-Safe Design for Structural Aluminum Alloys," NRL Report 6871, Naval Research Laboratory, March 1968.

(11)
    Leveau, C. W., "Aluminum and Its Use in Naval Craft, Parts
    I and II," Naval Engineers Journal, pp. 13-27, February 1965,
    and pp. 205 -219, April 1965.

(12)
    Little, R. S., "Introductory Notes on the Use of Aluminum
    Alloys for Vessel Structures," a paper presented at the
    Meeting of the Gulf Section, Society of Naval Architects
    and Marine Engineers, Houston, Texas, February 10, 1961.

(13)
    Masubuchi, K., "Integration of NASA-Sponsored Studies on
    Aluminum Welding," RSIC-670, Redstone Scientific Information
    Center, Redstone Arsenal, Alabama, September 1967.

(14)
    Metals Handbook, American Society for Metals, 1961.

(15)
    Muckle, W., "The Development of the Use of Aluminum in Ships,"
    Transactions, Eightieth Session of the North East Coast
    Institution of Engineers and Shipbuilders, Vol. 80, pp. 218-238,
    March 18, 1964.

(16)
    Rynewiez, J. F., Lockheed Missiles and Space Company, "Marine
    Corrosion Control for Deep Submergence Vehicles," Proceedings
    of the Air Force Materials Laboratory Fifthieth Anniversary
    Technical Conference on Corrosion of Military and Aerospace
    Equipment, Denver, Colorado, 23-25 May 1967, AFML-TR-329.

(17)
    Welding Handbook, Section IV, Fourth Edition, American
    Welding Society, 1960.

(18)
    Willner, A. R., "Aluminum Alloys," NRL Report 6167, "Status
    and Projections of Developments in Hull Structural Materials
    for Deep Ocean Vehicles and Fixed Bottom Installations,"
    U. S. Naval Research Laboratory, pp. 42-43, November 1964.

CHAPTER 5   TITANIUM AND ITS ALLOYS

5.1 Development of Titanium Alloys and Their Applications

Titanium is a relatively new metal that is just emerging from a lengthy period of development for aerospace. The use of titanium for marine applications is promising because titanium combines many of the needs of ocean materials: it is light-weight, has excellent corrosion resistance and potentially high strength.

Titanium and its Development

"Titanium comprises about 0.6% of the earth's crust. It is the fourth most plentiful structural metal; only iron, aluminum and magnesium occur in greater abundance. It is esti-mated that there is more titanium available in the earth than the combined amounts of chromium, copper, nickel, lead and zinc. Titanium is found in two important ores, ilmenite (TiO_2-FeO) and rutile (TiO_2). Significant ore bodies are located in Quebec (Canada), Virginia, New York, Wyoming, Florida, North Carolina and California. This natural abundance assures interest in and exploitation of titanium for many years to come. From the military viewpoint, it is a strategic metal that is independent of foreign sources."[14]

The density of titanium is approximately 60% that of plain carbon, alloy or stainless steels, and one and a half times that of aluminum. It is a silvery colored metal with a melting point of approximately 3035° F. [14]

In 1938, W. J. Kroll discovered magnesium reduction of $TiCl_4$ which produced titanium sponge suitable for melting. In 1946, the U. S. Bureau of Mines found the commercial practicalities of Kroll's process. And, later, the federal government became interested in the metal, partly because of its strength potential, and supported the development of titanium and its alloys with funds totalling $165 million.[8]

Applications of Titanium

Aerospace Applications. Titanium-based alloys were first used for aerospace application in 1957 when the builders of the Atlas Missile found that titanium vessels weighed 50 lbs., compared with 80 lbs. for aluminum and 125 lbs. for steel.[4] The suitability of its use in both aircraft and space structures is titanium's high strength to weight ratio, good corrosion resistance to aircraft structure environments and moderate modulus of elasticity. The government originally provided incentives in output and research in alloy development.

Today titanium alloys are used for various parts of aircraft structures and jet engines. Titanium alloys are planned for future use in major structural members of the supersonic transport (SST).

Marine Applications. It was not until 1963 that an active submarine hull plate program was started. "The immediate requirements for deep submergence hull materials and the then predicted growth of the marine market accelerated development and supported a rapid expansion of trial applications within the marine area."[4] The Alvin, a submersible diving to 6,500 feet, uses buoyancy spheres constructed from titanium. The <u>Autec I</u>, a research submersible, also uses titanium in the buoyancy sphere.[2]

The use of titanium alloys for marine application is in the stage of infancy. Their uses will probably grow **signifi-** cantly as the price of mill products drops. On the basis of characteristics of titanium alloys, general areas where titanium alloys are likely to be used include:

(1) A structure in which strength to weight ratio is critical. Examples include pressure hulls of deep submersibles and structural members of high-speed surface ships such as hydrofoil crafts and hydro-skimmers.

(2) A corrosion critical machine surface which cannot be painted and no corrosion can be tolerated, such as the ball valve for sea water use.

(3) A structure or a machine part which must have high corrosion-fatigue strength. Examples are propeller blades and a mast structure of a high-speed boat.[4]

Titanium alloys also may be used for machine parts or structural members where the following characteristics are important:

(4) Non-magnetic properties.

(5) Resistance to hot brine solutions and engine exhaust fumes.

(6) Cavitation resistance to high velocity sea water.[4]

5.2 Metallurgy of Titanium and Its Alloys

Pure Titanium

Pure titanium at room temperatures is in a hexagonal, close-packed form. At about 1620° F., allotropic transformation occurs and titanium takes on a body-centered cubic lattice. Only a 0.1% volume change takes place during this transformation. The low temperature phase of this process is designated alpha titanium. The high temperature phase is beta titanium.[14]

Pure titanium is not heat treated since the change from high temperature beta to the low temperature alpha on cooling from 1620° F. cannot be suppressed by very high cooling rates. However, the temperature at which transformation takes place may be decreased slightly by extremely high cooling rates. Different cooling rates result in different internal structure of the alpha phase. Rapid cooling gives an "acicular" structure which appears in a form similar to the martensitic steel structure. Slow cooling results in "equiaxed" alpha which appears to be like the very low-carbon annealed steels.[14]

Titanium Alloys

The addition of alloying elements to titanium either raises or lowers the transformation temperature and slows down or speeds up the transformation from the high-temperature beta phase to the room-temperature alpha phase. Titanium alloys are classified into alpha, alpha-beta, and beta types, depending upon the predominant phases in microstructure.[8] Furthermore, the type of structure that exists at room temperature determines if the titanium alloy can be heat treated and if it is sensitive to thermal embrittlement, which hinders weldability.[16]

An addition to titanium of an element such as aluminum, tin or oxygen raises the transformation temperature. These elements are called "alpha stabilizers." On cooling from a high temperature, the alloy with the above additions reaches the transformation stage sooner than pure titanium; therefore the alpha phase is formed faster.[15] An alloy with alpha stabilizers is referred to as a "non-heat treatable" alloy, since no further basic change takes place by subsequent heating in the low-temperature-phase region.

Elements such as iron, chromium, and vanadium, which lower the transition temperature, are called "beta stabilizers." Thus at a given cooling rate the transformation is retarded in time and takes place at a lower temperature than that of pure titanium. As a result, it is possible for the transformation at lower temperature to become much slower, permitting intermediate microstructures of alpha to be produced. These forms of alpha have different properties from the alpha formed at high temperatures. When added in sufficient amounts, certain beta-stabilizing elements will cause retention of the beta-phase indefinitely at room temperature. These alloys are metastable beta alloys at room temperature.

The relative amounts of alpha and beta stabilizers in an alloy (and the heat treatment) determine whether its microstructure is predominantly one-phase alpha, a mixture of alpha and beta, or the single-phase beta, over its useful temperature range.

Properties are directly related to microstructure. Single-phase alloys are usually weldable with good ductility. Some two-phase alloys are also weldable, but their welds are less ductile. Two-phase alpha-beta alloys are stronger than the one-phase alpha alloys, primarily because body-centered cubic beta is stronger than the close-packed hexagonal alpha. More important, two-phase alloys can be strengthened by heat treatments, because of the microstructure, which can be manipulated by controlling heating, quenching, and aging cycles.

Alpha Alloys. Alpha alloys usually have some beta-stabilizing alloy elements, which are balanced by a high aluminum content. Main features of alpha alloys are: 1) good weldability and (2) retention of their structures at high temperatures (due to the action of aluminum in the alloy).[8] Alpha alloys are insensitive to heat treatments and weldable except for alloys with over 12 atom per cent aluminum and tin combined with the interstitial elements oxygen and nitrogen.[15] Also, if the alloy contains more than 6% aluminum, hot working becomes hard; beta-stabilizers in small amounts may correct this.[8]

Alpha-beta Alloys. These alloys have enough beta-stabilizing elements to cause the beta phase to continue down to room temperatures. They are stronger than alpha alloys. The beta phase, strengthened by beta alloying addition in solution, is stronger than the alpha phase. If the alpha phase is strengthened by aluminum, then alpha-beta alloys are even stronger. These alloys can be strengthened about 35% with heat treatments which include quenching and aging phases.

Beta Alloys. These alloys are weldable in annealed and
heat treated conditions. Unlike alpha alloys, beta alloys
can be strengthened by heat treatments. Strength to 215,000
psi with 5% elongation is possible after heat treatments.
This is a 50% increase over annealed strength.[8]

5.3 Chemical Composition and Mechanical Properties
of Titanium and Its Alloys

Titanium Alloys and Their Properties

At present there are about 25 different titanium alloys
in commercial production. Table 5.1 contains a list of
these alloys with the appropriate specifications and mechanical
properties.[14] In some instances, cryongenic properties and
elevated temperature properties are given. Alloys for which
these properties are given are those which can be used
advantageously at the respective temperatures. The table also
notes the welding capabilities of particular alloys.

Titanium alloys which have been used and which may be
used for marine applications include:[4,14]

1. Commercially pure titanium of different grades
   depending upon differences in interstitial elements.
   For example, a 99.5% pure titanium contains the
   following elements:

0.1%	–	maximum oxygen
0.1%	–	maximum nitrogen
0.07%	–	maximum carbon
0.2%	–	maximum iron

   The ultimate tensile strength ranges from 38,000 psi

126

(a) TABLE 5-1   TITANIUM ALLOYS IN COMMERCIAL PRODUCTION

ALPHA-BETA ALLOY GRADES

Alloy	Weldability	Form	Condition	1	2	3	4	5	6	7	8	9	10	11	12	13
8Mn	W not recommended	S	Ann	16.4	137	125	15	14.4	98	75	13	800	80	59	15	
2Fe-2Cr-2Mo	W not recommended	b	Ann	16.7	137	125	18	14.7	95	65	19	800	75	55	30	12-15
		b	Aged	....	179	171	13	....	136	112	16					8-10
2.5Al-16V	W not recommended	S	SHT	....	105	45	16	13.5	155	140	8	800	140	125	10	
		S	Aged	15.0	180	165	6									
3Al-2.5V		S	Ann	15.5	100	85	20	13.0	70	50	25					
4Al-4Mn	W not recommended	b	Ann	16.4	148	135	15	13.9	110	90	17	800	100	85	21	10-15
		b	Aged	....	162	143	10	....	125	100	11					
4Al-3Mo-1V	Special conditions permit some W	S	Ann	16.5	140	120	15	14.0	152	120	.7	800	145	115	8	
		S	Aged	....	195	167	6	....								
5Al-1.25Fe-2.75Cr	Special conditions permit some W	b	Ann	16.8	155	145	15	15.5	122	102	20					10-15
		b	Aged	17.6	190	175	6	16.2	144	117	10					
5Al-1.5Fe-1.4Cr-1.2Mo	W not recommended	b	Ann	16.5	154	145	16	15.0	115	100	16	800	118	100	20	10
		b	Aged	17.0	195	184	9	14.6	150	125	14					
6Al-4V	Weldable	S, b	Ann	16.5	138	128	12	13.5	105	95	11	800	90	78	18	10-20
		S	Aged		170	155	8		130	105	7	800	130	100	8	
6Al-4V (low O)	Weldable	S	Ann	16.5	135	127	15	13.5	105	95	12	-320	220	205	13	10
6Al-6V-2Sn-1(Fe,Cu)	Special conditions permit some W	b	Ann	15.0	165	150	15	13.4	132	117	20	800	90	80	15	15
		b	Aged	16.5	190	180	10	14.5	150	132	15					
7Al-4Mo	Special conditions permit some W	b	Ann	16.2	160	150	16	14.2	127	108	18	800	117	94	18	18
		b	Aged	16.9	185	175	10	15.0	150	123	12					10

BETA ALLOY GRADES

Alloy	Weldability	Form	Condition	1	2	3	4	5	6	7	8	9	10	11	12	13
1Al-8V-5 Fe	W not recommended	b	Ann	16.5	177	170	8	14.7	128	115	19	800	108	85	32	
		b	Aged	16.5	221	215	10	14.5	140	123	12	800	120	100	30	
3Al-13V-11Cr	Weldable	S	Ann	14.2	135	130	16	13.2	175	145	-8	800	115	100	18	8
		S	Aged	14.8	185	175	8	13.8				800	160	120	12	
		S	CR+Aged	....	260	245	4									

(a) Other numbers T-12117 and WA-PD-76C (1) apply to all grades and all products; T-14557, T-14558, T-9046C, and T-9047C apply to all grades; and T-8884(ASG) applies to various grades
(b) Formerly Al-2Cb-1Ta. All data fgiven are for the 8-2-1 composition
(c) B-billet, b—bar, P—plate, S—sheet, s—strip. T—tubing, W—wire, E—extrusions
(d) AC—air cool, SC—slow cool, FC—furnace cool, WQ—water quench, CR½cold rolled.

(b) TABLE 5-1 TITANIUM ALLOYS IN COMMERCIAL PRODUCTION

Nominal Composition, percent	Other Designations		Forms Available[c]	Recommended Heat Treatments[d]		
	AMS No.	Military No.[a]		Stress-Relief Annealing	Annealing Treatment	Solution Treatment (Aging Treatment)
99.5			B,b,P,S,s,T,W,E	1000 to 1100 F, ½ hr, AC	1250 to 1300 F, 2 hr, AC	Not heat treatable
99.2	{4902 4941 4951}	T-9047B-1	B,b,P,S,s,T,W,E	1000 to 1100 F, ½ hr, AC	1250 to 1300 F, 2 hr, AC	Not heat treatable
99.0	4900A	T-7993B	B,b,P,S,s,T,W,E	1000 to 1100 F, ½ hr, AC	1250 to 1300 F, 2 hr, AC	Not heat treatable
99.0	{4901B		B,b,P,S,s,T,W,E	1000 to 1100 F, ½ hr, AC	1250 to 1300 F, 2 hr, AC	
98.9	4921		B,b, W,E	1000 to 1100 F, ½ hr, AC	1250 to 1300 F, 2 hr, AC	Not heat treatable
0.15 to 0.20Pd (Balance Ti)			B,b,P,S,s,T,W,E	1000 to 1100 F, ½ hr, AC	1250 to 1300 F, 2 hr, AC	Not heat treatable

ALPHA ALLOY GRADES

Nominal Composition, percent	AMS No.	Military No.[a]	Forms Available[c]	Stress-Relief Annealing	Annealing Treatment	Solution Treatment (Aging Treatment)
5Al-2.5Sn	{4910 4926 4953 4966}		B,b,P,S,s, W,E	1000 to 1200 F, ¼ to 2 hr, AC	1325 to 1550 F, 10 min to 4 hr, AC	Not heat treatable
5Al-2.5Sn (low O)			B,b,P,S,s, W,E	1000 to 1200 F, ¼ to 2 hr, AC	Same	Not heat treatable
5Al-5Sn-5Zr		In preparation	B,b,P,S,s	1100 F, ½ hr, AC	1650 F, 4 hr, AC	Not heat treatable
7Al-12Zr		In preparation	B,b,P,S,s	1000 F, ½ hr, AC	(1)[b]1600-1650 F, ½ to 4 hr, AC (2) 1300 F, 1 hr, AC	Not heat treatable
7Al-2Cb-1Ta[b]		In preparation	B,b,P,S, W,E	1100 to 1200 F, ½ hr, AC	1650 F, 1 hr, AC	
8Al-1Mo-1V	In preparation	In preparation	B,b,P,S,s, W,E	1100 to 1200 F, 1 hr, AC For sheet and plate — For forgings	(1) 1450 F, 8 hr, FC (Consult producers for other treatments) (2) 1450 F, 8 hr, FC + 1450 F, ¼ hr, AC (Duplex) (3) 1450 F, 8 hr, FC + 1850 F, 5 min; AC + 1375 F, ¼ hr, AC (Triplex) (4) 1850 F, 1 hr, AC + 1100 F, 8 hr, AC	

(c) TABLE 5-1 TITANIUM ALLOYS IN COMMERCIAL PRODUCTION

ALPHA-BETA ALLOY GRADES

Alloy	Spec. No.	Forms	Annealing	Stress-relief annealing	Solution treating and aging
8Mn	4908A	P,S	900 to 1100 F, ½ to 2 hr, AC	1250 to 1300 F, 1 hr, F C to 1000 F	Solution treatment not recommended
2Fe-2Cr-2Mo	4923	B,b, S,s	900 to 1000 F, ½ to 1 hr, AC	1200 F, ½ hr, AC	1400 to 1480 F, 1 hr, WO or AC (900 to 950 F, 2 to 8 hr, AC)
2.5Al-16V	T-8884(1)	B,b,P,S,s, W			1360 to 1400 F, 10 to 30 min, WQ (960 to 990 F, 4 hr, AC)
8Al-2.5V		s,T		1300 F, 1 hr, AC	Solution treatment not recommended
4Al-4Mn	4925A	B,b,P, W	1300 F, 2 hr, FC	1300 F, 2 to 4 hr, FC	1400 to 1500 F, ½ to 2 hr, WQ (800 to 1000 F, 8 to 24 hr, AC)
4Al-3Mo-1V	4912 4913	P,S,s	1000 to 1100 F, 1 hr, AC	1225 F, 4 hr, SC to 1050 F, AC	1625 to 1650 F, ¼ hr, WQ (925 F, 8 to 12 hr, AC)
5Al-1.25Fe-2.75Cr		B,b,P,S,	1100 F, 1 hr, AC	1450 F, 1 hr, SC to 1050 F, AC	1350 to 1500 F, 2 hr, WQ (900 to 950 F, 5 to 6 hr, AC)
5Al-1.5Fe-1.4Cr-1.2Mo	{4929 {4969	B,b,P	1200 F, 2 hr, AC	1200 F, 4 to 24 hr, AC	1600 to 1625 F, 1 hr, WQ (1000 F, 24 hr, AC)
6Al-4V	{4911 {4928A {4935 OS-10737 OS-10740	B,b,P,S,s, W,E	900 to 1200 F, 1 to 4 hr, AC (Usual: 1 hr, 1100 F, AC)	1300 to 1550 F, 1 to 8 hr, SC to 1050 F, AC	1550 to 1750 F, 5 min to 1 hr, WQ (900 to 1000 F, 4 to 8 hr, AC)
6Al-4V (low O)		B,b,P,S,s,T,W,E	Same	Same	Solution treatment not recommended
6Al-6V-2Sn-1(Fe,Cu)	T-46035 T-46038	B,b,P, W,E	1100 F, 2 hr, AC	1300 to 1400 F, 1 to 2 hr, AC	1600 to 1675 F, 1 hr, WQ (900 to 1100 F, 4 to 8 hr, AC)
7Al-4Mo		B,b,P, W,E	900 to 1300 F, 1 to 8 hr, AC	1450 F, 1 to 8 hr. SC to 1050 F, AC	1650 to 1750 F, ½ to 1½ hr, WQ (900 to 1200 F, 4 to 16 hr, AC)

BETA ALLOY GRADES

Alloy	Spec. No.	Forms	Annealing	Stress-relief annealing	Solution treating and aging
1Al-8V-5Fe		B,b,P	1000 to 1100 F, 1 hr, AC	1250 F. 1 hr, FC to 900 F, AC	1375 to 1425 F, 1 hr, WQ (925 F, 1000 F, 2 hr, AC)
3Al-13V-11Cr	4917	B,b,P,S,s, W			1400 to 1500 F, ¼ to 1 hr, WQ or AC (900 F, 2 to 96 hr, AC) 1450 F, ½ hr, AC + CR + 800 F, 24 hr, AC

(a) Other numbers T-12117 and WA-PD-76C (1) apply to all grades and all products; T-14557, T-14558, T-9046C, and T-9047C apply to all grades; and T-8884 (ASG) applies to various grades.
(b) Formerly Al-2Cb-1Ta. All data given are for the 8-2-1 composition.
(c) B—billet, b—bar, P—plate, S—sheet, s—strip, T—tubing, W—wire, E—extrusions.
(d) AC—air cool, SC—slow cool, FC—furnace cool, WQ—water quench, CR—cold rolled.

(d) TABLE 5-1 TITANIUM ALLOYS IN COMMERCIAL PRODUCTION

				Typical Tensile Properties												
Nominal Composition, per cent	Weldability Remarks	Form	Condition	Room Temperature				600 F				Extreme Temperatures				RT Charpy V Impact, ft-lb
				E, 10^6 psi	US, ksi	YS, ksi	EL, %	E, 10^6 psi	US, ksi	YS, ksi	EL, %	Test Temp, F	US, ksi	YS, ksi	EL, %	
99.5	All unalloyed grades are completely weldable (W)	S	Ann	14.9	38	27	30	12.1	20	10	50					
99.2		S	Ann	14.9	60	45	28	12.3	28	13	45	−423	175	...	...	25–40
99.0		S	Ann	15.0	75	60	25	12.5	33	19	33	−321	165	...	...	20–35
99.0 / 98.9		S / S	Ann / Ann	15.1 / 15.5	90 / 100	75 / 85	20 / 17	12.5 / 12.6	43 / 47	27 / 30	28 / 25	−321	175	...	...	11–15
0.15 to 0.20Pd (Balance Ti)	Completely W	S	Ann	14.9	62	46	27	12.3	28	13	30					

ALPHA ALLOY GRADES

Nominal Composition, per cent	Weldability Remarks	Form	Condition	E, 10^6 psi	US, ksi	YS, ksi	EL, %	E, 10^6 psi	US, ksi	YS, ksi	EL, %	Test Temp, F	US, ksi	YS, ksi	EL, %	RT Charpy V Impact, ft-lb
5Al-2.5Sn	Weldable	S / b	Ann / Ann	16.0 / 16.0	125 / 115	117 / 110	18 / 20	13.4	82	65	19	1000	75	56	18	19
5Al-2.5Sn (low O)	Weldable	S	Ann	16.0	110	95	20	13.4	78	60	20	−423	229	206	15	19
5Al-5Sn-5Zr	Weldable	S	Ann	16.0	125	120	18	14.2	94	74	20	1000	84	67	21	
7Al-12Zr	Weldable	S / b	(1) Ann / (2) Ann	16.0 /	135 / 165	130 / 159	15 / 14	14.3 /	109 / 130	86 / 119	21 / 18	1000	93	75	23	
7Al-2Cb-1Ta[b]	Weldable	b	Ann	17.7	126	120	17	15.1	100	81	25					
8Al-1Mo-1V	Weldable	S / S / S	(1) Ann / (2) Ann / (3) Ann	18.5 / 18.0 /	160 / 145 / 150	150 / 138 / 142	18 / 15 / 13					1000	85	70	20	
		b	(4) Ann	.....	141	130	18	.....	107	85	19	1000	88	71	20	

for 99.5% pure titanium to 100,000 psi for 98.9%
pure titanium. All grades have an alpha structure.

2. The Ti-7Al-2Cb-1Ta alpha alloy of about 125,000 psi
   ultimate tensile strength. The designation indicates
   that the principal alloying elements are 7% aluminum,
   2% columbium, and 1% tantalum.

3. The Ti-5Al-2.5Sn alpha alloy of about 120,000 psi
   ultimate tensile strength. The designation indicates
   that the principal alloying elements are 5% aluminum
   and 2.5% tin.

4. The Ti-6Al-4V heat treatable alpha-beta alloy. This
   has an ultimate tensile strength of 130,000 psi in
   the annealed condition and up to 160,000 psi in the
   heat treated condition. The designation indicates
   that the principal alloying elements are 6% aluminum
   and 4% vanadium.

5. The Ti-6Al-6V-2Sn heat treatable alpha-beta alloy
   with a tensile strength of 165,000 psi (annealed) and
   up to 190,000 psi (aged). The designation indicates
   that the principal alloying elements are 6% aluminum,
   6% vanadium, and 2% tin. Among these alloys, pure
   titanium and the Ti-6Al-4V alloy have been most
   commonly used for marine applications.

As an example of chemical compositions of actual titanium
alloys, Table 5-2 presents chemical compositions of some plates
one inch thick. These plates were used by Huber and others
for their experiments on fracture toughness and stress corrosion
cracking.[9]

TABLE 5-2 CHEMICAL COMPOSITION OF SOME 1 IN. THICK TITANIUM ALLOY PLATES (8)

ALLOY	CODE	Al	Mo	Sn	V	Fe	C	N	O	H	Other	Remarks
5Al-2.5Sn	T-18	5.18	...	2.56	...	0.08	0.03	0.013	0.07	0.008	...	...
6Al-4Sn-1V	T-20	6.0	...	3.90	1.0	0.024	0.024	0.01	0.062	0.003	...	...
6Al-2Mo	T-22	5.6	2.2	...	...	0.06	0.028	0.015	0.062	0.004	...	...
6Al-4V	T-27	5.85	...	...	3.85	0.07	0.042	0.06	0.06	0.005	...	...
6.5Al-5zr-1V	T-36	6.3	...	...	1.0	0.06	0.02	0.006	0.06	0.007	4.8Zr	...
6Al-2Sn-1Mo-1V	T-37	5.8	1.1	2.0	1.0	0.06	0.02	0.008	0.07	0.005– 0.010	...	...
7Al-2Cb-1Ta	T-39	Not Available										
7Al-2.5Mo	T-71	6.8	2.4	...	...	0.04	0.023	0.008	0.07	0.004	...	...
7Al-1Mo-1V	T-88	6.9	1.0	...	0.9	0.04	0.022	0.010	0.07	0.007	...	...
5Al-2Sn-2Mo-2V	T-90	4.9	2.1	2.2	1.9	0.06	0.022	0.009	0.06	0.012	...	Circular rolled
6Al-4V	T-91	6.0	...	...	4.0	0.05	0.023	0.009	0.07	0.007	...	Circular rolled
6Al-6V-2Sn	T-92	5.6	...	...	5.3	0.52	0.022	0.01	0.06	0.012	0.92 Cu	Circular rolled
6Al-3V-1Mo	T-93	5.9	1.0	...	3.1	0.05	0.022	0.007	0.07	0.006	...	Circular rolled
7Al-2.5Mo	T-94	6.9	2.6	...	...	0.04	0.023	0.008	0.073	0.006	...	Circular rolled
6Al-4V	T-95	5.9	...	...	4.0	0.01	0.022	0.013	0.114	0.005	...	...
6Al-2Cb-1Ta-0.8Mo	T-96	6.0	0.75	...	...	0.07	0.02	0.005	0.005	0.004	...	...
6Al-4V	T-100	5.7	...	...	4.0	0.07	0.022	0.01	0.07	0.007	...	...

Effects of Impurities and Alloying Elements
on Mechanical Properties

Pure titanium is very ductile and relatively low in strength. A small addition of impurities such as oxygen, nitrogen, and carbon causes increase in strength and a decrease in ductility. Other elements also can strengthen the metal with some reduction in ductility.

Impurities. The elements carbon, hydrogen, nitrogen, and oxygen form interstitial solid solutions with titanium. Figure 5.1 shows the individual effect of carbon, nitrogen, and oxygen on ultimate tensile strength and elongation of titanium. Hydrogen also causes an increase in strength and a reduction in ductility.

Alloying Elements. Figure 5-2 shows the individual effect of aluminum, tin, and zirconium, which are added to many titanium alloys, on the ultimate strength and elongation of titanium. These alpha alloys cannot be strengthened by heat treatment. Therefore, strengthening can be achieved either by increasing the alloy content or by cold working.

Fracture Toughness

Titanium and titanium alloys, like steel, are sensitive to a notch. Notch sensitivity can be evaluated by various impact tests (refer to Chapter 12 for general discussions of fracture toughness).

Figure 5-3 presents a summary of V-notch Charpy impact data for titanium alloys with different levels of yield strength. As the strength level increases, notch toughness

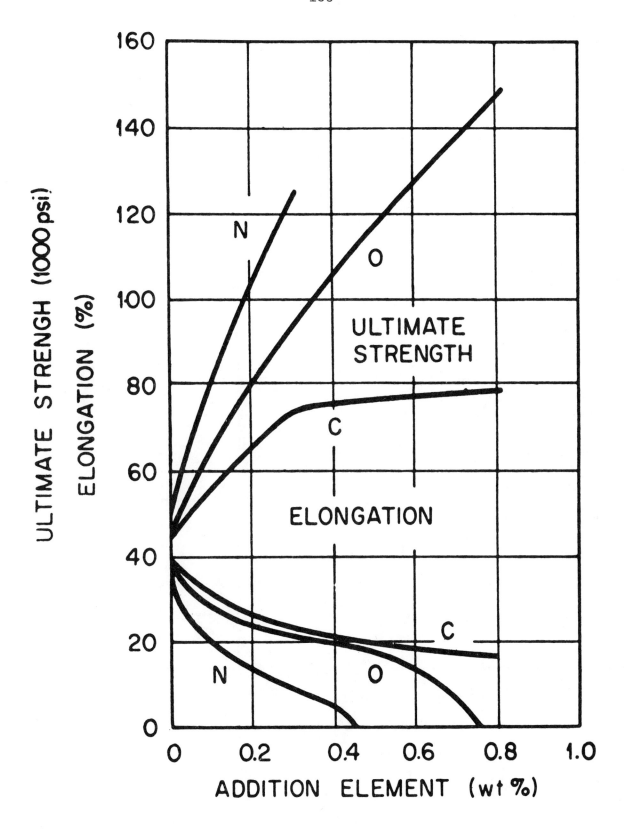

FIGURE 5-1   EFFECT OF IMPURITIES ON MECHANICAL PROPERTIES
OF TITANIUM[14]

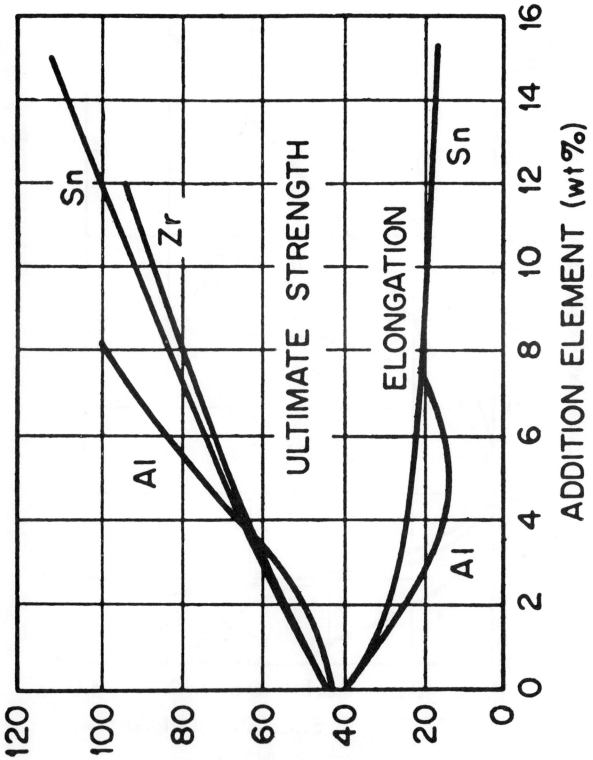

FIGURE 5-2 EFFECT OF ALLOY ADDITIONS ON MECHANICAL PROPERTIES OF ALPHA TITANIUM ALLOYS (14)

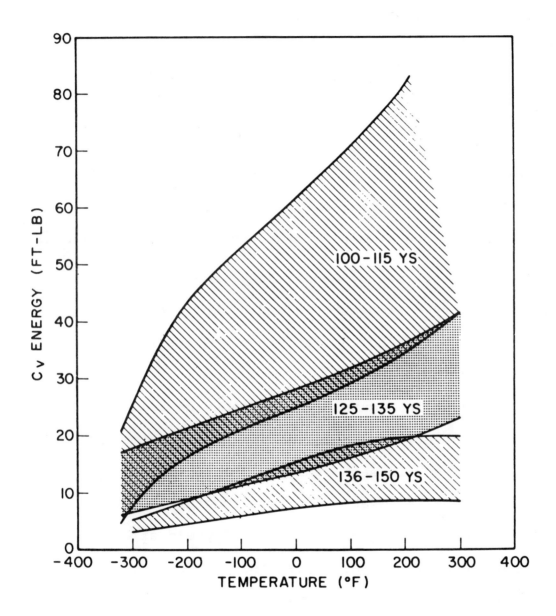

FIGURE 5-3   SUMMARY OF RELATIONSHIP BETWEEN V-NOTCH CHARPY
CURVES FOR TITANIUM ALLOYS AND DIFFERENT LEVEL
OF YIELD STRENGTH[5]

decreases. Compared to steels, titanium alloys do not show a sharp transition in fracture toughness with temperature changes and only a gradual change is noted over a relatively broad range of temperatures.

Optimum Material Trend Line (OMTL). Figure 5-4 shows the fracture toughness for all different generic families of titanium alloys. Here, fracture toughness is given in three different scales:

1. V-notch Charpy impact energy, Cv, ft-lb

2. Drop weight tear test energy, DT, ft-lb

3. Critical plane strain fracture toughness, K_{IC}, Ksi$\sqrt{in.}$

The results were obtained by Goode, Judy, and Huber of the Naval Research Laboratory.[5] Figures 5-5a and b show the DT test specimen and the single-notch fracture mechanics tensile specimen used in the NRL study.

The data shown in Figure 5-4 represents as-rolled material and a variety of processing and heat-treatment conditions for alloys in each family. The upper OMTL curve relates to the highest "weak" direction fracture toughness values determined for the related level of yield strength. The normal expectancy OMTL curve relates to the level of fracture toughness that can be expected with reasonable confidence if the chemistry, processing, and heat treatment are specified in the best way. The range of fracture toughness indicated for any given level of yield strength results from chemical composition, impurities, processing, and heat-treatment variables. Fracture toughness generally

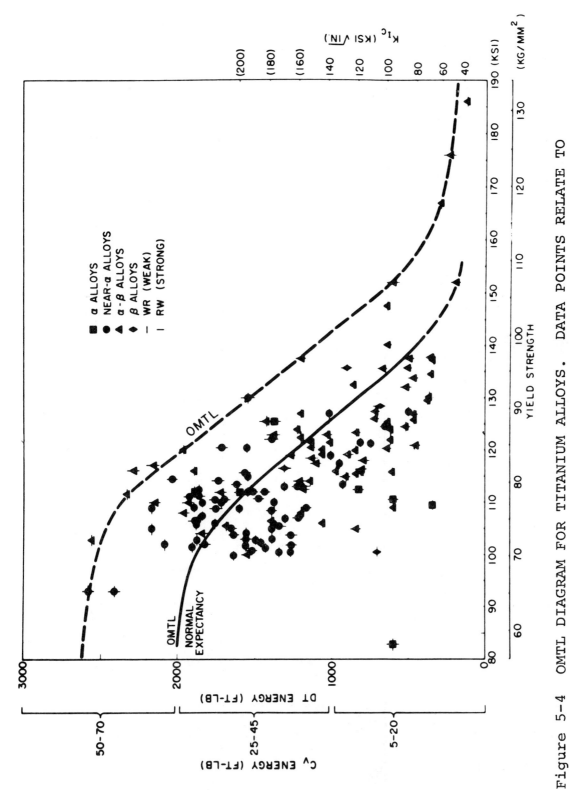

Figure 5-4   OMTL DIAGRAM FOR TITANIUM ALLOYS. DATA POINTS RELATE TO
1-INCH THICK DT TEST VALUES. THE OTHER FRACTURE TOUGHNESS
SCALES ARE INDEXED TO DT ENERGY BY CORRELATION.

138

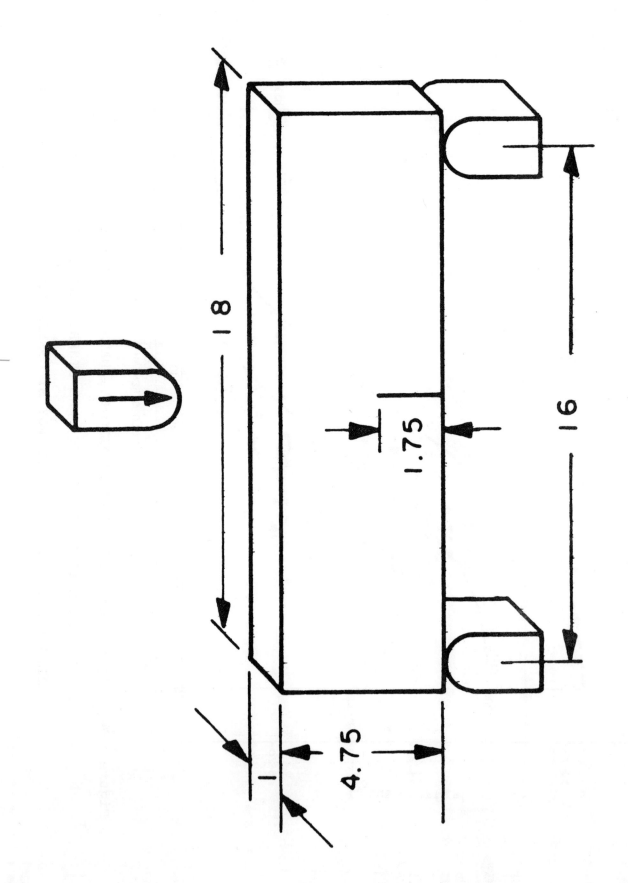

FIGURE 5-5 (a) DROP WEIGHT TEAR TEST SPECIMEN

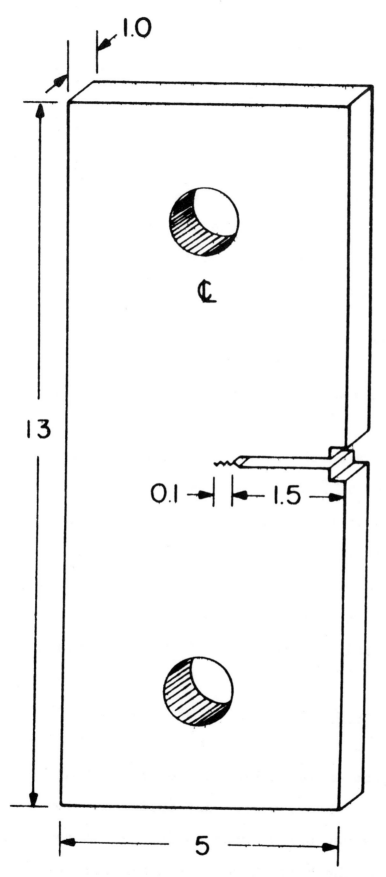

Figure 5-5 (b)   SINGLE-EDGE-NOTCH FRACTURE MECHANICS

TENSILE SPECIMEN

decreases as the yield strength increases.

Effects of Impurities. Interstitial impurities such as carbon, hydrogen, nitrogen, and oxygen are known to be detrimental to fracture toughness. Hydrogen is especially deterimental, particularly below 200°F. This is due to the precipitation of titanium hydride platelets in the alloy. A hydrogen level of 200 ppm can cause a significant loss of toughness. Consequently, commercial titanium and titanium alloys have a hydrogen level below 200 ppm.[14]

Section Size Effects

Section size effects on design properties can be estimated only through knowledge of the physical metallurgy of present alloys.

Alpha Alloys. With an increase in plate thickness in alpha alloys, the final product is subject to less working from ingot to plate than is the case in 1 in. plate. Probably increased thickness will result in a somewhat lower tensile strength without a decrease in toughness. Plates up to 4 in. thick may be subjected to the amount of working now given to 1 in. plate in the rolling operation if the roll is started with a thicker slab.[16]

Beta and Alpha-beta Alloys. Thermal cycling difficulties may curtail the use of this group of alloys when thick plates are formed. The problem is linked to the low thermal conductivity of titanium. When heavy plates are produced, slow cooling due to low heat loss may bring about a decrease in toughness through the aging of the beta phase. In a similar vein, welding of thick-plate alpha-beta alloys may require the weld metal and adjacent heat-affected zone to be at elevated temperatures for a relatively lengthy period of time due to the low thermal conductivity; this may lead to decreased toughness by aging of the beta. Quenching of the alloy prior to aging may be a difficulty in the event that heat-treatable, high-strength, alpha-beta or beta alloys find application as thick section.[16]

5.4 Corrosion and Stress Corrosion Cracking of Titanium Alloys

Corrosion Resistance

One of the great advantages of titanium is that it resists many types of attacks--sea water, animal life, high velocity, crevice corrosion, fatigue and temperature. Placed between Inconel and Monel in the galvanic series for various metals and alloys in sea water, titanium is cathodic to most of the other structural metals. Although titanium itself is not attacked when coupled with other materials, the corrosion of the more active metals is accelerated.[4]

Titanium has excellent corrosion resistance in elevated temperature water and hot brine. Such resistance is developed by the maintenance of a passive layer which also explains its good resistance to most environments. Effects of corrosion on titanium can be divided into three general categories divided easily by temperatures:

(1) Room temperature: environmental cracking and reduction of fatigue resistance.

(2) 250 - 450° F.: in solutions of hot brine--crevice corrosion or pitting corrosion.

(3) Above 500° F.: hot salt stress corrosion cracking.[4]

Environmental Cracking

The first line of protection in titanium against environmental cracking is the oxide film. If environmental cracking takes place, this film is broached and the restricted area of the crevice allows a shift in pH of the environment below 1.5. Chloride pitting in the crevice and a local cell reaction charging hydrogen into the lattice sustain the reaction only if sufficient stress is sustained to crack the hydride corrosion product extending the reacting surface area of the crack. Alloys most susceptible to this type of reaction are high aluminum alpha alloys where processing has precipitated a Ti_3Al compound and alloys which are prone to rapid hydrogen pickup and hydride precipitation.[4]

"In a practical design the concern of environmental cracking should be passed on the following check list:

1. The structure is fatigue load limited, i.e., above 20,000 psi under high cycle loading for all alloys, or above 60% of the tensile yield strength if low cycle loaded.

2. Plain strain conditions exist or could potentially exist in the structure. This condition is gage dependent, alloy dependent, and not configuration dependent. Surprisingly, in more marine structures a plain strain condition is not common except for massive welded devices of heavy gage.

3. If the above two conditions are controlling, the

following alloys are suggested:

a. Commercially pure titanium (Ti-50A)

b. Ti-6Al-4V (800 ppm oxygen maximum for heavy plate forgings and weldments and maximum toughness).

c. Ti-6Al-4V ELI (1200 ppm oxygen maximum for heavy plate forgings and weldments where higher strength is required).

d. Ti-6Al-4V (1600 ppm oxygen and sheet products, machined products, non-welded structures)."[4]

Crevice Corrosion

Crevice corrosion is a problem in heat exchanger equipment or engine exhaust fumes handling sea water where the operating temperature is over 250° F. Ti-50A has usually been used in the past. However, a new alloy, Ti-2Ni is promising in extending the temperature limit to 400° F. Elimination of the crevice, coupling titanium with nickel gearing alloys or nickel containing paint minimizes the probability of reactions. The use of Ti-2Ni seems to raise the threshold limit of the reaction to about 400° F.

Stress Corrosion Cracking

Hot stress corrosion cracking may be a problem in gas turbine engine components operating above 500° F. injecting hot sea air and having areas of stagnation where salt crystals can be retained on the surface. After a time, chloride pitting attack occurs under the salt particle and hydrogen is absorbed at the edge of the crystal.[4]

Corrosion Fatigue

Figure 5-6 shows data on Ti-6Al-4V presenting both

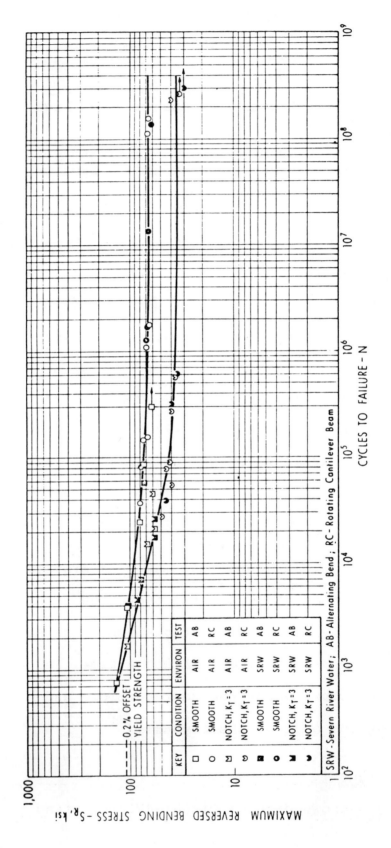

Figure 5-6 FLEXURAL FATIGUE CURVES FOR Ti-6A1-4V, 800 ppm OXYGEN MAX.

MEL REPORT 266/66 (4)

alternating plate bending and rotating cantilever beam specimens.
This work was performed by M. R. Gross of Naval Ships Research
and Development Center. Figure 5-7 illustrates plate bending
fatigue tests in which the rate of crack growth is measured
for both air and a 3.5% salt water environment. It is significant that there is no deterioration of fatigue resistance in
either test as the specimen is immersed in sea water.

Figures 5-8 and 5-9 compare various materials under the
same test condition to show the significant design advantage
of titanium over various steel alloys.[4]

"The use of fatigue data requires close interpretation
of testing technique best related to structure and the spectrum
of the loading cycle. In general, the more complex the hardware in structural shape fabrication, the lower the yield
strength and higher toughness required to develop adequate
service life. In the environment titanium provides one of
the highest levels of sustained or alternating strain in both
the area of crack initiation on smooth surfaces or corrosion
fatigue crack extension rates. Care must be taken in alloy
selection to provide adequate toughness to allow a sufficiently
large fatigue crack prior to fast fracture so that the crack
will be picked up during routine inspection of the components."[4]

5.5 Fabrication

Melting and Forming

The size of the final product is limited by the melting
capabilities. Wrought products can be worked from ingots on
steel-making equipment.[16] One of the largest ingots

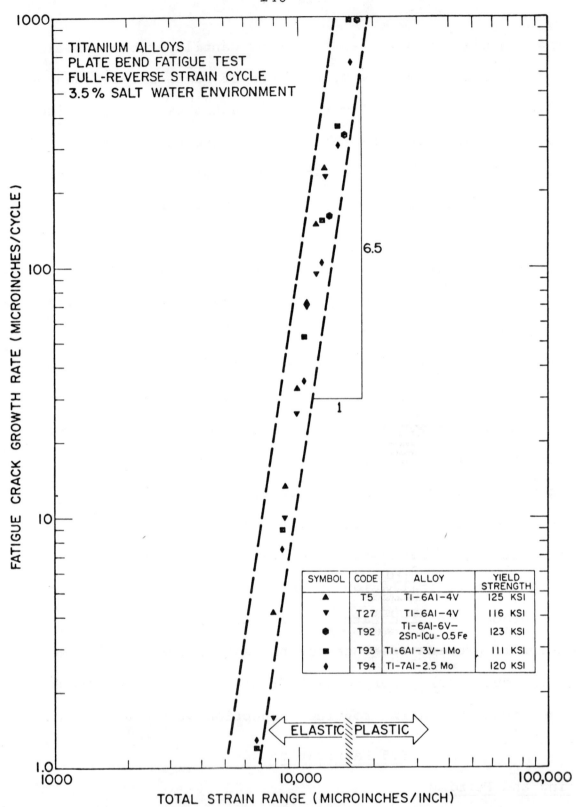

Figure 5-7   LOG-LOG PLOT OF FATIGUE CRACK GROWTH RATE VERSUS
TOTAL STRAIN RANGE FOR TITANIUM ALLOYS IN A 3.5%
SALT WATER ENVIRONMENT.   THE SCATTERBAND LIMITS
ARE REPRODUCED FROM THE AIR ENVIRONMENT DATA PLOT
FOR REFERENCE.(4)

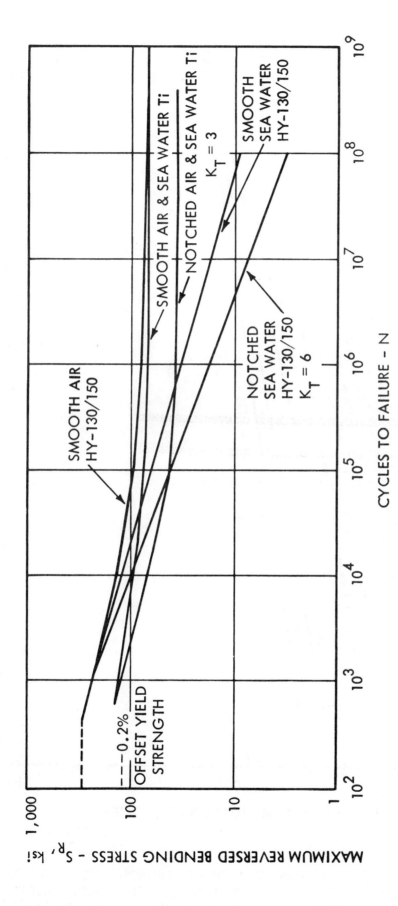

Figure 5-8   FLEXURAL FATIGUE CURVES FOR SPECIAL GRADE Ti-6Al-4V

AND HY-130/150 STEEL (4)

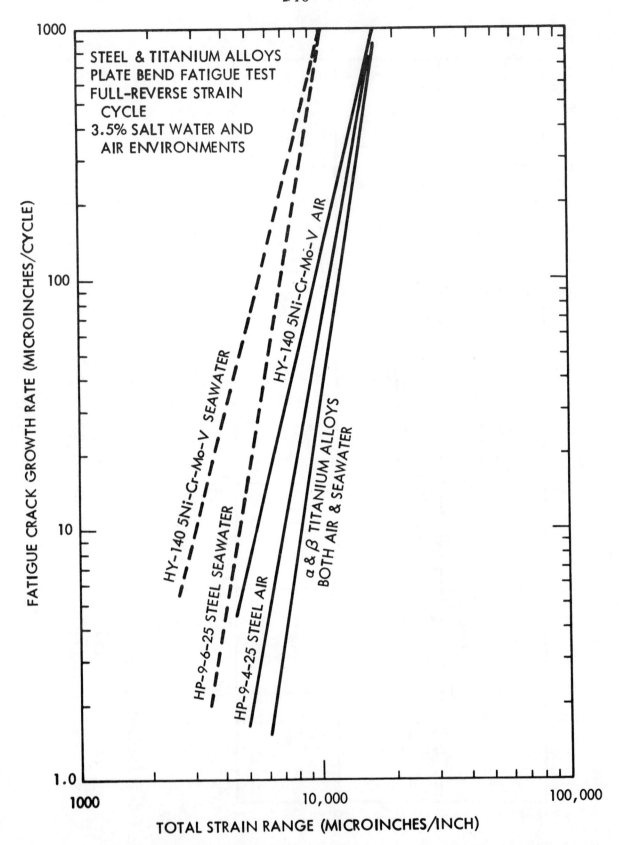

FIGURE 5-9   LOG-LOG PLOT OF FATIGUE CRACK GROWTH RATE VERSUS
TOTAL STRAIN RANGE FOR TITANIUM AND STEEL ALLOYS
IN BOTH AIR AND A 3.5% SALT WATER ENVIRONMENT.[4]

produced weighs 10,000 pounds. If there is a 60% yield from ingot to plate, the result is 30 sq. ft. of 2 in. plate or 60 sq. ft. of 1 in. plate.[16]

Although titanium can be formed with steel-forming equipment, it is important to remember that titanium has a lower modulus of elasticity than steel. Therefore, three times more energy is needed to form titanium at 120,000 psi yield strength than HY-80 steel. However, titanium can be heated to 400-600° F. to reduce the energy requirements.[16]

Welding of Titanium and Its Alloys

Shielding during the welding process is vital in preventing absorption of impurities at welding temperatures. For this purpose, inert gas-shielded arc welding processes are used most commonly.[14] The weld zones should be protected at 1200° F. or higher. The molten weld metal area is the most critical point for this protection. In this region, impurities diffuse into the titanium quickly and may result in severe weld embrittlement. Weld zones that are not molten but at temperatures of 1200° F. are subject to surface contamination from the air which may lead to early failures in service.[3]

Shielding fixtures required include: protection in open air welding and inert-gas filled welding chambers. Otherwise, conventional welding equipment may be used, either manual or automatic. Direct current power sources are best with tungsten-arc and consumable electrode welding. Straight polarity is used with tungsten-arc welding and reverse polarity with consumable electrode welding.[3]

Vacuum purging offers the best inert-gas atmosphere possible for welding titanium. Pre-weld cleaning with alkaline washes of dilute solutions of sodium hydroxide may also be required as well as an acid pickling treatment to remove the light oxide scale. This again stresses the purity of the material during welding and the purity of the gas used during the process.

Commercially Pure Titanium. A weld in titanium that is commercially pure is essentially a casting. The grain size is very large and the structure consists of beta dendrites. Oxygen and hydrogen in normal quantities have little effect on the microstructure, but extra carbon (over 0.2%) make TiC precipitate out of the melt into a network pattern which tends to lower ductility.[9]

Titanium Alloys. The structure of titanium alloy welds varies according to the alloy content. The weld is still a large grained casting which also has some transformation structures. A segregation effect during solidification occurs due to differences in composition between liquids and solids of the alloy. Dendritic structures, precipitated particles and some retained beta are present in certain microstructures.[14] Welding both commercially pure titanium and titanium alloys results in some changes in microstructure with no gross change in mechanical properties.[14]

"Two important factors in welding titanium and titanium alloys are:

1. The reactivity of the material at temperatures present

in the weldment and heat-affected zone, with air, and
with most elements and compounds (excluding the inert
gases), including all known refractories.

2. The mechanical properties of these materials are
   affected in an extreme manner by relatively minute
   amounts of impurities, especially nitrogen, oxygen,
   carbon, and hydrogen.[14]

Contamination. When hot, titanium will absorb easily or
combine with almost every other element. Most such combinations
result in greater strength and less ductility and toughness.
Consequently, when titanium is welded, it must be completely
protected from external elements, including not only the molten
weld puddle, but all metal that is hot. As a general rule,
"hot" is 1200° F. for the periods of time at temperature
associated with welding.

Normal fluxes are inadequate protection for titanium. In
fact, they may actually contribute to a loss of weld ductility
by alloying with the metal. Developmental work in the U.S.
shows the feasibility of fluxes and they are reportedly widely
used in Russia for joining titanium. However, it does not
necessarily follow that using fluxes will be practical or
economical. In general, a flux for titanium must be relatively
nonreactive and produce no harmful compounds in the metal.
Furthermore, it must have a greater free energy of formation
with oxygen than does titanium, and be oxygen free itself.
Only the halogen salts of the alkaline metals (specifically
calcium fluoride) come close to meeting these requirements.

The toxicity of this and similar compound always will restrict their use.[4]

Oxygen Effects. Oxygen has the greatest effect on titanium welds; it is ever present, readily absorbed by hot titanium and a small amount has a tremendous influence on mechanical properties. Titanium normally contains a thin, self-healing surface oxide film. This provides the corrosion resistance noted earlier and, in itself, does not cause any welding problems. This film thickens as titanium is treated in air, and the titanium dissolves its own oxide. Oxygen atoms diffuse from the surface oxide film into the metal, causing an increased oxygen content and reduced ductility toughness. Both processes are continuous; the surface oxidizes and the metal underneath dissolves the oxides. The surface oxidation proceeds generally at a faster rate than its solution, so that after a heating cycle the metal will have a heavier than normal surface oxide. If the surface oxide is not too heavy, it acts asa refractor of light and appears to the eye as colors. Different colors represent different thicknesses. They show the thermal cycle the metal has seen. However, they do not directly indicate the amount of oxygen in solution or weld contamination. Their only source of contamination is the diffusion of oxygen into the metal which is also dependent upon the thermal cycle, so the colors are an indirect measure of weld contamination. These colors on a weld always can indicate if a weld is bad, but they cannot prove that a weld is not contaminated.[4]

Problems of Porosity. Titanium welders are constantly
confronted with the problem of porosity. It is believed that
porosity is caused by hydrogen; however, the origin of this
hydrogen is not so certain. All titanium contains some hydrogen,
usually less than 100 ppm as an impurity. However, it is
unlikely that porosity is caused by this "dissolved" hydrogen.
Other possible causes of porosity include dirt, microscopic
particles, and water vapor. Whatever the cause, it has
been demonstrated that a fresh, clean joint edge is required
to produce minimum porosity.[4]

Mechanical Properties of Welds. Figure 5-10 shows
comparisons between tensile yield strength and drop-weight
tear energy of welds in various titanium alloys.[4] Alloy types
of the base metal and filler wire used are shown also.

154

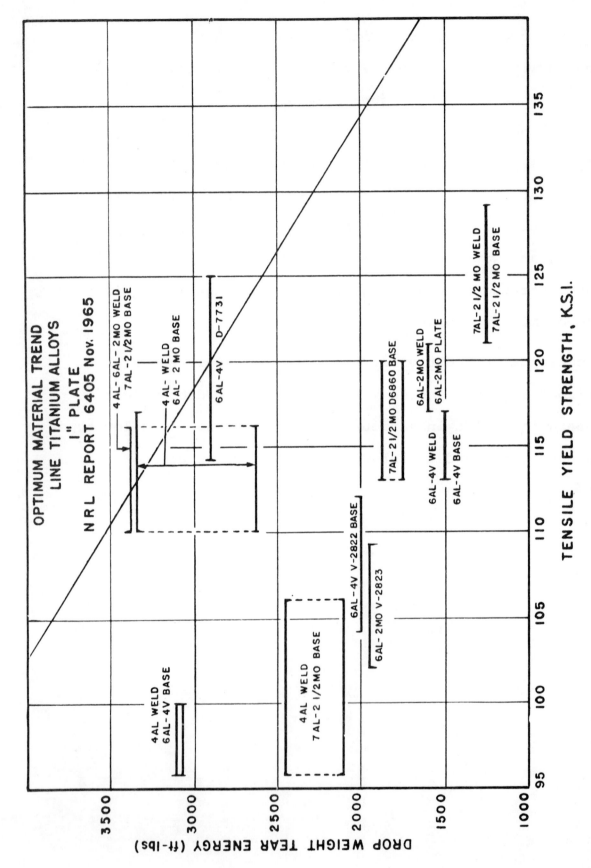

FIGURE 5-10   COMPARISON OF DROP WEIGHT TEAR TEST AND TENSILE YIELD STRENGTHS IN BASE PLATES AND WELDMENTS (4)

REFERENCES

(1) Battelle Memorial Institute, Defense Metals Information Center, "Residual Stresses, Stress Relief, and Annealing of Titanium Alloys," DMIC Report S-23, July 1, 1968.

(2) Cox, D. W., "Titanium Goes Undersea," Undersea Technology, pp. 31-32, February 1968.

(3) Faulkner, G. C., and Coldrich, C. B., "Interpetive Report on Welding Titanium and Titanium Alloys," Welding Research Council Bulletin Series, No. 56, pp. 1-20, December 1959.

(4) Feige, N. G., "Welding Titanium Alloys for Marine Applications," Lecture presented at a special summer session on "Welding Fabrication in Shipbuilding and Ocean Engineering" at M.I.T., August 20, 1969. Reprinted by Titanium Metals Corporation of America, 1969.

(5) Goode, R. J., Judy, R. W., Jr., and Huber, R. W., "Procedures for Fracture Toughness Characterization and Interpretation to Failure-Safe Design for Structural Titanium Alloys," NRL Report 6779, Naval Research Laboratory, December 1968.

(6) Huber, R. W., Goode, R. J., and Judy, R. W., Jr., "Fracture Toughness and Stress-Corrosion Cracking of some Titanium Alloy Weldments," Welding Journal, Research Supplement, pp. 439s-447s, October, 1967.

(7) "Marine Applications of Titanium and Its Alloys," Titanium Metals Corporation of America, November 1968.

(8) Metals Handbook, Properties and Selection, American Society for Metals, Eighth Edition, Vol. 1, 1961.

(9) Minkler, W. W., and Feige, N. G., "Titanium for Deep Submergence Vehicles," Undersea Technology, pp. 26-29, January 1965.

(10) Mitchell, D. R., and Feige, N. G., "Welding of Alpha-Beta Titanium Alloys in One Inch Plate," Welding Journal, Research Supplement, pp. 193s-202s, May 1967.

(11) Mitchell, D. R., and Kessler, H. D., "The Welding of Titanium to Steel," Welding Journal, Research Supplement, pp. 546s-552s, December 1961.

(12) Robelotto, R., Lambase, J. M., and Toy, A., "Residual Stresses in Welded Titanium and Their Effects on Mechanical Behavior," Welding Journal, Research Supplement, pp. 289s-298s, July 1968.

(13) Vorhis, F. H., "Titanium in Process Equipment," Metal Progress, pp. 105-114, February 1967.

(14) Welding Handbook, Section IV, American Welding Society, 1960 (Fourth Edition), and 1966 (Fifth Edition).

(15) Williams, W. L., "Metals for Hydrospace," Gillett Memorial Lecture, Reprinted in Journal of Materials, Vol. 2, No. 4, December 1967.

(16) Williams, W. L., and Lane, I. R., "Titanium Alloys," NRL Report 6167, "Status and Projections of Developments in Hull Structural Materials for Deep Ocean Vehicles and Fixed Bottom Installations," U. S. Navy Research Laboratory, pp. 31-41, November 1964.

CHAPTER 6 OTHER METALS

A wide variety of metals, other than steels, aluminum alloys, and titanium alloys, are used for various components of ocean engineering vehicles. The first part of this chapter (Section 6.1) covers metals used in hard water systems. The majority of this section is drawn from an article written by George Sorkin, Bureau of Ships, Department of the Navy, which covers metals which are used in hard water systems.[6]

The second part of this chapter (Section 6.2) covers materials used for various components of marine structures. This section is prepared from the information given in a booklet entitled "Guidelines for Selection of Marine Materials," published by the International Nickel Company, Ltd.[4]

6.1 Metals Used in Hard Water Systems

Systems and Components Included

Hard sea water systems are associated with nuclear main propulsion plants and include machinery equipment and components that are subjected to the effects of sea water pressure. The major components involved include:

Valve body and bonnet
Manifolds
Pumps
Pipe Fittings
Pipe End Connections
Sea Chests
Strainers

Pump Impeller

Pump Shafts
Bolting
Seamless Pipe
Welded Pipe
Condenser Tube
Condenser Tube Sheet
Condenser Shell (if heat exchange is located outside the hull)
Water Box

Wear Rings

Condenser Head
Valve Stem
Other Valve Trim

Flexible Connections (hose and end)
Weld Metal
Brazing Alloys [6]

In addition to the above, some auxiliary systems are exposed to sea pressure. These include brine overboard discharge, garbage disposal, ballasting, circulating water for jet propulsion and torpedo tubes. However, these will not be treated separately since they offer no greater difficulties than the main propulsion plant. [6]

Design Considerations

Hard sea water systems, when considered in relation to the thickness of metal, must be examined on the following basis:

Ultimate tensile strength

Tensile yield strength

Modulus of elasticity

Low-cycle fatigue strength

Coefficient of thermal expansion

Creep strength

Corrosion resistance (for corrosion allowance) [6]

"The general approach taken by designers of hard sea water systems is to calculate thickness of components based on allowable fiber stress at designated temperature. Figure 6-1 shows minimum wall thickness of hard sea water systems components using various allowable stress levels. The allowable stress is based on considerations of tensile strength, yield strength, and creep. Low cycle fatigue at specified number of cycles and shock requirements are also considered. Flexibility of the system is calculated so

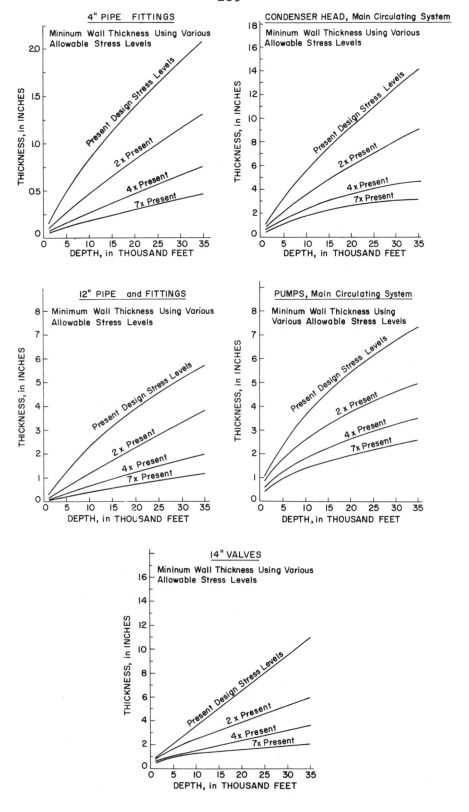

FIGURE 6-1   MINIMUM WALL THICKNESS OF HARD SEA WATER SYSTEM
COMPONENTS USING VARIOUS ALLOWABLE STRESS LEVELS[6]

TABLE 6-1:   METALS CONSIDERED FOR HARD SEA WATER SYSTEMS

METAL	COMPOSITION

COPPER ALLOYS

70-30 Cu Ni	68.9 Cu, 30 Ni, 0.6 Mn, 0.5 Fe
70-30 Cu Ni (Be)	69 Cu, 30 Ni, 0.5 Fe, 0.5 Be
70-30 Cu Ni (Cb)	66 Cu, 30 Ni, 1.5 Cb, 1.5 Mn, 0.75 Fe
70-30 Cu Ni (Hi Fe)	64 Cu, 30 Ni, 5 Fe
Cu Fenloy 40	55 Cu, 42 Ni, 2 Fe, 1 Mn
G-Bronze	88 Cu, 8 Sn, 4 Zn
M-Bronze	88 Cu, 6 Sn, 4 Zn, 2 Pb
Phosphor Bronze	95 Cu, 5 Sn, 0.25 P
Ni Al Bronze	79 Cu, 9 Al, 4 Fe, 5 Ni, 3 Mn
Mn Si Alloys	94 Cu, 2 Si, 1 Zn, 1 Mn, 2 Fe
Cu Ni Sil Alloy	97.5 Cu, 2 Ni, 0.5 Si
Superston Series	65-87 Cu, 0-14 Mn, 3-5 Ni, 3-5 Fe, 6.5-11.5 Al
Cu Be Alloy	98 Cu, 2 Be
Cu Al Si	91 Cu, 7 Al, 2 Si
Cu Zn Si	90 Cu, 9 Zn, 1.2 Si
Al Bronze (Hi Ni)	15 Ni

ALUMINUM

7079	4.3 Zn, 3.3 Mg, 0.6 Cu, 0.2 Mn, 0.2 Cr, Bal Al
5086	4 Mg, 0.5 Mn, 0.15 Cr, Bal Al
5456	5 Mg, 0.7 Mn, 0.15 Cu, 0.15 Cr, Bal Al
X7002	0.7 Cu, 3.0 Mg, 4.0 Zn, 0.2 Cr, Bal Al
X7106	2.0 Mg, 4.2 Zn, 0.11 Cr, 0.12 Zr, Bal Al
X7039	2.8 Mg, 4.0 Zn, 0.2 Cr, Bal Al

FERROUS

316 CRES	17 Cr, 10 Ni, 2.5 Mo, Bal Fe
17-4 PH	16.5 Cr, 4.25 Ni, 0.25 Cb, 3.6 Cu, 0.04C, Bal Fe
Maraging Steel	12 Ni, 5 Cr, 3 Mo, Bal Fe
Worthite	24 Ni, 20 Cr, 3 Mo, 3.2 Si, 1.8 Cu, Bal Fe
HY-150	5 Ni, Cr Mo V Steel

NICKEL ALLOYS

Monel	66 Ni, 1 Mn, 1 Fe, 32 Cu
K-Monel	66 Ni, 3 Al, 2 Fe, 1 Mn, 1 Si, 27 Cu
Inconel 718	53.5 Ni, 18 Fe, 19 Cr, 0.5 Al, 0.8 Ti, 5 Cb, 3 Mo, 0.2
Hastelloy C	16 Mo, 16 Cr, 5 Fe, 4W, Mn + Si + Co, Bal Ni
Inconel 625	62 Ni, 22 Cr, 9 Mo, 4 Cb, 3 Fe
Inconel X750	73 Ni, 16 Cr, 7 Fe, 2.5 Ti, 0.5 Al, 1 Cb
Ni-O-NEL	42 Ni, al Cr, 3 Mo, 2 Cu, 1 Ti, 31 Fe

TITANIUM ALLOYS

Pure	98.9 - 99.5 Ti
6-4	6 Al, 4 Va, Bal Ti
3-2-1	7 Al, 2 Cb, 1 Ta, Bal Ti
Alpha Beta Alloys	
Beta Alloys	

that the computed expansion stress will not exceed the allowable stress range specified in the ship specification."[6]

Material Selection

Table 6-1 shows metals considered for hard sea water systems.[6] Table 6-2 lists their desired characteristics.[6] In hard sea water systems, materials selected for use in submersible construction are considered on the basis of ease of fabrication, cost, strength/density ratio, weight and corrosion resistance. However, some non-corrosion resistant alloys are considered as appropriate load bearing materials to be developed as the outer portion of a composite structure. CuNi or Ni Cu, for example, could be utilized for corrosion resistance on the inside.

Material Characteristics

<u>Wrought 70-30 Copper Nickel Alloy</u>. This alloy, in an annealed form, is currently the wrought metal used most widely in hard sea water systems. Components constructed from 70-30 Cu-Ni include seamless and welded pipe, condenser tubes, tube sheets, welded water boxes, welded fittings and other similar components. A single phase solid solution alloy, 70-30 Cu-Ni has about 0.5% iron added to improve its resistance to corrosion errosion. This alloy may be hardened by cold working, but the improvement in strength of cold worked material is not used when welding or brazing are utilized to fabricate either the part or the system.

Difficulties in fabrication are increased by drawn tempers in pipe. 70-30 mill product specifications name a single strength

162

TABLE 6-2   DESIRED CHARACTERISTICS FOR METALS FOR HARD SEA WATER SYSTEMS

Density	Low
Tensile	High proportional limit, yield strength and ultimate tensile strength. Without need for heat treatment.
Toughness	Will not fracture in a brittle manner under severe plastic deformation at 0° or lower.
Modulus	High for bar, rod and plate; low for pipe so as to permit more compact expansion loops.
Fatigue	High in both high and low cycle with and without notches-axial, hoop and bending loads; in air and under sea water.
Corrosion-Erosion	Not susceptible to stress corrosion, low general and pitting corrosion rates under static, impingement, crevice, high and low velocity and cavitation conditions; no selective phase attack.
Fouling	Resists
Stability	No creep at 200° F.
Weldability	95% or better joint efficiency in as welded condition for yield strength, toughness and fatigue strength.
Formability	Hot formed or cold formed to shape without undue thinning or necking and without need for subsequent heat treatment.
Machinability	Good
Repairability	Weld repairable under service conditions.
Compatibility	Parts of system, electrically connected to have potential difference less than 0.060V in running sea water.
Castability (Castings)	Good
Weld Wire	Available in essentially base metal composition.
Availability	In large sizes and quantities at reasonable cost.

Screening tests will be on basis of yield strength, corrosion resistance and weldability. However, a complete material evaluation will include the following tests for base metal and weldments:

Density
Heat Transfer Coefficient (condenser alloys)
Yield Strength - tension, compression
Moduli - tension, shear
Ductility - elongation and reduction in area
Toughness - notch tensile, charpy
Weldability - resistance to cracking under restraint
Formability (wrought products) - hot and cold and accompanying properties
Fatigue - notched, un-notched, low cycle, high cycle
Corrosion - stress, fatigue, static in marine atmospheres and sea water velocities 3 - 30 foot/sec.
Creep - various loads and temperatures
Response to thermal cycling
Effect of residual stresses and stress relief conditions
Castability - for castings
Residual stress in weldments - relief
Location in galvanic series
Cavitation resistance

value, 18,000 psi for all thicknesses of annealed or soft material, with the exception that tube over 4-1/2 inches in diameter has a minimum yield strength of 15,000 psi. Mechanical requirements of seamless tube must be met by welded annealed tube.[6]

Wrought Nickel Copper Alloys. Components such as shafting, valve trim, bolting, lining of steel parts and fabricated valves are fabricated with Monel and K-Monel. K-Monel is precipitation hardenable due to its aluminum and silicon additions. Consequently, it has higher strength than Monel and usually is used in that condition. Monel does not resist pitting as much as does 70-30 Cu-Ni, although it is used widely in salt water. K-Monel can be welded but needs reheating treatments to avoid a degradation of the properties of the heat affected zone. Monel and K-Monel are approximately equal in corrosion resistance. Annealed Monel has a 25,000 psi yield strength. Although various processing or size variables can push the yield strengths to significantly higher levels, this advantage is not effective if welding is employed.[6]

Cast Valve Bronze and Gunmetal. These extensively specified casting alloys are used in sea water for such purposes as pumps, valves, fittings, heat exchangers, strainers and similar components. Cast valve bronze and gunmetal are both complicated alloys of tin and copper with zinc and lead added. Gunmetals are those bronzes composed of tin bronzes and zinc. Leaded gunmetals are the same material with lead added. 70-30 Cu-Ni, Monel and aluminum bronze have been developed as alternatives to tin bronze due to the serious difficulties in producing large castings which will pass radiographic inspection requirements as well as maintaining light weight. If improved pressure tightness is a factor, Ni modified tin bronze which is not heat treated is at times substituted for gunmetal without changes in allowable stresses.[6]

Aluminum Bronze. Aluminum bronzes have been investigated extensively and applied because of their relatively high allowable stress limit. These alloys are primarily copper-aluminum alloys which contain a maximum of 10% aluminum and appreciable quantities of iron, manganese, and nickel.

The alloys are of two groups: the first group contains up to about 7.5% aluminum and has a homogenous structure. They are applicable only in wrought iron form. The second group has more aluminum, a duplex structure, and can be used in both wrought and cast forms. "Due to eutectoid transformation, castings of the higher aluminum content alloys and those with less than 4 per cent nickel regardless of aluminum content can exhibit a microstructure containing a gamma-2 network and will be subject to catastrophic deterioration is sea water service by de-aluminization."[6]

These effects can be minimized by proper heat treatments; however, no non-destructive tests are available to determine whether castings have been properly heat treated. Aluminum bronzes are used to some degree in salt water systems.[6]

Clad or Lined Parts. Vital locations including hull and back valves and condenser waterboxes may involve steel, lined or clad with copper nickel alloy or nickel copper alloy. Designs for hull valves often require an HY-80 casting which is welded to the hull plating. The area exposed to sea water, the valve interior, is fitted with a Monel liner. With such a design, the necessity of a hull insert, separate sea chest and bolting is eliminated.[6]

Pipe and Tube Manufacturing. Sea water applications involving pipe and tube are either seamless or welded. Tube sizes under about 6 inches in outer diameter use seamless tubes drawn from hot extruded shells. Seamless tubes over about 6 inches in outer diameter are produced by a cold cupping and drawing practice using hot-rolled heavy gage circles as the starting stock. The American Brass Company, the only producer of large diameter seamless 70-30 Cu-Ni, is already unable to supply some of the sizes of seamless tube requested by fabricators of fittings who use seamless tube as base stock. An increase in wall thickness of any degree, required for higher pressure service, brings a serious facility problem for larger diameter seamless tube.[6]

Forming and seam welding fully annealed 70-30 Cu-Ni alloy flat stock results in welded tube in 12 to 20 foot lengths. All weld reinforcement is removed. 70-30 tube welded or seamless is furnished either bright annealed or annealed and acid pickled finish.

6.2 Materials Used for Various Components

This section discusses briefly a number of applications and mentions some of the more pertinent factors a designer would consider in the material selection process.

Pipe for Seawater[5]

The key considerations is selecting materials for pipes

to be used in seawater are a minimum life of twenty years, no
leakage, no clogging from marine growth and a six foot per second
design velocity. Materials most frequently used include galvanized
steel (6 to 9 years), copper (6 to 12 years), 90/10 copper-nickel
alloy with 1.5 Fe (10 years plus), and 70/30 copper-nickel alloy
with 0.5 Fe (22 years plus). Figure 6-2 shows the seawater velocity
for pipe and tube ranges. The tolerances of wrought alloys for
velocities commonly encountered in pipe and tubing are plotted.

Pumps for Seawater [5]

Primary considerations in selecting materials for seawater
pumps are that the impeller must resist turbulance and be more
noble than the body; the body must supply the impeller assembly
with substantial cathodic protection to reduce crevice corrosion
and pitting of the impeller assembly, particularly during down
periods; and the positive corrosion allowance must be added to the
case wall and straightening vanes where velocities over 20 feet
per second are encountered, unless more resistant case materials
are used. A cast iron body, bronze impeller or bronze shaft
often are materials used. These have an older standard, a frequent
impeller replacement and a short case life. Naval vessels and
merchant ships frequently use a G bronze body, a nickel-copper alloy
400 impeller, and a nickel-copper alloy 400 shaft. Designers of
coastal power plants probably would use an Austenitic cast iron
body, a type 316 impeller and nickel-copper alloys 400 or K-500
shaft. These materials are noted for their good service in a
marine environment.

167

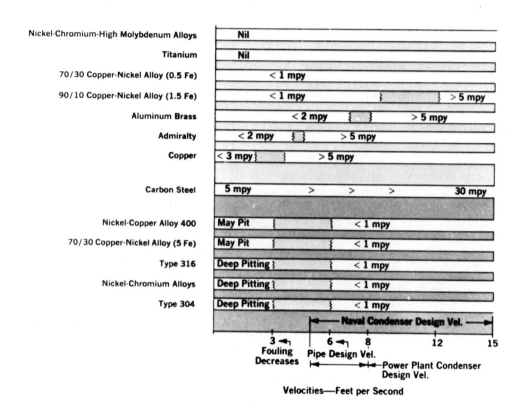

FIGURE 6-2 SEAWATER VELOCITY (PIPE AND TUBE RANGES) [5]

Valves for Seawater [5]

The overriding consideration with valves that must operate under a seawater environment is that the trim must resist turbulance and be more noble than the body. Cast iron I.B.B.M. (Iron Body Brass Mounted) gate valves have an older standard and high maintenance but exhibit some breakage. Ductile iron I.B.B.M. gate valves eliminate the breakage through a high degree of ductility. This material also demonstrates a high maintenance, if maintained. M bronze with a nickel-copper alloy 400 trim is standard for high performance bronze valves on both merchant and naval ships. A useful alternate to this bronze is an Austenitic nickel cast iron body with Type 316 trim. This combination, however, is used less frequently than the M bronze. Alloy 20 body and alloy 20 trim are good in seawater but must be insulated from copper alloy piping. And finally, a good selection for butterfly valves is an Austenitic nickel cast iron disc with nickel-copper alloys 400 or K-500 stem.

Tubing [5]

One of the materials often used for tubing is Admiralty which usually is limited to older, slow speed vessels with low velocity installations.

Aluminum brass is used widely in coastal and marine service and exhibits satisfactory results most of the time. In the areas where aluminum brass has shortcomings, 90/10 copper-nickel alloy (1.5 Fe) is used to better meet the seawater conditions. 70/30 copper-nickel alloy with 0.5 Fe is used where the greatest dependability and reliability are required. Nickel-copper alloy 400 is used to some extent in seawater evaporators. Type 316 is now being tried in coastal power plants with the most agressive and polluted waters. To date, the results of these experiments with Type 316 are mixed. When the tubes are kept thoroughly clean, generally good results are reported. In stagnant conditions, however, pitting has caused failures. A new product being evaluated for severe conditions is Type 316 over 90/10 copper-nickel alloy (1.5 Fe). This is a bimetallic tubing. See Figure 6-2 for information on the velocity of pipe and tubing.

Tubesheet[5]

The present trend in ocean engineering is toward welding tubes to the tubesheet. Galvanic considerations tend very strongly toward use of the same alloy in the tubesheet as in the tubes. This new system is generally preferred over the old system where the practice was to use muntz metal or naval brass tubesheets.

Water Boxes[5]

The primary material consideration is that the water box should not be more noble than the tubing. Carbon steel and cast iron are used because the heavy corrosion of the steel provides substantial cathodic protection to the copper alloy tube ends and the tubesheet. Carbon steel and cast iron lined with organic lining reduce the cathodic protection of tube ends. This combination also forces steel which is exposed at the pinholes and scratches to furnish the full amount of the current required by the galvanic couple with the copper alloy tubesheet and tube ends. Maintenance with this material can be high.

Austenitic nickel cast iron is considered a good material

selection. It has low corrosion but still is protective to the copper alloy tube ends and the tubesheet. Another good selection is 90/10 copper-nickel alloy with 1.5 Fe. This is used particularly when the tubes are 90/10 copper-nickel alloy with 15. Fe or 70/30 copper-nickel alloy with 0.5 Fe or nickel-copper alloy 400. If the tubes are 70/30 copper-nickel alloy (0.5 Fe) or nickel-copper alloy 400, the material used for the water boxes often is 70/30 copper-nickel alloy (0.5 Fe).

Water Lubricated Bearings [5]

The standard material used for water lubricated bearings in merhchant ships is a G bronze sleeve with Lignum Vitae bearings. The newer high performance ships report that the wear rates are so high that interim dry docking has been required. Substantially lower wear rates have been reported in comparative test work with 80/20 copper-nickel alloy sleeves with Lignum Vitae bearings. Early reports from shipboard installations show lower wear rates. Naval ships usually use a G bronze sleeve with rubber bearings. Wear apparently presents no problem with these materials although seal welding and welded repairs seem to present problems. Figure 6-3 shows general wasting for various materials which have been immersed in quiet seawater.

Topside Marine Hardware and Fasteners [5]

Chocks and Cleats. Cast iron is useful, inexpensive, and has a high maintenance. Formerly, cast brass and bronze were standard on large vessels and small boats. They require, however, frequent polishing. Cast aluminum has the drawbacks of pitting

171

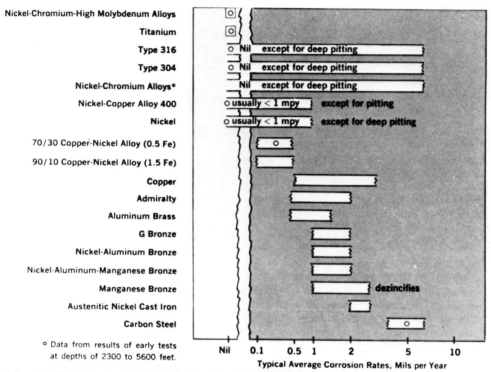

Nickel-Chromium-High Molybdenum Alloys
Titanium
Type 316 — Nil except for deep pitting
Type 304 — Nil except for deep pitting
Nickel-Chromium Alloys* — Nil except for deep pitting
Nickel-Copper Alloy 400 — usually < 1 mpy except for pitting
Nickel — usually < 1 mpy except for deep pitting
70/30 Copper-Nickel Alloy (0.5 Fe)
90/10 Copper-Nickel Alloy (1.5 Fe)
Copper
Admiralty
Aluminum Brass
G Bronze
Nickel-Aluminum Bronze
Nickel-Aluminum-Manganese Bronze
Manganese Bronze — dezincifies
Austenitic Nickel Cast Iron
Carbon Steel

° Data from results of early tests at depths of 2300 to 5600 feet.

Nil 0.1 0.5 1 2 5 10
Typical Average Corrosion Rates, Mils per Year

* Nickel-chromium alloys designate a family of nickel base alloys with substantial chromium contents with or without other alloying elements all of which, except those with high molybdenum contents, have related seawater corrosion characteristics.

FIGURE 6-3 GENERAL WASTING OF VARIOUS MATERIALS IMMERSED IN QUIET SEAWATER[5]

and losing its "sheen." Cast brass and bronze nickel-chromium plated hardware have good service when adequate thicknesses are used and properly applied. Cast stainless steel (CF-8) also has demonstrated good service.

Pulleys, Wrought Fittings, Trim. Aluminum, brass and bronze, and brass and bronze nickel-chromium plated are materials frequently used for these fixtures. They have the same characteristics as those mentioned in the preceding section. Type 304 also is used for these pieces of hardware; it has shown good service and it is easier to clean and restore the orginal "sheen" than with brass and bronze.

Wire Rope. Galvanized steel gives good service topside when used as the material for wire rope. Type 304 is also used and also gives good service except at deck fittings where crevice corrosion in swaged fittings has caused failures.

Fasteners. Brass and bronze are used widely and have a high maintenance. Type 316 also is employed but is subject to rusting on exposed threads and in crevices. Also, it may prove difficult to unscrew. Nickel-copper alloy 400 gives good service and is easy to unscrew. This material is standard for naval service.

Marine Wire Rope--Seawater[5]

Galvanized steel, including plow steel and improved plow steel, often is selected for a rope that must withstand the effects of seawater. However, this material does show some rusting in seawater and spools on deck winches. Stainless steel Types 304 and 305 are standard for many more rigorous naval applications. They also are used for rope, but crevice corrosion often limits the service to less than 30 days in some instances. When it is used as a working rope, failure from flexing is frequent. When Phosphor bronze is used as a wire rope material, it has about half the strength of galvanized plow steel or stainless wire ropes. Also, its availability is somewhat limited.

New products include galvanized steel with a plastic coating and aluminized steel. These products have some prospect of improvement over galvanized steel but results of testing are not yet conclusive. A 90/10 copper-nickel alloy-clad Type 304 stainless steel also is fairly new. It offers a promising improvement over bare stainless. The useful life of this material depends largely upon the thickness of the cladding.

Buoys and Floating Platforms [5]

Buoys and Floats. Coated carbon steel is the primary material used for buoys and floats. A great deal of effort has been expended in developing anti-fouling coatings for steel in seawater, all of which add significantly to the cost of steel. Unless the fouling is tolerable, cleaning maintenance is high.

Offshore Platforms. In the submerged zone, carbon steel is used either coated or bare with cathodic protection. Nickel-copper alloy 400 is used in the splash zone and is welded on steel. Above the splash zone, carbon steel witha coating is used.

REFERENCES

(1) Anti-Corrosion Methods and Materials, staff article, "Alloys Containing Large Amounts of Copper and Nickel for Use in Sea-Water," pp. 17-18, October 1966.

(2) Anti-Corrosion Methods and Materials, staff article, "Copper-base Alloys for Use in Sea-Water," pp. 19-20, June 1967.

(3) Bulow, C. L., "Use of Copper Base Alloys in Marine Services," Naval Engineers Journal, pp. 470-482, June 1965.

(4) International Nickel Company, Inc., company publication, "Guidelines for Selection of Marine Materials," 1966.

(5) Internation Nickel Company, Inc., company publication, "Nickel Alloy Steels and Other Nickel Alloys in Engineering Construction Machinery," 1960.

(6) Sorkin, G., "Hard Sea Water System Materials," NRL Report 6167, "Status and Projections of Developments in Hull Structural Materials for Deep Ocean Vehicles and Fixed Bottom Installations," U. S. Naval Research Laboratory, pp. 142-155, November 1964.

CHAPTER 7   FIBERGLASS REINFORCED PLASTICS AND OTHER FILAMENTOUS
COMPOSITES

7.1   Introduction

The bulk of this chapter is paraphrased from a paper written
by J. A. Kies and Irvin Wolock with some references to a paper
by Kenneth Hom.[5,6]

A composite material is one which is composed of dissimilar
materials bonded together by some method so that the component
materials act together in response to external conditions. The
final combination exhibits properties contained by neither of
the composite materials alone. Fiber reinforced composites are
used in many structures because of their high strength and
strength/density ratios. Filament-wound glass-fiber reinforced
epoxy composite is being applied to the Polaris and Minuteman
rocket chambers. This material in tension has a strength/density
ratio significantly greater than that of any metal currently
available. Glass reinforced plastic (GRP) compares well with other
materials on a specific strength or specific modulus basis. For
these reasons, these and similar composites will be examined with
an emphasis on their application to deep submergence vehicles.[6]

7.2 Materials for Filamentous Composites

Tables 7-1 through 7-5 show various properties pertinent to
discussions given in this chapter. They include:

Table 7-1: Specific Strengths of Glass Reinforced
Plastics and Some Metals.

Table 7-2: Buckling Strength and Stiffness of GRP
and Metals.

TABLE 7-1: SPECIFIC STRENGTH OF GLASS REINFORCED PLASTICS AND SOME METALS (6)

Material	Design Stress ($lb/in.^2$)	Density ($lb/in.^3$)	Stress/Density (10^6 in.)
Epoxy-Fiberglass	100,000 150,000 200,000	0.075	1.3 2.0 2.7
Steel	80,000 100,000 150,000 200,000	0.28	0.3 0.35 0.5 0.7
Aluminum	25,000 50,000 100,000	0.099	0.25 0.5 1.0
Titanium	100,000 125,000 150,000	0.17	0.6 0.75 0.9

TABLE 7-2: BUCKLING STRENGTH AND STIFFNESS OF G.R.P. AND METALS[6]

Material	Compressive Modulus (E) (10^6 lb/in.^{2})	Density (D) (lb/in.^{3})	$E/D^{2.5}$ $\left(10^9 \dfrac{\text{lb/in.}^2}{(\text{lb/in.}^3)^{2.5}}\right)$	$E^{1/3}/D$ $\left(10^2 \dfrac{(\text{lb/in.}^2)^{1/3}}{\text{lb/in.}^3}\right)$
Epoxy-Fiberglass				
(1964)	5.6	0.075	3.7	23.6
(1970)	8	0.075	5.3	26.7
Fiber Reinforced Composite				
(1980)	40	0.075	26.6	45.6
Steel	30	0.282	0.7	11.1
Aluminum	10	0.099	3.2	21.7
Titanium	15.5	0.17	1.3	14.6

Reinforcements

In this section, a variety of reinforcements will be discussed, along with their particular properties and composition.

Glass filaments. Glass reinforced plastics are of particular interest due to their high strength/weight ratio. This is particularly important in pressure hull applications on deep diving structures.[5]

Glass filaments now available on the commercial market have diameters of about 0.00038 inches. These involve zero compressive buckling strength in any useful length unless continuous lateral support is provided by the matrix, which is usually an epoxy resin for high strength filament windings. In spite of this, good lateral support is possible in composites. E glass and S or S994 glass are the most common types of commercial high strength filaments in use today. Both are sold with the "HTS" surface finish applied at the factory. Such a finish gives protection from mechanical damage which can be severe, especially if a clean bare glass surface touches another similar surface.[6]

E and S glass fiber strengths are not significantly affected by exposure to water or humid air for a number of months if no stress is imposed during the period of exposure. On the other hand, the presence of moisture during a typical short time tensile

TABLE 7-3: STRESS REQUIRED FOR EQUAL STRESS/DENSITY RATIOS FOR G.R.P. AND METALS[6]

Stress/Density (10^6 in.)	Required Stress (lb/in.^{2})			
	G.R.P.	Steel	Aluminum	Titanium
0.5	37,000	141,000	49,500	85,000
1.0	75,000	282,000	99,000	170,000
1.5	112,500	423,000	148,500	255,000
2.0	150,000	564,000	198,000	340,000

TABLE 7-4: PROPERTIES OF REINFORCEMENTS[6]

Material	Tensile Strength (lb/in.^{2})	Density (lb/in.^{3})	Strength/ Density (106 in.)	Modulus of Elasticity (106 lb/in.^{2})
S-glass (Single Filament)	650,000	0.091	7.1	12.5
S-glass roving (Strand test)	450,000	0.091	4.9	12.5
Berylium	180,000	0.066	2.7	44
Boron	500,000	0.094	5.3	55
Steel (Music wire)	600,000	0.282	2.1	30
Titanium	250,000	0.17	1.5	15
Tungsten	400,000	0.695	0.6	50

TABLE 7-5: SOME PROPERTIES OF CONTINUOUS FILAMENT OTHER THAN E GLASS AND S GLASS SOLID FILAMENTS(6)

Filament	Density D (lb/in.^{3})	Not Subject to Change Except*			
		E (lb/in.^{2})	$E^{1/3}/D$	Theoretical† σ comp/D	$E/D^{2.5}$
E glass, Hollow	0.06	6.7×10^6	33.0×10^2	5.6×10^6	7.59×10^9
SiO$_2$	0.083	11.0×10^6	38.3×10^2	6.6×10^6	5.56×10^9
Asbestos	0.072	25.0×10^6	43.0×10^2	17.4×10^6	18.0×10^9
Be	0.066	44.0×10^6	53.5×10^2	33.3×10^6	39.3×10^9
89% ZrO$_2$ 11% SiO$_2$	0.148	50.0×10^6	24.9×10^2	16.9×10^6	5.96×10^9
B	0.083	$50.0* \times 10^6$	49.7×10^2	15.0×10^6	25.2×10^9

*E expected to increase to 60×10^6 psi with R and D in 5 years.

†Theoretical strength = E/20.

test of 10 to 30 seconds results in a drastic decrease in strength. Therefore, most of the information currently applicable to the strength of glass filaments includes a large reduction caused by stress corrosion. This may be combated in two ways: 1) the development of glasses superior to S and E or 2) improved surface finishes for better moisture protection.[6]

Problems involved in applying GRP to submersible pressure hulls include:

```
least-weight design
hull penetrations
head closures
cylinder joints
head-cylinder junctures
structural fatigue
long-term exposure to a deep-sea environment.[5]
```

"Test results obtained with individual structural components indicate that it is possible to resolve problems associated with non-weldable material such as GRP for deep submergence application, that the overall weight of GRP hulls designed for a 15,000 foot operating depth is not prohibitive for oceanography, and that it is possible to design GRP hulls to stress levels of 50,000 psi at operating depth with no appreciable strength reduction after 10,000 cycles to test depth."[5]

Metals as reinforcements. The Polaris program involved an evaluation of metal wire as a reinforcement in filament wound structures. The conclusions from this experimentation with metal reinforcements are disappointing. The tensile strength to density ratio for glass filaments is much higher than that of any metal filaments now available.[6]

Whiskers. Whiskers -- single crystals in short stable

fiber form -- are extremely expensive. In order to make such
fibers applicable to submersible construction, production methods
would have to be revolutionized. The potential high strengths
indicated, however, make whiskers a possible contender for ocean
engineering use, should such production methods be found. Whiskers
offer no higher tensile strengths than glass fibers, however, in
equal gage lengths and diameters.[6]

Matrices

Currently, only epoxy resins are used as the matrix or
binder in military filament-wound vessels. They are preferable
to other possible materials, such as polyester resins, due to
their superior strength properties and resistance to water and
decreased shrinkage on curing.[6]

Such structures as the Polaris and Minuteman chamber employ
glass-fiber reinforced epoxy composites. The resins in these
laminates are virtually the same as those available ten years
ago, since no major advances in the field of epoxy resins have
been made since then. Epoxy resins, therefore, have served
satisfactorily in rocket chambers and improved resins would
yield probably only minimal improvements in internally loaded
vessels.[6]

Deep submergence structures, however, rely much more on
resins than do aerospace structures. The resin in ocean structures
must provide lateral support to the fibers for resistance to
buckling, resistance to interlaminar shear fractures, to transfer
stresses around discontinuities in the glass fibers and to exclude
water from the fibers. Not only is the role of the resin more

important, the conditions under which it must operate are more complex. Water, for example, results in a much greater degradative effect in the case of a deep submergence structure than does moist air in the case of a rocket chamber.[6]

With support from the federal government, however, substantial gains should be forthcoming in the field of resins for ocean engineering structures. Such federal support is vital since there is limited commercial demand for this premium type of resin at present. With the limited work done thus far, tensile strength has increased from 12,000 psi to 19,000 psi, compressive strength from 19,000 psi to 29,000 psi and Young's Modulus has gone from 4.5×10^5 psi to 7.0×10^5 psi.[6]

Surface Finishes

Chemicals applied in solution to fibers as they emerge from the drawing bushing are termed surface finishes. They protect the fibers from abrasion against one another and improve the wet strength retention and fatigue life of composites made with these fibers. The exact processes by which finishes provide protection are not yet fully understood. Such an understanding is fundamental for research into further improvements in finishes which will reduce the loss of strength of composites due to fatigue loading or exposure to water.[6] A program to accomplish such aims should include:

1. Studies of the reactions of the glass surface with typical finish functional groups.
2. Studies of wetting at the glass surface -- how it is affected by various chemical groups on the surface.

3. Investigation of the effects of various groups on the glass surface resin interaction.

4. Studies of the effect of water on the strength of glass fibers.[6]

7.3 Auxiliary Materials

Coatings

The development of new polymers is closely parallel to advances in coatings research. "Some idea of the versatility of polymers can be gained from many examples of polymers of the same general class finding application in the fields of coatings, plastics, adhesives, and sometimes elastomers. The field of coatings is the oldest and best developed of these technologies, and major furtherance of the state of the art is not foreseen in the near term..."[6]

Water permeability of organic polymers is not likely to be reduced below that rate now provided by coatings such as Saran F-120; Hypalon-30 and the copolymers of vinyl and vinylidene chloride. On the other hand, it may be possible to decrease the water permeability of some of the coatings that are rather poor in this respect. This would include the polyesters, polyurethanes, alkyds, epoxies and some others.[6]

In deep submergence vessels, there is a need for better exterior coatings. This would include work in anechoic coatings and epoxy coatings and sealants that will polymerize under water. The Navy's current sea based deterrents have demonstrated a need for internal coatings with special properties.[6]

7.4 Properties of Filamentous Composites

Advantages of Plastics

Deep submergence hulls may find the use of filament-wound plastics advantageous for a number of reasons, primarily because of the strength and weight characteristics.[6]

"Other advantages are that reinforced plastics are:

a) Non-corrosive
b) Non-magnetic
c) Non-critical (readily available supply)
d) Thermal insulative
e) Sound and vibration attenuative (in plane of lamination)
f) Dielectric
g) Easy to fabricate; F.W.P. process is automated, hence it is amenable to quality control; there are no foreseeable limitations to size; anticipated improvements in materials do not necessarily create new fabrication problems; no specialized labor requirements.
h) Amenable to high speed production and show a consistent trend of decreasing raw material prices point to potential finished product costs of $2/lb.
i) Capable of being used in F.W.P. - metal composite construction and sectionalized for ease of modification and repair.
j) Tough (Izod impact is in the neighborhood of 35-60 ft. lbs. per inch of notch; toughness and strength increase with decreasing temperature).
k) Proven in that models have been tested which meet arbitrarily established performance standards."[6]

Disadvantages of Plastics

"a) Non-metallic (Non-ductile)
b) Cannot be fabricated using conventional metal-working techniques (welding, bending and forming).
c) Absorb water and lose a fraction of 'dry' strength if no jackets or other external protective means are employed.
d) A new family of materials (date back ten years with regard to mine case applications and three years for deep-submergence applications - 12-year submarine use in fairings; 20 years entire history for glass reinforced plastics).

e) Need to develop background and basic knowledge with regard to materials, long-term performance, fabrication and design to raise acceptance and confidence level.
f) New design concepts, approaches and attitudes are required.
g) Abrasion resistance and cavitation erosion resistance may be problem areas."[6]

7.5 Applications of Plastics

Fiber-reinforced plastics can be successfully applied in a cylinder shape for pressure hulls of deep-submergence vehicles. The cylindrical shape is particularly applicable to filament-wound materials since it can be wound easily to almost any desired length-diameter ratio while aligning the fibers in the direction of the principal loads.[5]

"Little is known about filament-wound hulls in the following areas:

a) structural response to dynamic loading

b) structural response to long-term exposures to a deep-sea environment

c) repairs of damaged hulls made of non-weldable, low shear strength materials

d) the use of other reinforcements besides solid-glass fibers with circular cross-section

e) the use of other matrices besides epoxy resins."[5]

Other applications of reinforced plastics include the replacement of conventional metal in sea water-compressed air surfacing ballast tanks in the Alvin, a deep diving submersible, and an additional use in the outer hull construction enclosing the pressure tanks and aluminum frame.[7]

The Divar, an unmanned acoustical research vehicle employs

a reinforced plastic cylinder 16 inches in outer diameter, 3/4 inch in wall thickness, 12 1/2 inches in inner diameter, with nine ribs, 60 inches in length, weighing 180 lbs., diving to 6500 feet.[7]

7.6 Design Concepts for Fiber Reinforced Plastics

Past Efforts

In past years, composite materials have been investigated primarily on the basis of their properties and fabrication techniques. Only recently have efforts centered around the specific area of submarine hull design. Within this area, GRP have been of most interest due to their high strength/weight ratio. Models fabricated from GRP will act as the basis for the following conclusions.[5] Such conclusions, however, would apply also to applications and design of other fiber-reinforced plastics.

Design Considerations

"A study is now underway to demonstrate the capability of resolving problems associated with the overall design of GRP pressure hulls. These problems include:

1) Least-weight design

2) Hull penetrations

3) Head closures

4) Cylinder joints

5) Head-Cylinder junctions

6) Structural fatigue

7) Long term exposure to a deep-sea environment."[5]

The above design considerations were studied by Hom and Barnet using a ring-stiffened cylindrical pressure hull to which almost any length-diameter ratio can be wound while aligning the fibers in the directions of the principal loads. It also can be sealed easily with hemispherical end closures. In this study, these specifications were established:

1) Collapse depth of 30,000 feet.

2) 10,000 excursions to a depth of 15,000 feet with no loss in overall strength.

3) Opening in the closure head and cylinder equal to 1/5 the diameter of the pressure hull.

4) Overall length of the pressure hull equal to 5 diameters.[5]

Hydrostatic pressure tests and fatigue tests were conducted with small-scale models which incorporated the design specifications mentioned earlier. Results of the tests indicate that it is possible to resolve problems associated with non-weldable materials such as GRP in deep-submergence design and applications. It was also determined that the overall weight of GRP hulls designed for a 15,000 foot operating depth was not prohibitive for oceanography and that GRP hulls may be designed to stress levels of 50,000 psi at operating depths with no appreciable strength reduction after 10,000 cycles to test depth.[5]

Future Considerations

According to Hom and Barnet, ocean engineers have failed to take full advantage of the ultimate strength available in filament-wound composites. Analytical and experimental studies will be required to understand the failure mechanism of laminated, ortho-tropic structures more fully and to establish failure criteria which apply directly to hulls fabricated from such materials. Since such techniques are not now available, designers and other engineers have had to use engineering methods to modify existing criteria established for thin-walled hull structures fabricated from ductile, isotropic materials.

Protective Coatings in Design

In order to reduce the effects of water absorption for laminates having cut fibers exposed to the pressure medium, protective coatings or metallic jackets may have to be employed. Past experience has shown the effects of such water absorption to be serious. Besides coatings and metallic jackets, resins with greater resistance to water and greater affinity to the fiber reinforcements may also alleviate this problem. [5]

Future Hull Design

"Hydrostatic tests with GRP hull elements held together with a protective metallic jacket have shown that this type of construction is of practical consideration. Little is known about filament-wound hulls in the following areas:

a) structural response to dynamic loading

b) structural response to long-term exposures to a deep-sea environment

c) repairs of damaged hulls made of non-weldable, low shear strength materials

d) the use of other reinforcements besides solid-glass fibers with circular cross-section

e) the use of other matrices besides epoxy resins." [5]

REFERENCES

1. Aerospace Technology, staff report, "Glass reinforced Plas-
    tic Hull Set for Delivery Soon to DSRV," pp. 39-41,
    November 6, 1967.

2. Freund, J.F., and Graner, W.R., "Filament Wound Plastics for
    Deep Submergence Pressure Hulls," BuShips Journal,
    pp. 2-9, November 1965.

3. Henton, D., "Glass Reinforced Plastics in the Royal Navy,"
    Read in London at a meeting of the Royal Institution
    of Naval Architects, March 23, 1967.

4. Hom, K., Buhl, J.E., and Willner, A.R., "Glass-Reinforced
    Plastics for Submersible Pressure Hulls," Naval
    Engineers Journal, pp. 827-834, October 1963.

5. Hom, K., Barnet, F.R., and Zable, J.J., "Applications Con-
    cepts for F.R.P.," NRL Report 6167, "Status and Pro-
    jections of Developments in Hull Structural Materials
    for Deep Ocean Vehicles and Fixed Bottom Installations,"
    Naval Research Laboratory, pp. 84-99, November 1964.

6. Kies, J.A. and Wolock, I., "Fiberglass Reinforced Plastics
    and Other Filamentous Composites," NRL Report 6167,
    pp. 54-83, November 1964.

7. Rosato, D.V., "Plastics in Oceanography," Plastics World,
    pp. 14-20, September 1965.

8. Siegrist, F.L., "Reinforced Plastics in Engineering Applica-
    tions," Metal Progress, pp. 93-99, May 1965.

9. Spaulding, K.B., "Fiberglass Boats in Naval Service,"
    Naval Engineers Journal, pp. 333-340, April 1966.

CHAPTER 8 GLASS AND CERAMICS

Since glass is one of the world's oldest materials, a great deal of information has been compiled concerning its characteristics and applications. In this chapter, the discussion of glass and ceramics will be restricted to properties most applicable to ocean engineering structures. The bulk of this chapter is paraphrased from a paper by H. A. Perry.[5]

8.1 Properties of Glass

Although several types of glass are today available commercially, one general grouping known as "E" Glass composes the major proportion of all glass produced in continuous filaments. Major families of glass are shown in Table 8-1, which outlines chemical requirements for various types of glass.[3]

Characteristics of "E" Glass

This type of glass is a lime-alumina-borosilicate glass which exhibits both high bulk electrical resistivity and high surface resistivity as well as good fiber forming characteristics.[3]

Chemical Composition. The primary components of "E" glass are SiO_2, Al_2O_3 and Ca. The alkali content ($Na_2O + K_2O$) of the glass is kept lower than 2% to provide for corrosion resistance and high surface resistivity. Boric oxide may be substituted as a network former for SiO_2 and MgO may be substituted for CaO. Table 8-2 shows the compositional range of "E" glass and a typical analysis in weight percent.[2]

TABLE 8-1: APPROXIMATE COMPOSITIONS OF COMMERCIAL GLASSES[3]

NO.	DESIGNATION	PER CENT								
		SiO_2	Na_2O	K_2O	CaO	MgO	BaO	PbO	BrO_2	Al_2O_3
1	Silica glass (fused silica)	99.5+	....	....	....	....	....	....	....	....
2	96% silica glass	96.3	0.2	0.2	....	....	....	....	2.9	....
3	Soda-lime-window sheet	71–73	14–15	....	8–10	1.5–3.5	....	....	....	....
4	Soda-lime-plate glass	71–73	12–14	....	10–12	1–4	....	....	....	0.5–1.5
5	Soda-lime-containers	70–73	Na_2O 13–16	K_2O	CaO 10–13	MgO	....	....	....	1.5–2.5
6	Soda-lime-electric-lamp bulbs	73.6	16	0.6	5.2	3.6	....	....	....	1
7	Lead silicate-electrical	63	7.6	6	0.3	0.2	....	21	0.2	0.6
8	Lead silicate-high lead	35	....	7.2	....	....	....	58	....	....
9	Aluminoborosilicate (apparatus)	74.7	6.4	0.5	0.9	....	2.2	....	9.6	5.6
10	Borosilicate-low expansion	80.5	3.8	0.4	....	....	....	....	12.9	2.2
11	Borosilicate-low electrical loss	70.0	....	0.5	0.1	0.2	....	Li_2O 1.2	28.0	1.1
12	Borosilicate-tungsten sealing	67.3	4.6	1.0	....	0.2	....	....	24.6	1.7
13	Aluminosilicate	57	1.0	....	5.5	12	....	....	4	20.5

In commercial glasses, iron may be present, in the form of Fe_2O_2, to the extent of 0.02 to 0.1% or more. In infrared-absorbing glasses it is in the form of FeO in amounts from 0.5 to 1%. The numbers listed in this table may be identified with commercial glasses as follows: 2, Corning glasses 7900; 791–; 7911; 6, Corning glass 0080; 7, Corning glass 0010; 8, Corning glass 8870; 9, Kimble glass N5la; 10, Corning glass 7740; 11, Corning glass 7070; 12, Corning glass 7050; 13, Corning glass 1710, 1720.

TABLE 8-2: COMPOSITIONAL RANGE OF "E" GLASS AND A TYPICAL ANALYSIS IN WEIGHT PERCENT[2]

	SiO_2	Al_2O_3	CaO	MgO	B_2O_3	$Na_2O + K_2O$	TiO_2	Fe_2O_3
RANGE	52	12	16	0	8	0	0[a]	0.05[a]
	56	16	25	6	13	3	0.4	0.4
TYPICAL	54.4	14.4	17.5	4.5	8.0	0.5	0	0.4

[a]The low values are for glasses of low color level.

Since the term "E" glass is a general category of glasses composed of the $CaO-Al_2O_3-SiO_2$ system, the composition of any particular glass may vary from any other particular glass within the general category of "E" glass.

General physical properties. Table 8-3 shows the physical properties of "E" glasses. Since "E" glass is not a single composition, the values in Table 9-3 are representative of the range of composition and property values are those at room temperature unless otherwise stated. Most of the physical properties relating to ocean engineering structures are given in Table 9-3.

Special Purpose Glasses

"C" glasses. This glass was developed for applications which required a greater corrosion resistance to acids than that available with "E" glass. The typical composition in weight percentage is:

SiO_2	65.0%
CaO	14.0
MgO	3.0
Na_2O	8.0
K_2O	0.5
B_2O_3	5.5
Al_2O_3	4.0 (2)

Besides its greater resistance to acids, "C" glass has a tensile strength (of virgin glass fibers) of about 450,000 psi and the Young's modulus of "C" glass fibers is 10×10^6.[2]

"D" glass. This glass was developed particularly to have

TABLE 8-3: PHYSICAL PROPERTIES OF E-GLASS FIBER[2]

Property	Condition	Value
Density		2.54 $\pm$ 0.03 gm/cm³ (bulk glass 2.58 $\pm$ 0.03)
Viscosity (bulk glass b)		
Strain point	(10 poise)	1140 F (616 C)
Annealing point	(10 poise)	
Softening point	(10 to 10 poise)	1555 F (846 C)
Young's modulus	73 F (23 C)	10.5 x 10⁶ psi
	73 F (23 C)	12.3 x 10⁶ psi (heat compacted)
	1000 F(538 C)	11.9 x 10⁶ psi (heat compacted)
Shear modulus		4.4 x 10⁶ psi (heat compacted 5.1 x 10⁶)
Bulk modulus		5.8 x 10⁶ psi (heat compacted 6.8 x 10⁶)
Poisson's ratio		0.20 (heat compacted 0.20)
Tensile strength c	73 F (23 C)	500,000 psi
	-45 F (-43 C)	745,000 psi
	-90 F (-68 C)	775,000 psi
	-310 F(-190 C)	820,000 psi
	1000 F(538 C)	250,000 psid
	1000 F(538 C)	120,000 psid
Yield strength		
Elastic elongation	73 F(23 C)	4.8
	10 cps	6.43
	10 cps	6.32
	10 cps	6.12
Specific heat	73 F (23 C)	0.19 Btu lb F or cal gm C
	500 F (260 C)	0.25
	1000 F(538 C)	0.29
Coefficient of thermal expansion		
80 to 212 F (25 to 100 C)		2.7 x 10⁻⁶ F (4.9 x 10⁻⁶ C)
80 to 1100 F (25 to 600 C)		3.3 x 10⁻⁶ F (6.0 x 10⁻⁶ C)
Thermal conductivity		6.7 Btu in hr ft F (8.3 cal cm sec C)

a low dielectric constant for use in randomes; its composition
has not been released. Fibers have a low density of 2.16 gm/cm^3
and an index of refraction of 1.47.[2] The modulus of elasticity
is 7.6 x 10^6 psi with a strength of 350,000 psi.

"M" glass. "M" glass was developed at Owens-Corning Fiberglas
under an Air Force contract to investigate the possibility of
fiberizing high-modulus glasses. As this glass has been developed
and used, it has shown excellent resistance to attack by water
and fair acid resistance. Young's modulus of as-drawn "M" glass
fibers is 15.9 x 10^6 psi, 50% greater than the "E" glass level.
The modulus of a thermally compacted fiber is 16.7 x 10^6 psi at
room temperature. The average strength of "M" glass fibers is about
500,000 psi, or about the same as "E" glass.[2]

"S" glass. The common variety of this glass is AF-994
or S-994 glass which was developed as a commercially fiberizable
glass with a higher strength and elastic moduli than "E" glass
and with better retention of strength at high temperatures. The
elastic modulus of as-drawn "S" glass fibers is 12.4 x 10^6 psi,
which is an 18% improvement in "E" glass in modulus and a 20%
improvement in modulus-to-density ratio. Thermally compacted
"S" glass fibers have an elastic modulus of 13.5 x 10^6 psi at
room temperature and a strength at 1200° F of 216,000 psi.

Strength

The strength of glass depends, of course, upon the specific
type of glass. In high-silicate glass, the high yield strength
is in the glass's random structure and high energy bonds. All

dⁱslocations are blocked. Silica tetrahedra are bonded randomly and strongly into a 3-dimensional network. These bonds must be broken for creep or flow.[4]

"The intrinsic compressive strength of a high-silicate glass is that of its random networks. It does not depend on the size, or shape or mass of the part."[4]

Tests of unstiffened glass cylinders and small hemispheres under pressure demonstrated that all failures were due to buckling. The highest stress obtained was 300,000 psi.[4]

Depth hardening effect. Ordinary glass, in a hollow structure, becomes increasingly resistant to damage by mechanical impacts or underwater shockwaves at great depths. This is the opposite of the "depth softening" effect of external pressure occurring with hulls composed of ductile alloys.[4] (See Figure 8-1).

Thermochemical strengthening. Tension and bending properties are low in abraded glass because of the high concentration of stress arising at the tips of cracks when the part is under load. Cracks do not seem to originate from the body of a glass part but from the surfaces and interfaces. This is due perhaps to the fact that the random networks of glass are as strong in tension as in compression. Surfaces of glass parts can be left in permanent states of compression by thermochemical treatments.[4]

This treatment involves chill-tempering, swelling all surfaces by an iron exchange reaction with a molten ionic salt bath and superficial phase changes which may occur in some glass compositions.[4]

"Small, transparent, low expansion crystals are generated

in all surfaces during a heating cycle. These crystals reduce the coefficient of thermal expansion of the outer layers which then shrink less, during cooling, than the rest of the part. All of these treatments lead to an intense permanent compression of the surface layers of the part, to varying degrees. They strengthen a part by keeping its surfaces from getting into tension while under a load."[4]

With no treatment, abraded glass parts may be as weak as 2,000 psi under bending or tension. After thermochemical treatment, the strength may reach 120,000 psi in tension or bending. "The modulus of rupture of a thermochemical-treated glass part is that of its surfaces. It does not depend on the size, shape or mass of the part."[4]

Stability

High-silicon glass does not creep under stress or become permanently deformed at ordinary temperatures. Glass parts do, however, go through a time of delayed elastic deflection or recovery after loading or unloading them.

Durability

In saline solutions, high silicon glass is resistant to early deterioration. The strength retention of thermochemically-treated glass in saline solutions is not well known. The leaching rate of essential elements from the compressed surface layer and its thickness determine the behavior. The time of treatment alone determines the depth of treatment of glass surfaces for compression.[4]

Non-Ductility

A ductile yield point over 10^6 psi - about E/10 - is shown
in high-silicate glass. Further, it does not creep at usual
temperatures. However, attempts at making high-silicate glass
ductile at lower stresses have been unsuccessful. If this
could be accomplished, the glass would lose some of its high
compression strength and creep resistance.[4]

8.2 Properties of Ceramics

Attempts have been made since the end of World War I to
combine metal and ceramics in order to achieve an end product
that combines the ductility of metal and the high temperature
strength and creep resistance of ceramics. Thus far, it has
been difficult to get the best properties of the two components
at once.[1]

"In general, starting from a metal and adding a ceramic
material in increasing proportion, where the ceramic particles
are relatively large there is at first little effect on the
properties of the metal. There may be some improvement in creep
resistance in special cases where the ceramic is finely dispersed.
With increasing addition of the ceramic material, there is
transition from a continuous metal phase to a continuous ceramic
phase, and the composition then has properties little different
from those of the ceramic."[1]

Types of Ceramics

Although ceramics may include a wide range of types, we will
discuss those most applicable to ocean engineering vehicles.

Ceramics with high thermal conductivity. An example of a ceramic with a high thermal conductivity is 1000°C. beryllia with a conductivity of 0.046, and silicon carbide 0.072, both of which have values as high as many metals at high temperatures. Beryllia is very toxic and therefore very little work is being carried out on its application. Silicon carbide is very good in some ways, but limited in use due to its tendency to oxidation and reluctance to sinter. Both are sometimes used in turbine blades.[1]

Low expansion ceramics. Fused silica is a good ceramic, with especially good resistance to thermal shock. This resistance is due to its low coefficient of thermal expansion (0.5×10^{-6} deg. c.). However, it has bad creep resistance and loses thermal shock resistance at temperatures approaching 1200° C.[1]

Silicon nitride is manufactured by a new process called reaction sintering. Compacts of silicon powder are heat-treated in a nitrogen atmosphere and the compacts convert to silicon nitrate and simultaneously sinter. The thermal shock resistance of silicon nitride is excellent, approaching that of metals. Also, it is dimensionally stable and resistant to oxidation to 1400° C.[1]

8.3 Design Considerations

Designs with Thermochemically Treated Glass

For years, the use of syntatic foams containing glass microspheres, hollow glass fibers and hollow glass balls of differing sizes as supplementary flotation materials has been studied.

"In brief, the rules for designing with an unyielding

material are: avoid buckling; avoid principal stresses in tension; avoid gross or sudden variations of wall thickness. Fortunately, the 'Hookian' or elastic behavior of glass simplifies all design computations. Also, photoelastic stress analyses can be performed on actual structures instead of models."[3]

Currently, most designs call for the use of a simple sphere or cylinder. More advanced work is being done experimenting with connected spheres and rib-stiffened cylinders with ogival or hemispherical closures.[4] Large size presents no technical obstacles since a glass hemisphere up to 56 inches in diameter is being cast, ground and polished for use. Tools are available also for structures even larger in size.[7]

8.4 Applications of Glass and Ceramics

The Naval Ordnance Laboratory has suggested the use of clad-glass hulls for deep diving submersibles because of:

1. new research and development in glass and ceramics, especially improvements in bending strength, resistance to impact and resistance to stress-corrosion cracking.

2. advances in the designs of nonmetal external pressure hulls.

3. great buoyancy of glass and ceramics.[7]

Flotation Units

Hollow glass spheres, made by welding pressed hemispheres together, is used in deep ocean applications. The greater the depth, the more the glass is resistant to damage by explosions.[4]

A relatively new borosilicate glass sphere with buoyancy/

displacement ratios of 0.55 withstands pressures up to 29,500 psi. At this pressure, the compression stress in walls of these spheres is 200,000 psi. Pressed and welded glass flotation systems may be more resistant to external pressures than is actually necessary for marine structures. In such cases, the safety margin is significantly increased.[4]

Applications of Brittle Materials

Designs and applications of brittle materials must recognize that:

1. glass and ceramics are strong in compression

2. glass and ceramics fail in tension, not in shear, and cracks begin only at flawed surfaces or flawed internal interfaces.

3. glass or ceramic that is highly surface compressed can "dice," that is, break itself up into small bits due to its stored elastic energy, once a crack begins to move.

4. dicing cannot occur if the center tensions are eliminated.[4]

REFERENCES

1. Aldred, F.H. and Hinchliffe, N.W., "General Developments
    in Ceramics for Marine Engineering," Transactions of
    the Institute of Marine Engineers, vol. 73, pp. 81-88,
    1961.

2. Broutman, L.J. and Krock, P.R., Modern Composite Materials,
    Addison-Wesley Publishing Co., Inc., 1967.

3. Mantell, C.L., editor, Engineering Material Handbook, McGraw-
    Hill Book Company, Chapter 27, 1958.

4. Perry, H.A., "The Argument for Glass Submersibles," Naval
    Engineers Journal, pp. 77-82, February 1965.

5. Perry, H.A., "Marine Applications of Glasses and Ceramics,"
    NRL Report 6167, "Status and Projections of Develop-
    ments in Hull Structural Materials for Deep Ocean
    Vehicles and Fixed Bottom Installations," Naval
    Research Laboratory, pp. 100-119, November 1964.

6. Perry, H.A., "Massive Glass for Deep Submergence," NRL Report
    6167, "Naval Research Laboratory, pp. 100-119, November
    1964.

CHAPTER 9   CONCRETE

Concrete is used extensively for civil engineering structures such as dams, piers and bottom foundations of various other structures. Because of its high compressive strength, concrete can be used for some ocean engineering structures, especially stationary structures on the ocean bottom. The majority of the material presented in this chapter is drawn from an article by Paul Klieger and Armand H. Gustaferro.[3]

9.1 Properties of Concrete

Consideration of concrete for use in ocean engineering structures should take into account the following properties:

1) strength, particularly compressive strength, of the concrete

2) permeability

3) resistance to attack by sea water

4) structural design

5) construction procedures and processes

Concrete meeting the above criteria can be used in deep ocean structures for housing personnel and equipment for long periods of time.

Strength

Compressive strengths of concretes currently in use range from 5,000 to 10,000 psi at an age of 28 days, depending upon the type of cement used, aggregate properties and degree of control of concrete making operations. Strength of up to 15,000 psi may be attained under laboratory conditions. If tensile strength is a consideration, it may be 7 to 10% of the compressive strength, and perhaps greater in the lower compressive strength ranges.

The density of the concrete material is affected by the type of aggregate used. Sand and gravels or crushed stone produce concretes weighing about 150 lb. per cu. ft. Portland cement clinker, magnitite, ilmenite, limonite and steel punchings yield densities up to about 300 lb. cu. ft. Table 9-1 shows the approximate maximum 28 day compressive strengths now possible under controlled conditions.

Section size influences the strength of concrete only slightly. A 6 inch diameter by 12 inch high cylinder, for instance, is the usual test specimen used for the strengths just mentioned. A 12 inch diameter by 24 inch high cylinder of the same concrete would have strengths about 9% lower. A reduction of strength up to a maximum of 5% is possible for more massive sections.

According to Klieger and Gustaferro, "It is probable that the level of strengths cited above will prevail for the next 5 to 10 years, with a probable increase of about 20 percent at the end of that period. By 1980-1985, it is probable that a further increase of 25 percent in strength will be possible."[3]

Permeability

Concrete is permeable to liquids and gases. This permeability may be diminished, however, by proper proportioning of low water-cement ratio, use of dense aggregates and adequate moist curing. If a high quality/high strength concrete is made with

TABLE 9-1   APPROXIMATE MAXIMUM 28-DAY COMPRESSIVE STRENGTHS[3]

Aggregate Type	Concrete Density in lb./cu. ft.	Approximate Maximum 28-day Compressive Strength, psi
Lightweight aggregate	100	8-10,000
Sand and gravel or crushed stone	150	12-15,000
Heavy aggregates	200 to 300	9-11,000

dense, relatively impermeable aggregates, and is cured moist and kept moist until submerged to a depth of 1,500 feet, the passage of sea water will be about 0.001 to 0.005 lb. of water/ day/ft.^{2}/ft. thickness (0.004 to 0.020 for 6,000 ft. depth). Maintaining the humidity inside the structure at a level below 100% increases the effective force driving water through the concrete. Fresh water permeability would be approximately 1/3 greater than that of sea water. Concrete constructed above the water should be saturated until submerged since drying increases permeability markedly. An impermeable sheath on the exterior surface is required if no passage of water can be tolerated.

Resistance to Sea Water Attack

Protection against attack from sea water (sulfates in solution in sea water) and leaching of lime from the concrete during the passage of sea water is provided through the use of high quality, high strength concrete demanded by structural and other considerations. Limiting the tricalcium aluminate content of the cement can further increase resistance to chemical attack in an ocean environment. The threat of sea water attack and leaching is eliminated if sheathing to avoid passage of water through the concrete is provided.

In the case of steel encased in concrete, corrosion of the steel is not highly probable if the concrete cover is at least 2 1/2 to 3 inches thick. Prestressing is not usually indicated in this application.

9.2 Design Considerations

In this section, the discussion is concentrated on two simple shapes likely to occur with concrete ocean engineering structures - spheres and cylinders.

Spheres

The axial (or tangential) stress in psi is:

$$s = 0.444 \, \frac{A_o}{A_c} \times \text{(Depth)}$$

where the depth is in feet. The coefficient 0.444 is the hydrostatic pressure in psi of a head of one foot of sea water weighing 64 lb. per cu. ft. The stress, s, is the compressive stress in the concrete if there is no reinforcement, or it is the weighed average stress in the concrete and steel if the concrete is properly reinforced. A_o is the area of a circle having a diameter equal to the outside diameter of the sphere. A_c is the cross sectional area of the concrete shell. Both A_o and A_c are in the same units, e.g., square feet. Figure 9-1 shows values of stress in spheres under various conditions.

Cylinders

In this type of design, the tangential stress, s, in psi in the shell is:

$$s = 0.444 \, (\frac{D_o}{2t}) \times \text{(Depth)}$$

D_o is the outside diameter of the cylinder in feet, and t is the shell thickness in feet. Figure 9-2 shows values of s in cylinders under various conditions.

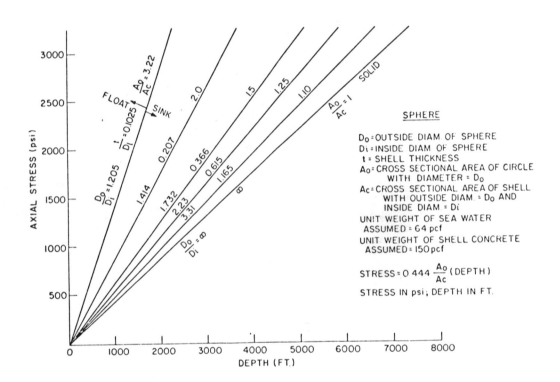

(3)

Figure 9-1. Stress values in spheres.

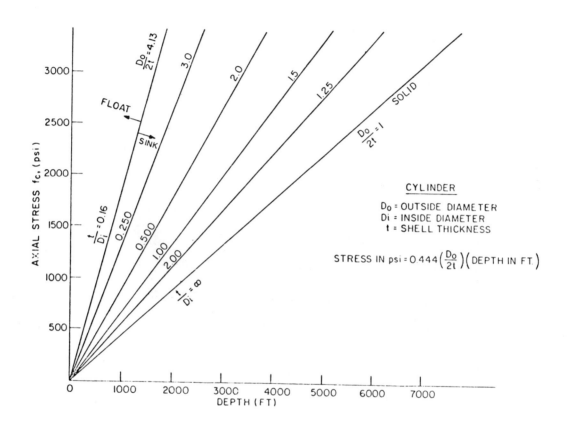

(3)

Figure 9-2. Stress values in cylinders.

General Considerations in Design

In design, any bending stresses in the shell should be minimized where ever possible. Such stresses are due to non-uniform loading of the shell and changes in the shell thickness. Non-uniform loading could be the consequence of the use of ballast if the structure were buoyant. The shell should be thick enough to result in submergence of the entire structure. Another consideration, however, is that the thinner the shell is, the better it is able to accommodate bending stresses, especially bending stresses at the openings. Therefore, the best thickness for deep submergence vehicles would be one that is just thick enough to cause submergence.

Figure 9-3 shows the material illustrated in Figures 9-1 and 9-2 in terms of thickness. Figures 9-4 and 9-5 are magnified views of Figure 9-3. The latter figures show the influence of reinforcement and increases in the unit weight of the concrete on the buoyancy of the structure.

Figures 9-1 through 9-5 illustrate the stresses due to hydrostatic pressure. An adequate factor of safety may be assured by a concrete strength that is at least 2 to 3 times as high as the calculated compressive stress.

Figure 9-6 shows the influence of unit weight of concrete on the buoyancy of the structure.

To further minimize bending stresses, the ends of cylindrical structures should be hemispherical in shape. Reinforcement may be provided with two equal mats of bars or welded wire fabric placed symmetrically within the shell. Small diameter bars are

213

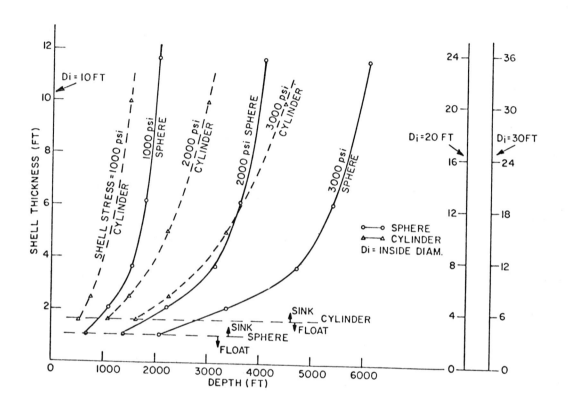

(3)

Figure 9-3. Thickness shown for spheres and cylinders.

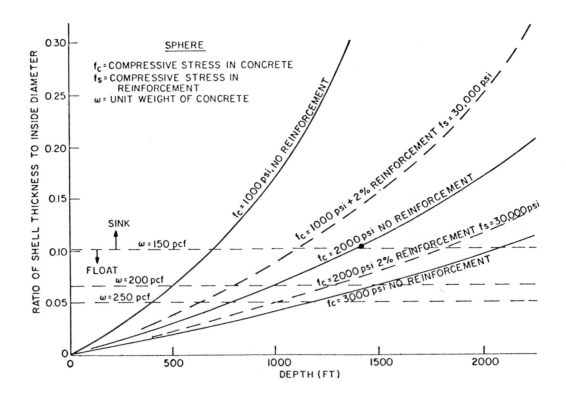

Figure 9-4. A magnified view of Figure 9-3, showing the influence of reinforcements.[3]

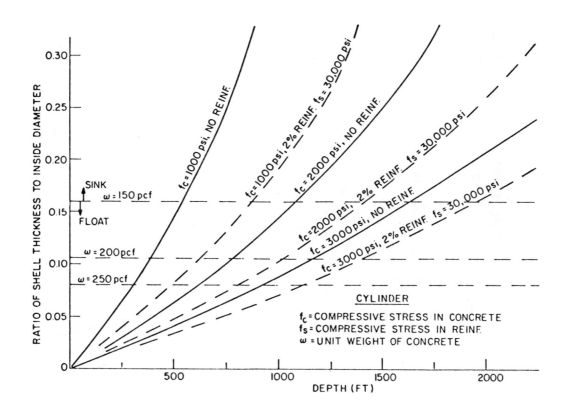

Figure 9-5. A magnified view of Figure 9-3, showing the influence of reinforcements and increases in the unit weight of concrete on the buoyancy of the structure.[3]

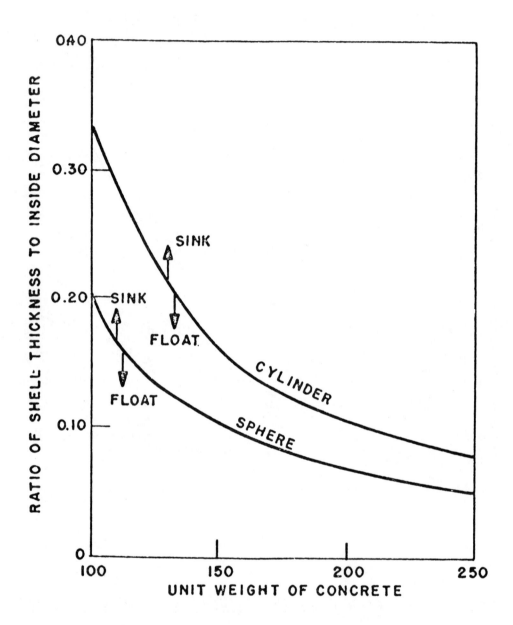

Figure 9-6. Influence of unit weight of concrete on the buoyancy
of the structure.[3]

preferred to large bars, but the openings in the mats should be no less than approximately 4 x 4 inches.

9.3 Ferro-Cement in Marine Applications

Ferro-cement is a composite material made of portland cement mortar and wire mesh. This material has received attention in recent years, including its application to a marine environment. The following is a brief discussion on some of the problems and characteristics of ferro-cement.

Characteristics

Ferro-cement exhibits many of the characteristics of other types of cement in its relationship to marine applications. Therefore, the discussion in this section concentrates on those areas of particular difference.

Reinforcement. Fibrous reinforcement of ferro-cement tends to enhance compressive strength whereas mesh reinforcements have little or no effect. Design compressive strengths in either case depend mostly on the unreinforced mortar compressive strength.[1]

Tensile loading. The characteristics of ferro-cement under tensile loading are not yet fully understood. However, there seems to be a great dependence on the bond strength between mortar and wire. An acceptance value of tensile yield stress is 1,000 psi for steel content which is above 4.5% by volume.[1]

Shear strength. Ultimate shear strength of both mesh and fibrous reinforced mortar seems dependent on the unreinforced mortar shear strength, increasing slightly with an increase in the steel content.

Fatigue. The fatigue limit for mesh and fibrous reinforced mortar is somewhat better than unreinforced concrete. It has been determined that a nominal design value of 60-70% of pre-fatigue static strength is a relatively realistic estimate.[1]

Impact damage. Impact damage may have a noticeable effect on ferro-cement. Therefore, design should incorporate a protective rub rail in the vicinity of expected impact points. Impact loads which do not cause visible damage may result in the disruption of the internal bond between mortar and wire and consequently reduce the strength of the material.

9.4   Research and Development

Research and development of concrete and a submersible structure should consider the following areas of interest:

1.   Permeability of concrete to sea water. Water-cement ratio, curing, aggregate type, sea water pressure and temperature, varied temperature and humidity on the interior surface would be areas to be considered. The leaching of lime from concrete would also fall into this area.

2.   New processes and techniques for significant increases in concrete strength.

3.   Corrosion resistance of reinforcement under the various exposure conditions possible.

4.   Construction practices needed to obtain the specified concrete properties and structural design features.

219

References

1. Collins, J. F., and Claman, J. S., "Ferro-Cement for Marine Application -- An Engineering Evaluation," a paper prepared for presentation before the New England Section of the Society of Naval Architects and Marine Engineers, M.I.T., March 1969.

2. Goode, R. J., and Judy, R. W., "Addendum to Concrete for Underwater Structures," NRL Report 6167, "Status and Projections of Developments in Hull Structural Materials for Deep Ocean Vehicles and Fixed Bottom Installations," U.S. Naval Research Laboratory, pp. 127-129, November 1964.

3. Klieger, P. and Gustaferro, A. H., "Concrete for Underwater Structures," NLR Report 6167, "Status and Projections of Developments in Hull Structural Materials for Deep Ocean Vehicles and Fixed Bottom Installations," U.S. Naval Research Laboratory, pp. 120-126, November 1964.

CHAPTER 10   BUOYANCY MATERIALS

Buoyancy materials will be considered in this chapter with
an emphasis on their application to submersibles. In this context,
types of buoyancy materials are discussed, as well as their
particular properties, their function in submersibles, and their
fabrication processes. The bulk of this chapter is drawn from
a paper by Bukzin and Resnick.[1]

10.1 Requirements for Buoyancy Materials

Density

The primary requirement for a buoyancy material is that its
density be significantly lower than that of water. At a specific
gravity range from 0.67-0.72, gasoline provides about 1/2 lb. of
buoyancy for each pound of fluid. However, the volume of tanks
necessary for such a ratio puts severe limitations on the structure's
operating capabilities. The propulsion unit, maneuverability and
speed of the submersible vehicle are all related to the buoyancy
tank size required. Based on the design requirements and operating
experience with the "Trieste," which used gasoline, the density
range of gasoline is an upper limit for any buoyancy material in
large applications.[2]

Cost

The expense of a particular buoyancy material would of
course rest on the material and the application. However, it
would be different to match or excel the cost of gasoline - 9¢/lb
of buoyancy.[2]

"An excellent standard would be $1.00 per pound of buoyancy.
As a basis of comparison, a 44 pcf syntactic for deep submergence
would cost $2.50-$5.00/lb. of buoyancy and lithium a minimum of
$15.00/lb. of buoyancy."[2]

Other Requirements

Besides the major demands listed above, a buoyancy material
should also prove non-corrosive, non-flamable, non-toxic and shock
resistant. Furthermore, its variation in compressibility over a
range of pressures and temperatures is very important. Buoyancy
systems closest to matching the variation of sea water with changes
in temperature and pressure are most desirable. Buoyancy depends
on the differences in density between the buoyant material and the
sea water medium. Any change in the temperature or pressure will
affect the buoyancy of the system.[2]

10.2 Various Types of Buoyancy Materials

Various types of buoyancy materials are used for different
applications. Buoyancy materials may be gases, liquids, or solids.
They are discussed in this section. Syntactic forms, which are
composite, light materials made up of low density hollow microsphere filler embedded in a resin matrix, also are used. Syntactic
forms are discussed in Section 10.3.

Gases

Maximum buoyancy at any submergence pressure could be provided
by hydrogen gas. Generated from a solid such as lithium hydride
or sodium hydride reacting with water, it could provide a "one
shot" buoyant force for the recovery of a vehicle or structure in

the event of an emergency. The hydrogen filled structure must be vented as it rises from the depth.[2]

The specific volume of hydrogen at 680 atmospheres and 0°C. is about 24.3cc/gm, which would provide a buoyancy of over 63 lbs. per cubic foot of hydrogen. This is the maximum buoyancy with no reductions for hydride weight or hydroxide by-product.[2]

Since hydrides are difficult to handle, store and keep stable, the application of such gases for buoyancy systems is limited. Other gases, with the exception of helium and neon, do not approach the buoyancy values of hydrogen.[2]

Liquids

Many liquids are used in submersibles, including naptha, gasoline, concentrated amonia solutions and silicone oil. Table 10-1 presents comparisons of properties of low density liquids.[1] All of the above are less dense than water and consequently provide buoyancy for submersibles. Enclosed in a thin-walled, light-weight shell, these buoyancy materials are under conditions of equalized internal and external pressures.[1]

Advantages. Liquids may still be considered for submersible buoyancy systems if disadvantages can be overcome. Positioning the material to obtain the best streamlining and trimming of the structure possible is relatively simple with a liquid buoyancy material. Releasing a liquid for an increased rate of descent is more practical with a fluid than a solid. Furthermore, equipment already available for loading and unloading ordinary gasoline or diesel oil on the high seas could be used for liquid buoyancy materials.[2]

TABLE 10-1   COMPARISON OF PROPERTIES OF LOW DENSITY LIQUIDS

PROPERTY	GASOLINE & NAPHTHA	AMMONIA SOLUTIONS	SILICONE OIL
Cost	Low	Low	High
Compressibility	High, causing a loss of buoyancy at great depth, thereby causing a trim stability problem	Low, nearly the same as sea water	High
Bulk Modulus x 10^5 psi	1.4	4	1.5
Fire Hazard	High, both aboard the vehicle and the mother ship	None	None
Thermal contraction	High, causing a loss in buoyancy	----	----
Relative buoyancy, pcf. (at 20° C) in 20° water	17	17.5	3-7
Effect on aluminum and copper base alloys	None	Highly corrosive	None

Disadvantages. Liquids display many disadvantages as buoyancy materials. Among these is the fact that they require delicate, thin shells which can be damaged easily with disastrous results. Increases surface handling of the vessel and greater depths make damage even more likely. Furthermore, many of the liquid buoyancy materials mentioned above have potentially hazardous properties. For example, gasoline is flammable, amonia solution is corrosive, and silicon oil is high in cost and low in net buoyancy.[1]

Solids

Low density solids most appropriate for submersible buoyancy systems include lithium metal, woods, organic polymers (such as polyethylene and polypropylene), expanded plastics, inorganic foams and syntactic foams. The latter will be discussed in a separate section. The basic properties of the above materials are outlined in Table 10-2.[1]

Potentials of low density solids. Lithium metal or wood have extremely limited potentials for buoyancy materials in submersibles. Lithium is very reactive and requires a container while the woods have low strengths and absorb water quickly. Foamed plastics and inorganic foams are also limited due to their low strength and permeability. As a group, the low density plastics exhibit limited buoyancy but may be applied as thick walled spheres or sections for buoyancy at moderate pressures. Syntactic foams show the best potential as buoyancy materials.[1]

Glass as a buoyancy material. Hollow glass spheres have a very high compressive and shear strength that exhibits a favorable

TABLE 10-2   COMPARISON OF PROPERTIES OF LOW DENSITY SOLIDS

MATERIAL	DENSITY	BULK MODULUS x 10^5 psi	COMMENTS AND LIMITATION
Lithium metal	0.53	22	Reacts vigorously with water releasing hydrogen. It must be contained in suitable non-corrosive metal containers which reduce buoyancy.
Wood	0.4-1.3	--	Has low compressive strength, absorbs water rapidly at pressures greater than 500 psi.
Solid Polyethylene, Polypropylene	0.9-0.95	2.6	Marginal compressive strength. Low buoyancy.
Expanded plastics	wide range	--	Have low strength. Are permeable to water and must be packaged in a barrier material.
Inorganic foams	wide range	--	Have low strength or open celled structure.
Syntactic foam	0.65-0.75	5.5	Offer exceptional promise for use at 10,000 psi.

weight/displacement ratio. Sympathetic "implosions" have
caused some difficulty in the application of glass spheres for
buoyancy. Currently, laboratory work is being carried out on
strengthened glass, that is, glass with a surface compression
layer. One company is offering glass spheres and cylinders of
various sizes and weight/displacement ratios for pressures down
to 10,000 psi.[2]

Glass reinforced plastics. This type of material has a
potential application as a buoyancy material due to its excellent
weight/displacement ratio. While it is being developed extensively
as a basic hull material, it could be used in a spherical structure
for flotation.[2]

10.3 Syntactic Foams

Syntactic foms offer the most promising applications as
buoyancy materials. The particular type of syntactic foam
depends upon the type of resin component, microsphere component,
ratio of each component and fabrication techniques.[4]

Applications

"Most of the current applications for syntactic foams are
for deep-submergence buoyancy. In addition, syntactic foam applied
as a structural, as well as a buoyancy material should also be
recognized. Typical applications of syntactic foam include the
following:

Buoyancy
Deep Sea Buoys
Oceanographic Instrumentation Capsules

Deep Diving Research Vehicles

Structural and Buoyancy

Submarine Void Filling

Fixed Deep Ocean Floor Installations

Sandwich Core Materials."[4]

Deep Sea Buoys. Buoys of this sort mainly offer buoyancy for diverse oceanographic installations. An electrical or acoustic device, for instance, may be moored by an anchor and buoy at some predetermined distance off the ocean floor. The NASL and Woods Hole Oceanographic Institution have developed six buoys of this sort. These are composed of a solid syntactic foam cylinder with an outside diameter of 11 inches and an overall length of 18 inches, with a net buoyancy of 22 lb.[4]

The Naval Applied Science Laboratory has also used syntactic foams as a part of a deep sea materials exposure array recovery system. In such a system, the foam is a spherical subsurface buoy which gives lift to a taut titanium monofilament line. Since such a buoy is not subjected to corrosion or water penetration, recovery of the buoy is highly reliable and use could extend to fairly significant depths.[4]

Oceanographic instrumentation capsules. The Woods Hole Oceanographic Institution has designed and constructed a sub-surface, acoustic beacon replacing the metallic container housing electronic circuits. This beacon uses a syntactic foam buoyancy collar which is placed around an acoustic transponder unit.[4]

In this and similar applications, the syntactic foam displays
the additional advantage of excellent acoustical transmission.

"For example, it was found that an acoustic signal generated
at 1MHz, at 1 atmosphere, is attenuated at a rate of approximately
3.5 db/in. in a block of syntactic foam having a nominal density
of 40 pcf. Measured attenuation, investigated in a frequency
range of 3 to 8 KHz at hydrostatic pressures to 6,000 psi, was
found to be less than 0.05 db/in."[4]

Deep submergence research vehicles. Syntactic has the advantages in deep diving submersibles of being strong, stable and
versatile. Gasoline is limited as a buoyancy material at great
depths due to its flammability and vertical depth control problems
from high compressibility and thermal contraction of gasoline
at great depths. A number of deep submergence structures use or
plan to use syntactic foams for buoyancy. These applications
include Lockheed's "Deep Quest," Woods Hole's "Alvin," Electric
Boat's "Star" and Westinghouse's "Deep Star."

Fixed deep ocean installations. The riser-syntactic foam
concept involves deep off-shore oil drilling operations. Such
an application was considered for the recently terminated project
MOHOLE, which would have used syntactic foam as a buoyancy
material to support the riser pipe. The drilling shaft, contained
within the pipe, was to be used in exploratory drilling into the
earth's crust and mantle through 14,000 ft. of water. The pipe
was to be supported by syntactic foam collars along its length.[4]

Properties of Syntactic Foams

Table 10-3 shows the typical properties of a family of syntactic foams which were made with hollow glass microspheres and epoxy resin. Different grades of hollow glass microspheres produce differences in the density and strength of each material. Differences in the ratio of hollow microsphere filler to resin matrix also accounts for variations in the final material. Different properties also result from the use of different component materials such as hollow, phenolic microspheres or various resin systems.[4]

Syntactic foams of high density are usually stronger than the low density foams. For a specific ocean depth and use, syntactic foams can be made to provide the maximum density and strength required.

Used as a deep ocean buoyancy material, syntactic foams exhibit density, compressive strength, bulky modulus, and water absorption to an advantage. In applications such as core material or void filling, the physical properties such as the compressive modulus of elasticity, shear strength, flexural strength and impact strength are important. These characteristics can be improved or adjusted within limits by the election of resin, microsphere or other components for the syntactic foam.[4]

TABLE 10-3: REPRESENTATIVE PHYSICAL PROPERTIES OF SYNTACTIC FOAMS PRESENTLY AVAILABLE, NAVAL APPLIED SCIENCE LABORATORY AND COMMERCIALLY DEVELOPED.

PRINCIPAL PROPERTIES

Nominal density, pfc	44
Net buoyancy, pcf	20
Initial compressive strength	
2% offset yield, psi	15,000
Ultimate, psi	18,000
Compressive modulus, psi	550,000
Compressive strength after hydrostatic immersion *	
2% offset yield, psi	14,000
Ultimate, psi	17,000
Compressive modulus, psi	520,000
Water absorption % **	1.5

ADDITIONAL PROPERTIES

Shear strength, psi	6,000
Tensile strength, psi	5,000
Tensile modulus, psi	600,000
Bulk modulus, psi	550,000
Impact strength, ft. lbs/in.	0.25
Fatigue, No. of cycles (between atmos. and 10,000 psi hydrostatic pressure) that material will withstand more than	10,000
Creep, no permanent deformation after 10 days hydrostatic immersion at pressures up to	12,000 psi

* 1000 cycles, each 20 min., between atmospheric and 10,000 psi hydrostatic pressure.
** Specimen size 1 x 1 x 1 in. Material shows lower percentage of water absorption as specimen size increases.

References

1. Bukzin, E. A., and Resnick, I., "Buoyancy Materials," NRL
    Report 6167, "Status and Projections of Developments in
    Hull Structural Materials for Deep Ocean Vehicles and
    Fixed Bottom Installations," Naval Research Laboratory,
    pp. 130-141, November 1964.

2. Hobaica, E. C., "Buoyancy Systems for Deep Submergence
    Structures," Naval Engineers Journal, pp. 733-741,
    October 1964.

3. Hobaica, E. D., and Cook, S. D., "The Characteristics of
    Syntactic Foams Used for Buoyancy," Journal of Cellular
    Plastics, pp. 143-148, April 1968.

4. Resnick, I., and MaCander, A., "Suntactic Foams for Deep Sea
    Engineering Applications," Naval Engineers Journal,
    pp. 235-243, April 1968.

CHAPTER 11   SEALING MATERIALS

The bulk of this chapter is drawn from an article
by Watt V. Smith.[1]

11.1   Requirements

The purpose of submersible seals is to provide closure
against sea pressure for various hull penetrations. Since there
are different sizes, speeds and operations of submersibles, the
design of seals has to be widely diversified. Since the type of
seal plays a significant role in the material selection, seals
will be discussed according to their applications.

Shaft Seals

Sealing materials intended for propeller shaft seals must
be functional within two limits--the lower leakage limit set by
the heat generated at the sealing surface and the upper limit
set by the maximum flow tolerable into the ship. The first depends
on the shaft diameter, shaft speed and depth of operation as well
as on the materials and surface condition of the seal surface.

"The propeller shaft seal is complicated by the necessity
for accepting a variety of shaft displacements. In the axial
direction the seal must accommodate uncertainties in the
positioning of the shaft with respect to the after bulkhead,
the displacement of the shaft resulting from changing operation from
ahead to astern, and from surface to fully submerged. These relatively slow axial displacements are also accompanied by axial
motions of small amplitudes occurring at shaft rotational

speeds as a result of out-of-squareness of seal rotor and thrust bearing rotor."

Figure 11-1 shows a very approximate arrangement of the submarine stern. The propeller and shaft are supported on flexible water-lubricated bearings.

Currently, fleet service involves an outboard inflatable safety seal, an installed spare face-type seal, an operating seal and emergency packing. The first type of seal, the outboard inflatable seal, is made of a tube of rubber with piping to the ship's interior which allows inflation of the rubber. When the rubber is inflated against the shaft, sealing is provided. Such sealing occurs only when the shaft is stopped while the vehicle is motionless on the surface or at a depth to provide required concealment. Since many of the inflatable tubes have been discovered to be torn or punctured, this type of sealing has not been considered highly successful.

The installed spare face-type seal is identical to the operating seal. Which of the two is sealing is determined by a valve placed in the cooling water supply to the seals. The seal rotor, a rotating component, is keyed to the shaft sleeve providing axial location and positive drive. Between the seal rotor and the shaft sleeve, an 0-ring gives a stationary secondary seal along the leakage path between the rotor and shaft. This rotor is made from monel and hard-faced with Stellite in the seal region except in the deepest diving submersibles where an insert of tungsten carbide with a nickel binder may be used. An example of a submersible using the

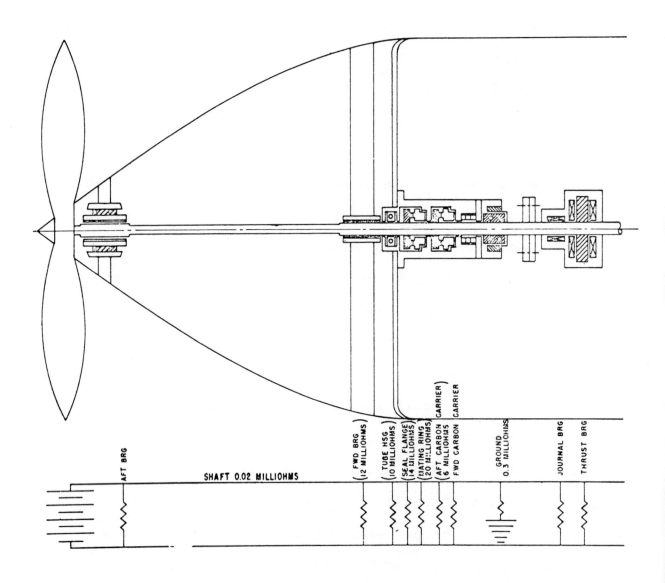

FIGURE 11-1. A SUBMARINE STERN WITH PROPELLER AND SHAFT
SUPPORTED ON FLEXIBLE WATER-LUBRICATED BEARINGS.

deepest diving submarine, AGSS 555.

Radio Antenna Seals

Radio antenna seals used for floating antenna must be able
to pass up to 3,000 feet of floating wire into the sea against
water pressures. Furthermore, the antenna seal must provide for
retrieving the antenna, cutting it off if necessary and automatically sealing the hull if the antenna should break and pass
out of the pass through.

This seal system involves an outboard ball valve welded into
the structure's hull, a ball check valve able to close it if the
antenna pulls out of the seal, an elastomer packing with external
hydraulic compression control, and a drain space and wiper.

In early antenna sealing systems, the chief failure occured
in worn places in the jacket which developed from slippage of
the rolls on the jacket. An endless track carrying the plastic
grippers over pairs of geared sprockets driven by hand or motors
was developed, however, minimizing this problem. The most likely
failures in such systems today is the possibility of cutting the
inner surface of the seal rubber by dirt trapped on the surface
of the antenna jacket, allowing the accidental entry of salt
water into the ball bearings of the drive mechanism.

Rod Seals

Operator rod seals must be present to transmit rotary or
translatory motion of small diameter shafts at low slow speeds
through the pressure hull. Grease packed, standard design
0-quad or Vee-ring packings are usually used. These packings

are self-acting, that is, increased pressure tends to reduce leakage. Periodic relubrication from inside the hull is provided for in all these seals.

The greatest limitation of rod seals is seal friction. A grease lubrication may reduce this friction while extrusion of the elastomer into the space between the rod and the housing may increase the torque.

Hatch Seals

Flexible packing seals are also used in hatches. In submersibles, hatches may be one of two types: 1) those in which pressure closes the hatch and 2) those in which pressure tends to open the hatch. As Smith explains,

"The hatches are basically flat-to-flat seals with an elastic lip-type seal between the surfaces. They may contain one lip for sealing in one direction or two-lip seals for sealing in two directions. As with most flexible seals, the elastomers are compressed about 10% of their free height to provide initial sealing. Even in large diameter missile seals, an initial out of flatness of 10 mils is of no consequence, since the pressure will flatten out the hatches. Such is not the case where the pressure tends to open the seal."

The greatest difficulty with modern hatch seals is protection against explosion.

Electrical Seals

To provide for an insulated path for a number of conductors through the pressure hull, electrical pass-throughs are needed.

Other than pressure stress, the mechanical stresses present are small. Glass-to-metal seals on the pin connectors are currently widely used. The glass bead is kept in a heavy compression by shrinking on an outer stainless steel plate. Many connectors may be held in the same plate. Standard O-rings seal against pressure in the fitting, which is held by a hull fitting welded into the pressure hull. An O-ring seal is also used between the electrical fitting and the hull fitting.

No failures up to 10,000 psi have been demonstrated by current testing.

11.2 Future Developments

Most research and development efforts should concentrate on the shaft seal since this seal involves the highest risk and is shortest lived of the seals presently used on submarines and similar submersibles. According to Smith, such an R&D program should emphasize the following:

"1. Analytical - Seek a solution of the coupled navier-stokes, elastic and possible energy equations for the face-type seal under rotation with periodically varying face waviness (or elastic modulus). This solution would be the complete analytical solution of face-type seals with real surfaces in contact.[2] The magnitude of the required solution will probably dictate using simplified solutions for the immediate future.

2. Experimental - Seek experimental verification of the analytical solutions proposing new models to the analysts as indicated by the experimental observations....

3. Development - Develop seals to meet specific requirements

for shaft size, speeds, pressure, and lives imposed by merging machinery concepts."

REFERENCES

1. Smith, W.V., "Submarine Seals," NRL Report 6167, "Status and Projections of Developments in Hull Structural Materials for Deep Ocean Vehicles and Fixed Bottom Installations," Naval Research Laboratory, pp. 231-239, November 1964.

2. Snapp, R.B., "Theoretical Analysis of Face-Type Seals with Varying Radial Face Profiles," M.S. Thesis, George Washington University, 1962.

PART II   SELECTED MATERIAL PROBLEMS

CHAPTER 12 FRACTURE TOUGHNESS

Since extensive publications exist on fracture tougness of
engineering materials, this chapter covers only subjects that are
important in understanding materials for ocean engineering structures.
For those who are interested in further information on fracture
toughness, the following textbooks and reports are recommended:

a. General Textbooks

1. Parker, E. R., Brittle Behavior of Engineering Structures,
   John Wiley and Sons, Inc., New York, 1957.

2. Biggs, W. D., The Brittle Fracture of Steel, Pitman
   Publishing Corporation, New York, 1960.

3. Hall, W. J., Kihara, H., Soete, W., and Wells, A. A.,
   Brittle Fracture of Welded Plates, Prentice-Hall, Inc., 1967.

4. Hayden, H. W., Moffat, W. G., and Wulff, J., The Structure
   and Properties of Materials, Volume III, Mechanical
   Behavior, John Wiley and Sons, Inc., New York, 1965.

b. Microfracture Mechanisms

5. Tetelman, A. S., and McEvilly, A. J. Jr., Fracture of
   Structural Materials, John Wiley and Sons, Inc., New
   York, 1967.

6. Averbach, B. L., Felbeck, D. K., Hahn, G. T., and Thomas, D. A.,
   editors, Fracture - Proceedings of an international
   conference on the Atomic Mechanisms of Fracture, held in
   Swampscott, Massachusetts, April 12-16, 1959, The Technology
   Press of M.I.T. and John Wiley and Sons, Inc., New York (1959).

7. Low, J. R. Jr.,"The Fracture of Metals," Progress in
   Materials Science, Pergamon Press, 12 (1), 1-96 (1963).

The following NRL report, although it is not available commercially,
is highly recommended:

8. Pellini, W. S., "Evolution of Engineering Principles for
   Fracture-Safe Design of Steel Structures," NRL Report
   6957, Naval Research Laboratory, Washington, D. C.,
   September 23, 1969.

12.1   Elementary Concepts of Fracture

Fracture is the separation of a body into two or more parts. The nature of fracture differs with materials and is often affected by the nature of the applied stress, geometrical features of the sample, and conditions of temperature and strain rate.[21] The differing types of fracture produced in ductile materials and brittle materials, under repeated stresses or at high temperatures, arise from differences in the modes of crack nucleation and propagation, which vary for each of these conditions.

Transgranular and Intergranular Fractures

Fractures in polycrystalline materials may be classified into transgranular and intergranular types depending upon the crack path.[43] The transgranular fracture traverses the grains of a polycrystalline aggregate, as shown in Figure 12-1a. In the intergranualr fracture, separation takes place between grain boundaries or along grain boundaries, as shown in Figure 12-1b.

Grain boundary separation is characteristic of the behavior of materials at elevated temperatures. Creep fractures, which occur under prolonged loading at elevated temperature, generally occur along grain boundaries. Intergranular hot cracks often occur in the weld metal during solidification and in the heat-affected base metal near the fusion zone where incipient melting occurs along grain boundaries. Intergranular cracks also occur in steel at room temperature under the simultaneous action of stress and of certain environments. This phenomenon, called stress corrosion cracking, is covered in Chapter 14.

Other fractures which include brittle and ductile fractures

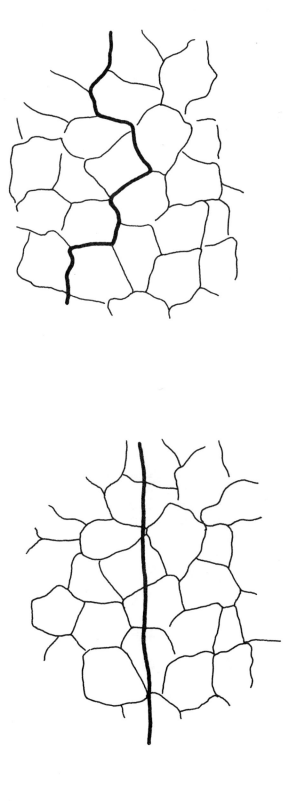

a. TRANSGRANULAR FRACTURE

b. INTERGRANULAR FRACTURE

Figure 12.1 Transgranular and Intergranular Fracture

(covered in this chapter) and fatigue fractures (covered in Chapter 13) are transgranular.

Cleavage and Shear Fractures

Two modes of fracture are of importance to the present discussion: cleavage and shear modes (both are transgranular).

The crystallographic nature of these failures can best be described with reference to the crystal structure of iron.[43] Figure 12-2 shows the unit cell of the body centered cubic (BBC) lattice, consisting of an atom located at each corner of the cube and another at its corner. Slip, or plastic flow, takes place by the shearing of certain crystallographic planes over one another. The important feature of the slip is that translation always occurs in the direction having the minimum interatomic distance and usually on the planes having the greatest atomic density.

There are three known types of slip planes in iron, all having common slip directions--the cube diagonals, such as the (110) plane shown in Figure 12-2a. Shear type fractures are promoted by the action of shear stresses and may be compared to the separation of cards in a deck when one part of the deck slides over the other until the translation separates the deck into two stacks.

The cleavage mode of fracture, on the other hand, is caused by normal tensile stresses and is typified by the fracture of mica when sheets are peeled apart. This type occurs in iron on a different set of crystallographic planes, of which (010) plane is shown in Figure 12-2b.

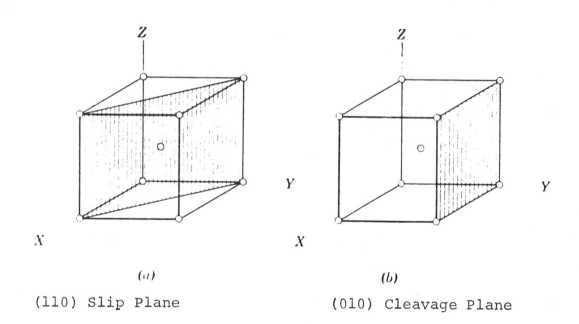

(a)

(b)

(110) Slip Plane

(010) Cleavage Plane

FIGURE 12-2 Body Centered Cubic Iron Showing Slip and Cleavage
Planes

Table 12-1 shows cleavage and shear planes for various structures and materials. The three most common crystal structures in metals and alloys are:

Body centered cubic (BCC)

Face centered cubic (FCC)

Hexagonal close packed (HPC)

All metals except those with an FCC structure have cleavage planes; therefore they show a ductile-to-brittle transition which is covered in this chapter. Aluminum alloys and austenitic stainless steels are two ordinary structural metals which do not fracture in a cleavage mode. These metals are commonly used for cryogenic applications.

Macroscopic Fracture Appearance. The two basic types of separation produce fractures that differ radically in macroscopic appearance. Figure 12-3 presents photographs of fractured steel specimens showing the fracture appearance for (a) shear, (b) cleavage, and (c) mixed shear and cleavage. Occasionally, different parts of the same specimen fail in different ways, resulting in a fracture with a mixed appearance.

The part failing by shear appears gray and silky, while the part failing by cleavage appears bright and granular.

Microscopic Observations of Fracture Surfaces. Cleavage and shear produce fracture surfaces which are distinctively different under the electron microscope. Figure 12-4 is a cleavage fracture surface of a low carbon steel broken by impact at 78° K. A cleavage fracture surface is characterized by what is called the "river pattern." It is believed that the river pattern is caused

TABLE 2-1   CLEAVAGE AND SHEAR PLANES FOR VARIOUS STRUCTURES
AND MATERIALS[60]

Crystal Structure	Example	Cleavage Plane	Primary Shear Planes
BCC	Li, Na, K, Fe, most steels, V, Cr, Mn, Cb, Mo, W, Ta, Ti (B)	{100}	{112}, {110}
FCC	Cu, Ag, Au, Al, Ni, brass, 300 series stainless steels	None	{100}
HCP	Be, Mg, Zn, Sn, Ti (α), U, Cd, graphite	{1000}	{1122}, {1010}, {1000}
Diamond	diamond, Si, Ge	{111}	{111}
Rock salt	NaCl, LiF, MgO, AgCl	{100}	{110
Zinc blend	ZnS, BeO	{110}	{111}
Fluorite	CaF_2, UO_2, ThO_2	{111}	{100}, {110}

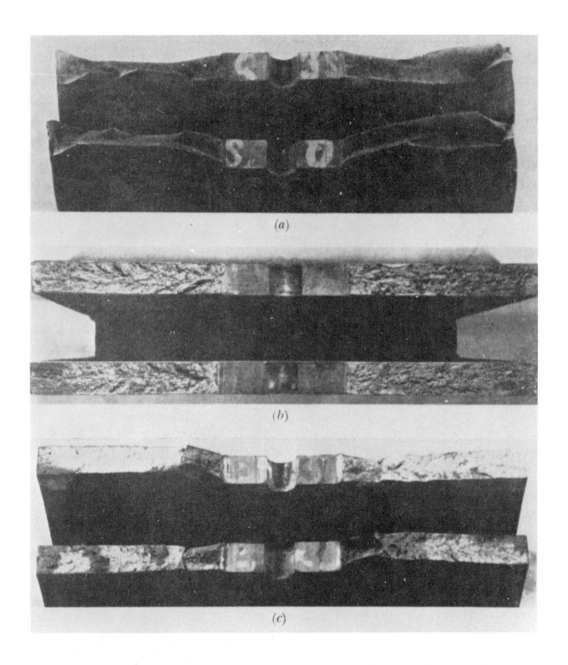

FIGURE 12-3   Photographs of Fractured Steel Specimens
Showing the Fracture Appearance for the
Following Modes of Fracture: (a) Shear,
(b) Cleavage, (c) mixed shear and cleavage.[43]

FIGURE 12-4   Cleavage Fracture Surface of a Low-Carbon
            Steel Broken by Impact at 78 K.[10]

ABC is a grain boundary; BD is a typical river marking.
The long arrows indicate local crack-propagation directions
and the circled arrow idicates a cleavage step. The facet
F is due to cleavage along a deformation twin.   Direct
carbon replica. (Reprinted from Ref. (2)).   5500 x reduced
in reproduction.

by the fact that transgranular cleavage fracture propagates on more than one level.[10,60]

Figure 12-5 shows a shear type fracture surface of a 4340 steel. The presence of "equiaxed dimples" on the fracture surface is the characteristic feature of a normal shear type fracture.[10]

Various Technical Terms Used

The literature has been somewhat confused by the use of conflicting and ambiguous terminology. For example, cleavage type fractures have been called cleavage, brittle, crystalline, or granular; and shear type fractures have been described as shear, ductile, silky, or fibrous.

While the crystallographic mode of fracture, i.e. cleavage or shear, is specific in meaning, the terms "brittle" and "ductile" are not. Brittle fracture literally means that separation occurred without plastic flow. However, even with the most brittle specimens, small amounts of plastic flow have been detected at the fracture surface by means of X-ray diffraction.

Ductile fracture, conversely, is considered to be associated with a substantial amount of plastic flow. Obviously, then, the dividing line between ductile and brittle fracture is an arbitrary one, depending upon the judgement of the investigator. Since there is no fixed boundary between ductile and brittle behavior, there is always room for differences in opinion among investigators. In general, a specimen having less than a few per cent reduction in area is called "brittle."

The following table, given in a textbook by Parker,[43] summarizes the terms commonly used in connection with discussions

FIGURE 12-5   Typical Shear-Rupture Dimples on Surface
of a Shear Lip in a Steel Specimen
3000 x Reduced in Reproduction.

of fractures:

Behavior Described	Term Used	
Crystallographic mode	Shear	Cleavage
Appearance of fracture	Fibrous	Granular
Energy or strain to fracture	Ductile	Brittle

Granular appearance is associated with cleavage fractures, and fibrous appearance is characteristic of shear.

12.2 Brittle Fracture of Welded Structures

Brittle fractures of structural steels have plagued engineers ever since the 1850's, when steel first became available in quantities large enough for structural use.[43] Brittle fractures are not uniquely associated with welded construction, and failures occurred in a number of riveted structures. Serious failures are, however, likely to occur in welded structures more than in riveted structures, because:

(1) A welded structure does not have riveted joints which can interrupt the progress of a brittle crack.

(2) Welds may have various defects including cracks, slag inclusions, etc.

(3) High tensile residual stresses in carbon steels and low-alloy high-strength steels have experienced catastrophic brittle failures.

Failures in Welded Ships. The most extensive and widely known failures are those of cargo ships and tankers which were built in the U.S.A. during World War II. To meet the urgent demand for a large number of ships needed for the war, the

United States entered for the first time in history the large-scale production of welded ships. At that time, the technique of welding steel plates had been well established. However, there had not been enough knowledge and experience regarding design and fabrication of large welded structures and their fracture characteristics.

A number of ship failures first occurred in the winter of 1942-43. Figure 12-6 shows a T-2 tanker, "Schenectady," which failed on January 16, 1943 at her fitting-out pier at Portland, Oregon. The failure occurred with no prior warning. The sea was calm, the weather mild, her computed deck stress was only 9,900 psi.

The fracture extended across the deck just aft of the bridge and about miship; the break extended down both sides and around the bilges, but did not cross the bottom plating.[43] The fracture transversed all girders and plating, thus almost completely severing the ship.

In April 1943, the Secretary of the Navy established the Board to Investigate the Design and Method of Construction of Welded Steel Merchant Vessels. The board issued a very comprehensive report in 1946.[43,67]

Among approximately 5,000 merchant ships built in the United States during World War II, about 1,000 ships have experienced about 1,300 structural failures in varying degrees before April 1946, when most of the ships were less than three years old.[7,67] Serious failures, such as complete fracture of deck and bottom plating, occurred in about 250 ships. These numbers do not include casualties resulting from war damage or from external causes such

FIGURE 12-6 A T-2 Tanker that Fractured a Pier[43]

as grounding or collision. About 20 ships were broken in two or
abandoned due to structural failures. Details of these failures
have been covered in several reports and books.

Broadly, it was concluded from these early investigations
that the fractures occurred due to the brittle behavior of the
steel in the presence of notches in the structure, either geometric
notches or defects in welds. It was also realized that conventional
factors of safety, based on the ultimate tensile properties of the
steel as usually measured, which had proved satisfactory, did not
cover this type of failure.[22]

Recommendations made as a result of these early investigations
were mainly directed toward minimizing all forms of notch effects
and toward improving the toughness of steel. For instance, the
American Bureau of Shipping Specifications included in 1948 notch
toughness requirements for hull steel by specifying various grades
and steel making practices. At the same time a great deal has been
done to improve welding technique and standards to minimize defects
in welds.

As a result of these improvements in design, materials, and
fabrication, the number and extent of brittle fractures that have
occurred in post-war welded or partially welded ships have decreased
drastically. However, brittle failure has not disappeared
completely. For example, between 1951 and 1953, two comparatively
new all-welded cargo ships and a tanker broke in two. In the
winter of 1954, a longitudinally framed welded tanker, in which
improved design, weld quality, and steel had been embodied, also
broke in two.[22]

256

<u>Structures Other Than Ships.</u> Brittle fractures in large structures made of medium carbon structural steel are far from uncommon. Shank[54] conducted an extensive survey of brittle fractures in structures other than ships. His report, published in 1953, covers 64 structural failures in both riveted and welded structures, including tanks, bridges, pressure vessels, power shovels, gas tansmission lines, a smoke stack, and a penstock.

In October 1886, a failure occurred during the hydrostatic test required for acceptance of a 250-foot high standpipe fabricated by riveting in Long Island, New York. A vertical crack about 20 feet long appeared at the bottom and the tower immediately collapsed. In January 1919, a riveted molasses tank in Boston, Massachusetts fractured, killing 12 persons and injuring 40 others.[43,54]

An early example of brittle failures in a welded structure is the failure in March of 1938 of a Vierendeel truss bridge over the Albert Canal in Hasselt, Belgium. The bridge was only about a year old, when it broke into three pieces and fell into the canal. Two other welded bridges over the Albert Canal suffered structural damages in January 1940, although they did not collapse.[43]

In January 1951, the Duplessis Bridge in Quebec, Canada, suddenly collapsed and fell into the river. Failures of many storage tanks and pressure vessels have been reported. A large number of failures have also been reported for transcontinental natural gas transmission lines. Many of these failures occurred during pressure testing, and often were due to defective welding.[43]

King's Bridge in Melbourne fractured in July (winter in Australia) 1962, when the bridge was only one year old. One span collapsed as a result of cracks which developed in a welded girder made in a high-tensile steel BS968 grade.[11]

Failures also have been reported in a number of high-strength steel members in military ships, aircrafts, and rockets. A recent example is the failure of a 260-inch-diameter rocket motor case. The motor case was constructed of 250 grade maraging steel plate joined mostly by submerged arc welding. The case fractured into pieces at a pressure which was about 56 per cent of proof pressure. A comprehensive report on this accident has been issued by NASA.[57]

Characteristics of Brittle Fractures

Brittle fractures which occurred in welded ships and other structures have the following characteristics:

1. Fracture appearance. A brittle fracture surface is approximately perpendicular to the plate surface, and it has a granular appearance. The reduction of thickness near the fracture surface is very small, ordinarily less than 3 per cent.

However, it is important to mention that steel plates which fracture are not brittle when tensile tests are conducted on ordinary specimens such as round bars and flat plates with no notch. Fracture generally occurs in shear mode after a considerable amount of plastic deformation has taken place. A question here is why a structure made of materials that are proved to be ductile under ordinary test conditions fails in a brittle manner.

A brittle fracture surface ordinarily contains chevron markings as shown in Figure 12-7. The chevron markings point back to the site of the flow responsible for fracture initiation. In the investigation of failures in an actual structure, chevron markings

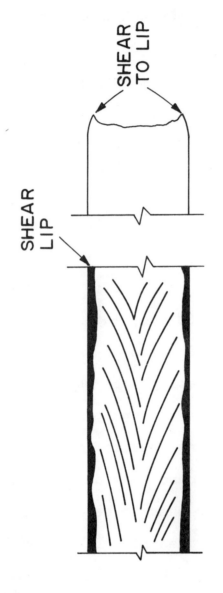

Figure 12.7 Brittle Fracture Surface

offer important information concerning where the fracture originated and how it propagated.

The right figure in Figure 12-7 shows a cross section of a plate near the fracture surface. Considerable deformation usually takes place in areas near the surfaces of the plate, and these deformed areas are called shear lips.

2. <u>Temperature.</u> Most brittle failures have occurred during cold winters. In the cases of brittle failures of World War II ships, failures did occur at temperatures as high as 70° F. However, according to the analysis by Acker, the probability of serious failures increases as the temperature is lowered.[43]

3. <u>Stresses at failures.</u> According to the statistics of brittle failures of World War II ships, more failures occurred in heavy sea than in moderate and calm sea. However, a number of failures occurred when the average stress in the structure was well below the yield stress of the material. For example, a T-2 tanker, "Schenectady," fractured completely when the calculated stress at the deck was only 9,900 psi.[43]

It is characteristic of brittle fracture that a catastrophic fracture can occur without general yielding under a low average stress.

4. <u>Origins of Failures.</u> According to a statistical investigation of failures of American ships built during World War II, about 50 per cent of the failures originated from structural discontinuities, including square hatch corners, cutouts in sheer strakes, ends of bilge keels, etc.[67] About 40 per cent of failures started from weld defects including weld cracks, undercuts,

and lack of fusion. The remaining 10 per cent of the failures
originated from metallurgical defects such as the weld heat-affected
zones and notches in flame-cut plate edges. In other words, all
failures originated from notches which created severe stress
concentrations.

A number of serious failures started from unimportant,
incidental welds such as tack welds and arc strikes made on a
major strength member. For example, a tanker, "Ponagansett," broke
in two in Boston Harbor in December 1947.[43] The fracture initiated from a tack weld between a small clip and the deck plate.

5. Fracture propagation. In most cases fracture propagated
in the base plate. Fractures seldom propagated in the weld metal
or along the heat-affected zone in carbon steel structures.

Fractures in deck or bottom plates in ships ordinarily propagate in the transverse direction because major stresses in a ship
hull are in the longitudinal direction.

Brittle fracture in a long pipe propagates in a rather unique
manner. In most cases, fractures travel in a fairly regular sinusoidal pattern, with a wave length of 3 feet or so, down the longitudinal axis of the pipe, as shown in Figure 12-8.[46]

Brittle fractures usually propagate at a very high speed.
In fracture tests of pipelines, propagation speeds around 3,000
feet per second have been observed.[46]

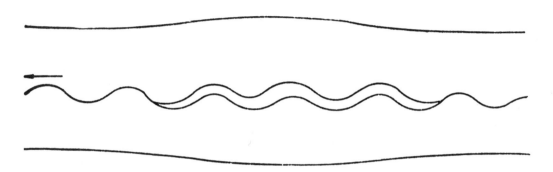

FIGURE 12-8 Sinusoidal Pattern of Cleavage Fracture
            of Pipeline

12.3 Notch Toughness of Steel

Notch Sensitivity and Transition Temperature

It has been established that to avoid brittle fracture of a welded structure, the materials used must have adequate notch toughness.

Low carbon steel, when it is tested with an ordinary tensile specimen, exhibits good ductility. However, when the steel contains a sharp notch and the temperature is low, a crack may initiate from the notch and the plate may fracture in a brittle manner. Such a phenomenon is called "notch brittleness."

The temperature at which the mode of fracture changes from shear to cleavage is called the "transition temperature." The transition temperature is often used as a parameter to express the "notch sensitivity," or the sensitivity of a material to notch brittleness. The term "notch toughness," which indicates the resistance of a material against notch brittleness, also is often used as an antonym of notch sensitivity. The following sentences show how these terms are used:

"In order to avoid brittle fracture of a structure, it is necessary:

    a. 'That the transition temperature of the material used is
       lower than the lowest anticipated service temperature of
       the structure.'

    b. 'That the material does not show notch brittleness under
       the service temperature.'

    c. 'That the material is not notch sensitive.'

    d. 'That the material with good notch toughness is used.'"

Although the transition temperature is widely used to express the notch sensitivity of a material, it is not a definite and fixed temperature for a certain material, such as the melting temperature and a transformation temperature. In most cases, brittle-to-ductile transition takes place over a range.

Furthermore, the transition temperature of a material varies depending upon a number of factors including size and shape of the specimen used, type of loading (tensile, bending, etc.), and loading speed (static or dynamic).

A material may exhibit a sharp transition over a narrow temperature range in one test, while the same material may exhibit a gradual transition over a wide temperature range in another test. Transition temperatures of a certain material determined in different tests may differ considerably. Even when the same test is used, the transition temperature may differ depending upon the criteria used (absorbed energy, mode of fracture, etc.).

Tremendous efforts have been made during the last 25 years by a number of investigations all over the world in search of a simple test which is most suitable for evaluating the adequacy of a material to avoid brittle fracture of a welded structure. Although many tests have been proposed, no test has yet been selected by the majority of investigators as the most appropriate test. Consequently, is presenting the transition temperature of a material, it is very important to describe the test method and the criterion used.

Among many tests that have been proposed to evaluate notch toughness, Charpy impact tests have been used most widely, especially for commercial applications. The U. S. Navy uses extensively the drop-weight test and the drop-weight tear test (recently renamed

the dynamic tear test), both developed at the Naval Research
Laboratory.

Charpy Impact Tests [34,36]

The details of the Charpy impact-testing procedures are specified by the American Society for Testing Materials. The Charpy
impact test uses a notched bar 10 mm. square and 55 mm. long (0.394
inch square and 2.165 inches long). ASTM specification E23-60 covers
three types of notches: (1) V-notch, a 45 vee-shaped notch 2 mm.
(0.079 inch) deep, (2) keyhole notch, and (3) U-notch, 5 mm. (0.197
inch) deep. Figure 12-9 shows the V-notch and the keyhole notch
Charpy specimens and the method of supporting and striking the
specimen.

A V-notch is smoothly machined with a special milling cutter,
a carbide-tip cutter, or a specially prepared grinding wheel. It
is important to prepare the root of the notch in an exact shape in
order to avoid scattering of experimental data.

The Charpy impact tests, especially with the V-notch specimen,
are widely used in the United States, Western European countries,
and Japan. In Russia and other Eastern European countries, the
Mesnager specimen is used. This is essentially a Charpy-type
specimen with a 2 mm. (0.79 inch) deep V-notch. [34]

Figure 12-10 shows three common ways of preparing Charpy
V-notch specimens from a plate. In Case 1, the specimen axis is
parallel to the rolling direction of the plate and the notch is in
the thickness direction. In Case 2, the specimen axis is perpendicular to the rolling direction and the notch is in the thickness
direction. In Case 3, the notch is parallel to the plate surface.

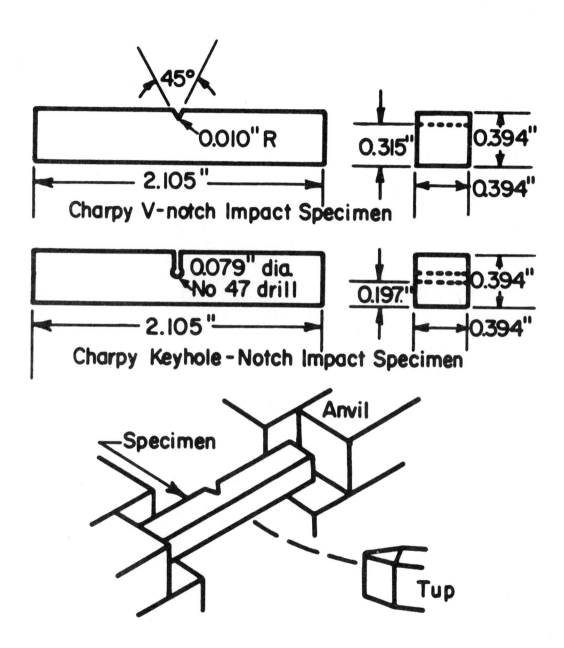

45°

0.010" R

2.105"

Charpy V-notch Impact Specimen

0.315"

0.394"

0.394"

0.079" dia.
No 47 drill

2.105"

Charpy Keyhole-Notch Impact Specimen

0.197."

0.394"

0.394"

Specimen

Anvil

Tup

FIGURE 12-9   Charpy Impact Test

266

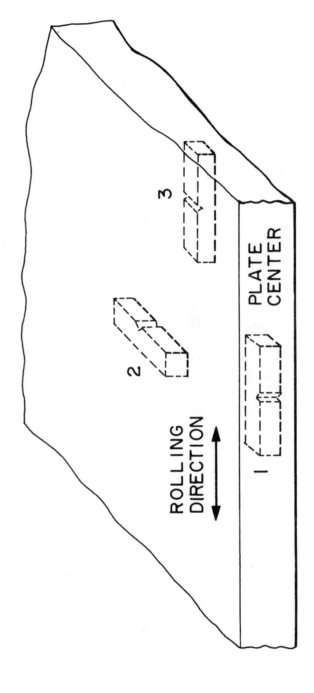

Figure 12.10 Methods for Preparing Charpy V-Notch Specimens from a Plate

Specimens may be taken from the midthick part of the plate as shown in Figure 12-10, or they may be taken in areas close to the plate surface. Case 1 is most commonly used for evaluating notch toughness of steel.

Standard Charpy specimens cannot be prepared from a plate thinner than 0.4 inch. Subsized specimens are often used for evaluating notch toughness of a plate thinner than 0.4 inch.

To obtain a Charpy curve, specimens are tested over a range in temperature. In conducting tests at other than room temperature, specimens are usually placed in a liquid bath to attain the desired temperature and then placed in an impact testing machine and struck with a pendulum. The energy absorbed by the specimen in fracturing is measured, and the percentage of the fracture area that is cleavage or shear can be determined.

Figure 12-11 shows an example of tests results on low carbon steel (carbon 0.23 per cent, manganese 0.46 per cent, silicon 0.05 per cent). Shown here are:[55]

(1) Absorbed energy, in foot lb.*

(2) Lateral contraction at the notch root, per cent

(3) Percentage of fracture surface area in fibrous appearance

---

*The Charpy V-notch absorbed energy is given by ft-lb or kg-m. The specific absorbed energy per square centimeter of sectional area (0.8 cm^2), $kg-m/cm^2$, also is often used.

$$1 \text{ ft-lb} = 0.138 \text{ kg-m} = 0.174 \text{ kg-m/cm}^2$$

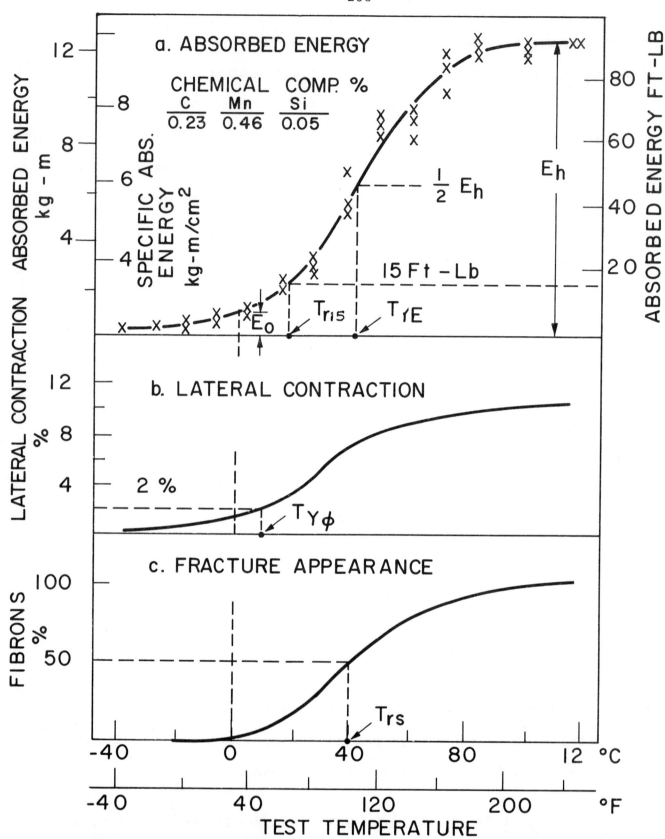

Figure 12.11 Example of Charpy V-Notch Impact Test Data

There are several criteria that have been used to interpret the results and determine the transition temperature:

(1) Temperature at which the amount of energy absorption decreases to one-half (or some other fraction) of the maximum amount at high temperature. TrE in Figure 12-11.

(2) Temperature at which lateral contraction at the notch root rapidly decreases or temperature which corresponds to a certain amount (2 per cent, for example) of the lateral contraction. Trϕ in Figure 12-11.

(3) Temperature which corresponds to 50 per cent fibrous fracture appearance. Trs in Figure 12-11.

(4) Temperature which corresponds to a certain energy level (15 ft-lb, for example). Tr15 in Figure 12-11.

The value of absorbed energy at a certain temperature (32 F, for example) also is often used as a parameter to express the notch toughness of the material.

Use of Charpy Impact Test Data for Evaluating Notch Toughness of Steel

What is the best criterion to evaluate the adequacy of a material for use in a welded structure is still a matter of controversy among investigators. Many studies have been conducted to determine correlations between service behaviors of welded structures and results of laboratory tests which include Charpy V-notch impact tests.

NBS Study on Ship Failures. In an extensive study carried out at the National Bureau of Standards on fractures of World War II ships, a correlation was obtained between Charpy V-notch impact test data and characteristics of ship failures.[66,67]   Figure 12-12

shows the spread of Charpy impact test data of ship fracture steels.

Charpy V-notch impact tests show that plates in which fractures

originated (source plates) were generally more sensitive than plates

which did not contain a source of fracture (through and end plates).

The correlation showed:

(1) Only 10 per cent of the fracture source plates absorbed
    more than 10 ft-lb at the failure temperature.

(2) 33 per cent of the fracture through plates absorbed more
    than 10 ft-lb at the failure temperature.

(3) 73 per cent of the fracture end plates absorbed more
    than 10 ft-lb at the failure temperature.

No relation was found between service performance and the

tensile properties of the steel.

Based upon the above finding, Williams[66,67] proposed that

the 15 foot-lb transition temperature (Tr15 in Figure 12-11) be

used to evaluate the adequacy of carbon steel for welded ships.

NRL Studies. Pellini and associates conducted a series of

studies on the usefulness of the Charpy V-notch impact test, from

a rather critical point of view.[44]

Figure 12-13 illustrates a typical correlation of the fracture

performance in the explosion crack starter test with the charpy V-notch

transition curve for steels taken from fractured ships. The

explosion crack starter test, which will be discussed in a later

part of this chapter (Figure 12-23), was used as a method to

simulate the service performance of steels. The transition from

ductile fracture to brittle fracture (or flat break) occurred at

temperatures which correspond to 10 to 20 ft-lbs of absorbed energy

in the Charpy V-notch test. The results indicate the adequacy of the

271

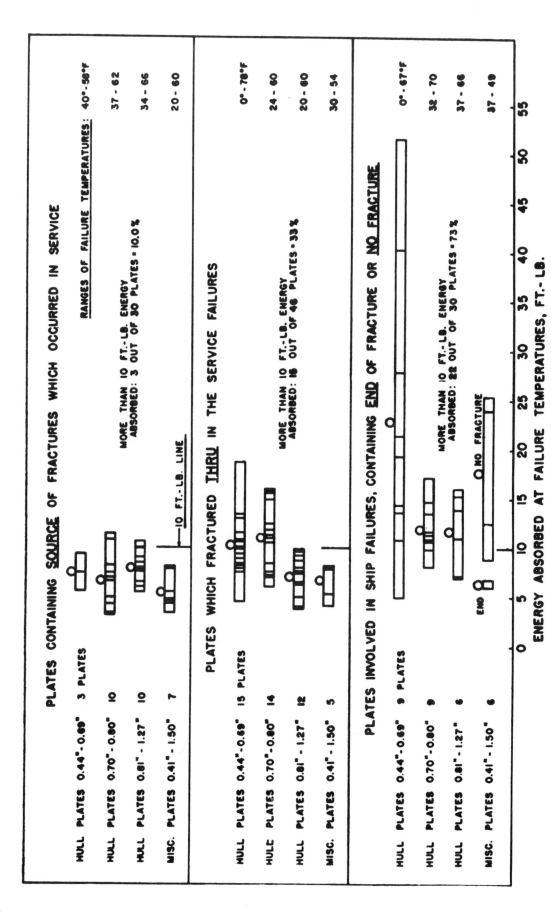

FIGURE 12-12 Relation of Energy Absorbed by Charpy V-Notch Specimens at the Temperature of the Ship Failures to the Nature of the Fracture in the Ship Plates. Vertical Lines in the bars Indicate Values for Individual Plates. Circles above Bars Indicate Average Value for each group of Plates.(66)

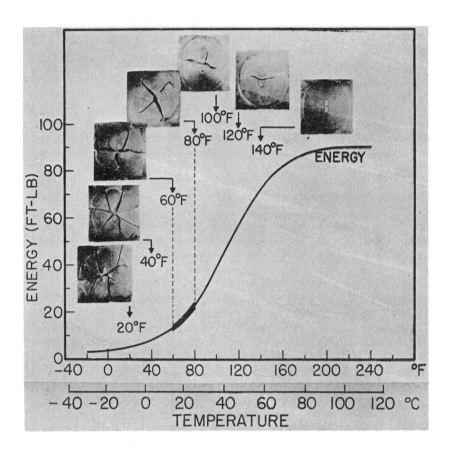

FIGURE 12-13 Typical Correlation of Explosion Crack Starter
Test Performance with Charpy V-Notch Transition
Curves for Ship Fracture Steels. Flat Break (NDT)
Fractures are obtained at temperatures below the
10 ft-lb transition temperature.[44] Arrest
Characteristics are developed in the range indicated
by the solid band which spans the 10 to 20 ft-lb
transition temperature range. Full ductility is
attained on approach to shelf temperatures.

Charpy V-notch 15 ft-lb transition temperature for evaluating notch toughness of low carbon steels.

Pellini and associates expanded their studies to cover different steels including improved carbon steels, high-strength quenched-and-tempered steels, and intermediate strength low-alloy bainitic steels.[44] Results are summarized in Figure 12-14. It was found that NDT temperatures correspond to Charpy impact test temperatures at energy values much higher than 15 ft-lb. Plus the corresponding energy values vary greatly depending upon steel types.

On the basis of the above findings, Pellini and associates at the Naval Research Laboratory have developed the drop-weight test and the drop-weight tear test (recently renamed the dynamic tear test), which will be covered in the following pages.

NRL Drop-Weight Test

The drop-weight test was developed by Pellini et al. of the Naval Research Laboratory. The type of specimen and the method of test are shown in Figure 12-15. The test procedure consists of dropping a weight on a rectangular flat-plate specimen containing a crack starter of a notched, brittle, hard-surfacing weld bead. This procedure is employed over a range of specimen temperatures. The specimen support is provided with a stop such that the maximum angle of bend of the specimen is 5 degrees. The crack starter weld develops a cleavage crack when the bend deformation reaches 3 degrees, which corresponds to incipient yielding. The additional 2 degrees of bend (dynamic bend) allowed by the stop provides a test of the ability of a metal to develop deformation in the presence of an extremely sharp notch.

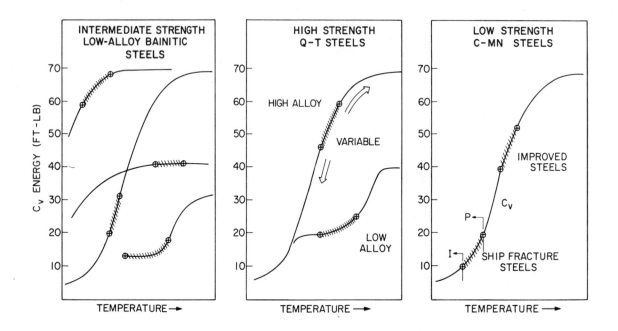

FIGURE 12-14   Complications in the Interpretation of Charpy V-Notch
Impact Test TRansition Temperature Range Curves for
Different Steels.   The relative positions of the
initiation to arrest correlation bands (NDT to FTE)
are shifted widely in comparison to the toe region
position of the Charpy V-Notch curve which was typical
of the ship fracture steels.[44]   The correlations for
improved ship steels (c-Mn steels) are displaced to
higher relative position of the Charpy V-Notch curve.
More complex relationships are indicated for steels
of over 50 ksi (35 kg/mm^2) yield strength.

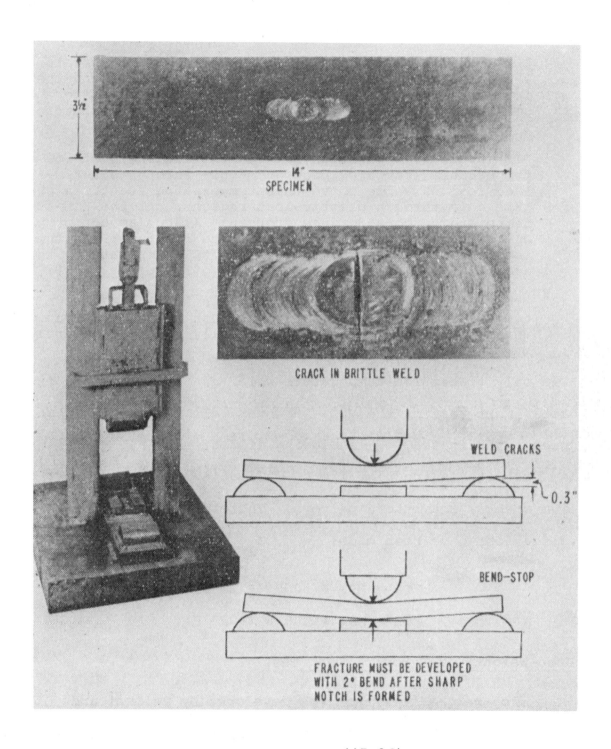

FIGURE 12-15   NRL Drop Weight Test [47,36]

The drop-weight test results evaluate the performance of a steel in the presence of a sharp notch in terms of three critical fracture transition temperatures[48] proposed by Pellini:

(1) NDT (nil-ductility transition). Below this temperature the steel does not deform prior to fracturing, and the fracture occurs immediately upon reaching the yield point. The fracture propagates easily through elastic-load regions.

(2) FTE (fracture transition for elastic loading). Below this temperature fractures propagate only through elastic-load regions while above this temperature fractures propagate only through plastic-load regions. At the FTE temperature, considerable "forcing" (large amount of plastic deformation) is required to initiate fracture. As the temperature is lowered to approach NDT temperature, the amount of "forcing" required falls off rapidly.

(3) FTP (fracture transition for plastic loading). Above this temperature cleavage fractures cannot propagate, even through material which is plastically deformed severely by high overstressing.

The only transition temperature which can be determined from the drop-weight test is the NDT. However, all the materials which have been investigated,[48] including mild steels, high-strength steels, cast steels, nodular irons, malleable iron, and 12 per cent chromium steels, fit into a pattern. Therefore, the following generalizations have been made:

$$\text{FTE} = \text{NDT} + 50\ \text{F}\ (\pm 10\ \text{F})$$

$$\text{FTE} = \text{NDT} + 100\ \text{F}\ (\pm 20\ \text{F}).$$

NRL Drop-Weight (or Dynamic) Tear Test

The drop-weight tear test was developed by Pellini and associates[44] at the Naval Research Laboratory. The first version featured a notch brittle bar welded to a test specimen. By 1964, the DWTT was redesigned to eliminate the brittle crack starter bar and the test was redefined as the Dynamic Tear (DT) test.

Figure 12-16 illustrates the feature of 5/8 in. and 1-in. thick DT specimens. The original standard version involves a deep sharp crack introduced by the use of an electron beam weld which is embrittled metallurgically by alloying. For example, a titanium wire added to the site of the weld results in a brittle Fe-Ti alloy. The narrow weld is fractured easily in loading and thus provides a reproducible sharp crack. It has now been established that equivalent results may be obtained by the use of a deep sharp crack produced by fatigue or by slitting, and then sharpening a deep notch by a pressed knife edge. DT specimens featuring a deep flaw produced by any of these methods are tested over a range of temperature using a pendulum-type machine.

Ductility Transition, Fracture Appearance Transition and Crack Arrest Temperature

By using different type specimens and different criteria, a variety of transition temperatures have been used by different investigators. These transition temperatures may be classified into three groups: ductility transition temperature, fracture (appearance) transition temperature, and crack arrest temperature.

Ductility Transition Temperature.
In order that brittle fracture occur, it is first necessary to initiate a brittle crack from the pre-existing notch. Consequently, whether or not a crack initiates under a small amount of plastic strain, or absorbed energy is an important criterion. A transition temperature can be determined on the basis of the amount of deformation or energy absorption required to initiate a crack. This type of transition

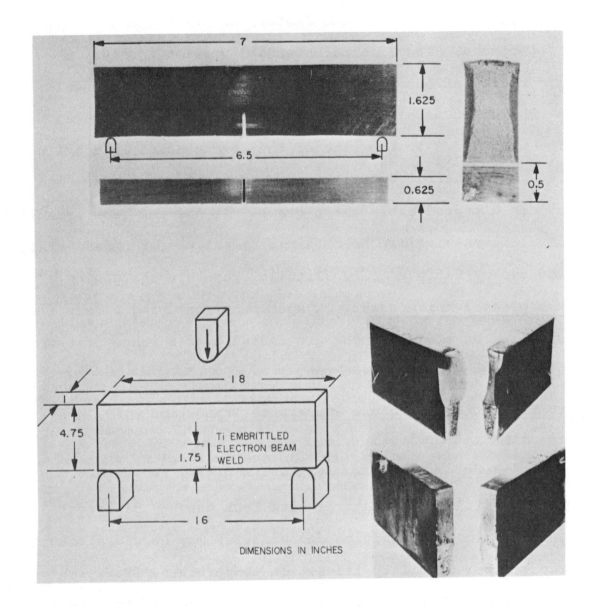

FIGURE 12-16   Features of   5/8 in. and 1 in. DT Test Specimens.[40]
The 5/8 in. DT specimen (top) features a machine slit,
with a knife-edge-sharpened notch tip. The 1 in. DT
specimen (bottom) features the brittle electron beam
weld, which is also used for the 5/8 in. DT, as desired.
The broken halves of the 1-in. DT specimens illustrate
brittle and ductile type fractures.

temperature is called the "ductility transition temperature, T_d ."
At a temperature below the ductility temperature, fracture occurs
in a completely cleavage mode.

Transition temperature determined by (1) deformation at the
notch root, (2) deformation at the maximum load, and (3) energy
absorption before crack initiation can be considered as "ductility
transition temperatures." The NDT temperature determined by the
NRL drop-weight test, and the dynamic tear test is a typical example
of ductility transition temperature. In case of the Charpy V-notch
impact test, $Tr\phi$ and $Tr15$ shown in Figure 12-11, can be considered
to be ductility transition temperatures.

Fracture Appearance Transition Temperature. When a test is
made at a temperature somewhat above the ductility transition
temperature, a shear type crack initiates from the notch root.
When the test temperature is sufficiently high, the crack continues
to grow in the shear mode. However, when the temperature is not
high enough, the crack becomes unstable when its length reaches a
certain critical size and the mode of fracture changes to cleavage
resulting in a rapid fracture propagation. The amount of energy
absorbed during the growth of the shear crack and the size of
critical crack change with temperatures. Transition temperatures
determined by these criteria are called the "fracture appearance
transition temperature, T_f," because in this transition temperature
range, the fracture appearance changes significantly. For a
given material, T_f is always higher than Td.

Transition temperatures determined by percentage of fibrous
fracture appearance, energy absorption after the maximum load,

and other parameters which indicate the amount of shear fracture are considered to be Tr_f. In the case of the Charpy V-notch test, for example, TrE and Trs shown in Figure 12-11 are fracture appearance transition temperatures.

Crack Arrest Temperature. The crack arrest temperature is determined by the Robertson test, ESSO test, and the double-tension test, which are designed to study fracture propagation characteristics of a material. These tests are described later in this chapter. Figure 12-17 shows a typical result which shows the relationship between the test temperature and the critical stress necessary for fracture propagation. When the temperature is below a certain value, brittle fracture can propagate at a fairly low stress. However, when the temperature is above a certain temperature, brittle crack is arrested even under fairly high stress. This limiting temperature is called the "crack arresting (transition) temperature, Ta."

12.4 Various Tests for Evaluating Notch Toughness[34,36]

Many investigators have used various types of tests for evaluating notch toughness of steel. Table 12-2 lists typical tests for evaluating notch toughness, which can be classified into the following four groups:

(a) Impact tests with small specimens

(b) Static tests with small specimens

(c) Fracture tests of weldments by dynamic loading

(d) Wide-plate tension tests.

Some of the laboratory tests appear to emphasize the initiation of cleavage cracks and others in the stopping of a propagating

281

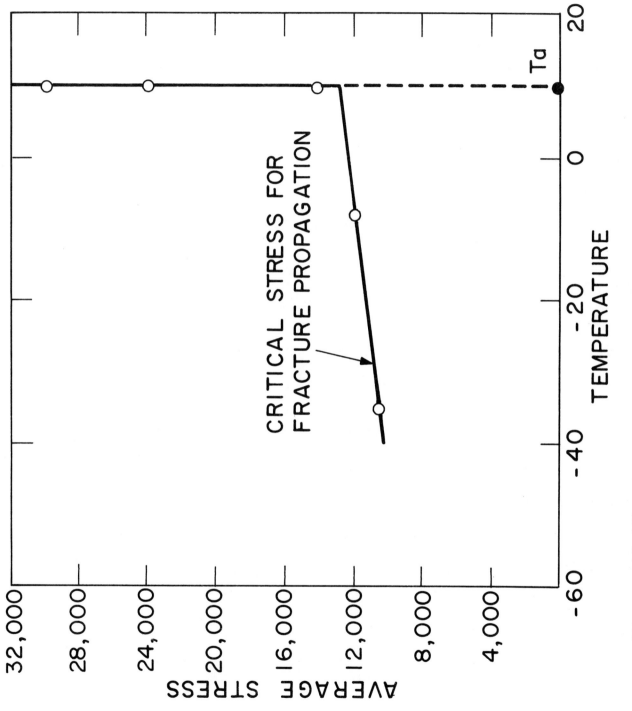

Figure 12.17 Crack Arrest Temperature

TABLE 12-2   VARIOUS TESTS FOR EVALUATING NOTCH TOUGHNESS OF STEEL
AND BRITTLE-FRACTURE CHARACTERISTICS OF WELDS*

A.   Impact Tests with Small Specimens

      1. Standard Charpy impact tests and modified tests
      2. NRL Drop Weight test
      3. NRL DT test

B.   Static Tests with Small Specimens

    B-1 Static Fracture Tests on Notched Specimens
      4. Tipper test
      5. Navy tear test
      6. Van der Veen test
    B-2 Bend Tests on Welded Specimens
      7. Lehigh test and Kinzel test
      8. Kommerell test

C. Fracture Tests of Weldments by Dynamic Loading

      9. Crack-starter explosion test
    10. Explosion bulge test

D.   Tensile Tests of Wide-Plate Specimens

    D-1 Fracture Propagation Tests
     11. Robertson test
     12. ESSO (or SOD) test
     13. Double tension test
    D-2 Welded-and-Notched Wide-Plate Tensile Tests
     14. Wells-Kihara test

---

*This table does not include those tests which are aimed at determining the critical stress intensity factor, K_c. Fracture toughness tests are shown in Figures 12-36 and 12-37.

cleavage crack, but almost all of the tests involve the introduction of a notch and the observation of brittle behavior as the
test temperature is lowered. Each of these tests emphasizes
different features of the brittle-fracture process, and it is not
surprising that thay may rate the ability of a material to resist
cleavage fracture differently.

The tests almost always define a transition temperature below
which cleavage fracture occurs under the test conditions.

Impact Tests with Small Specimens

Impact tests using small specimens have been developed primarily
to evaluate notch toughness of the base plate. The standard Charpy
impact tests, the NRL drop-weight test, and the NRL drop-weight
tear test (or dynamic tear test) have been covered previously.
Consequently, the following pages cover tests other than the above
three.

Modified Charpy Tests. There are a number of tests which may
be classified as modified Charpy tests, since they either (1) use
the standard specimen but employ different test procedures or (2)
use slightly different specimens.

The "double blow" or "low blow" technique has been proposed
and used by Orner and Hartblower[41] to evaluate separately the
energy required to initiate a crack and that required to propagate
the crack. A blow with the pendulum is applied twice; the first
blow with low energy is applied to initiate a crack, and the
second blow is applied to propagate the crack. The low-blow
transition temperature is believed to be the maximum temperature
at which an initiating crack can become self-propagating in a thick

plate (where the energy required to produce shear lips is small compared with the elastic energy available for crack propagation).

Several investigators have proposed the use of pressed-notch specimens instead of specimens with machined vee notches.[69] A pressed notch can be made by pressing against the specimen surface a tool in hard material with a vee-shaped edge.

Subsized specimens are often used to evaluate notch toughness of plates thinner than 3/8 inch, while specimens larger than the standard size also have been used for heavy plates. A problem involved in the use of specimens with sizes different from the standard size is how to interpret the results. For example, the transition temperature obtained with subsized specimens is shifted to a lower value because of a decreases in the degree of triaxiality of the stresses near the root of the notch. However, the shift is not the same for all regions of the curve.

Schnadt Test. Schadt developed a series of notch impact specimens which can be tested with a Charpy machine.[53] There are five standard specimens with V-notches of varying root radii. In the specimens, where the notch-root radii are larger than 0.5 mm, the notches are made with milling cutters of the requisite root radii. In the "coheracic" specimen, the notch is made by pressing a tungsten carbide knife into a previous notch. A pin made of tungsten carbide is inserted in a cylindrical hole on the compression side of the specimen. This is done in order to prevent formation of a compression zone in the portion of the specimen where the crack propagates; all of the cross section of the specimen remains in tension.

285

Static Tests with Small Specimens

Static tests using small specimens include:[34]

(A) Static fracture tests on notched specimens

    (1) Notched-specimen tension tests, such as the Tipper test[1] and the Noren nominal cleavage (N-C) strength test[40]

    (2) Tear tests, such as the Navy tear test[25]

    (3) Notched-specimen bend tests, such as the Van der Veen test[4,61]

(B) Bend tests on welded specimens

    (4) Longitudinal-bead-on-weld notched-bend tests, such as the Lehigh test[58,59] and the Kinzeltest[32]

    (5) Longitudinal-bead-on-weld bend tests, such as the Kommerell test[33]

Tests in Group (A) have been developed primarily for evaluating the notch toughness of unwelded base plates. Tests in Group (B) have been developed for evaluating the fracture characteristics of weldments or the effects of welding on the base plate, i.e., the weldability of steel.[59]

Tipper Test. Tipper and Baker[1] at Cambridge University developed a side-notch tension specimen as shown in Figure 12-18. A specimen is prepared in the full thickness of a test plate and then tested over a range of temperatures. The transition temperature is determined either from the reduction in thickness (at the middle point of the fractured surface between the notches) or from the fracture appearance.

Navy Tear Test. The specimen utilized in the Navy Tear Test, originated by Kahan and Imbembo[25] at the New York Naval Shipyard, is shown in Figure 12-19. The test specimen is flame cut from the

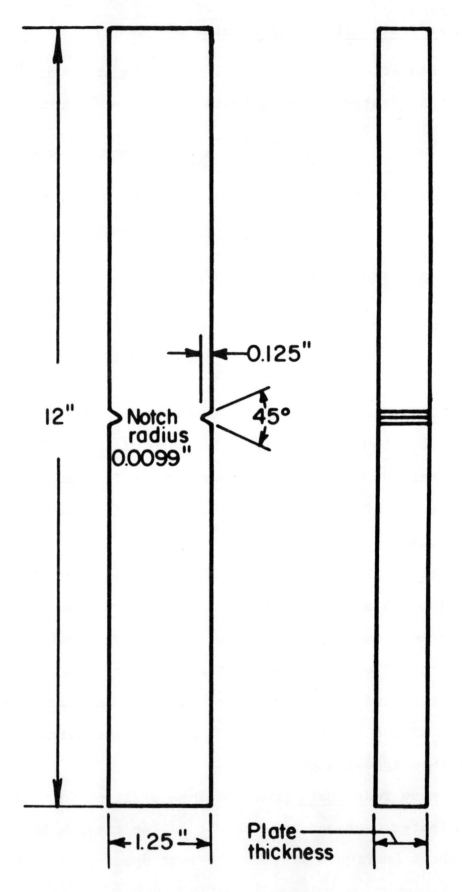

FIGURE 12-18 Tipper-type Side-notched tension-test Specimen.

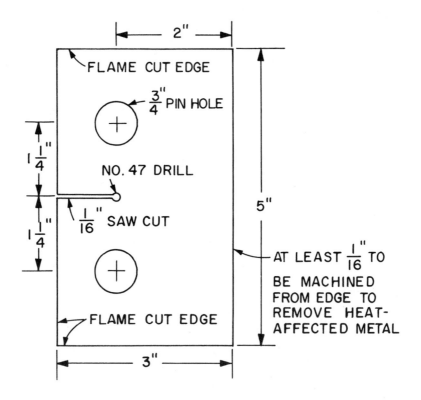

Figure 12.19 Navy Tear Test Specimen

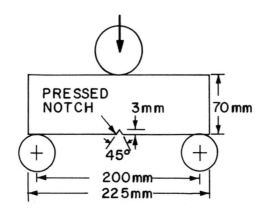

Figure 12.20 Van der Veen Test Specimen

full plate thickness and is machined only on the edge opposite
the notch. Static asymmetric tensile loading is applied to a
specimen through pins inserted in the pinholes. A transition
temperature is determined from the following values:

    (1) energy to propagate fracture (energy absorbed by a
        specimen after maximum load until fracture)

    (2) Fracture appearance (percentage of fibrous fracture).

Van der Veen Test. Figure 12-20 shows the static bend test
developed by Van der Veen.[4,61]

Lehigh Test and Kinzel Test. Longitudinal-bead-on-weld notch-
bend tests were developed by Stout, et al.[58,59] at Lehigh
University, and Kinzel, et al.[32]. In both tests, a bending load
is applied on a specimen having a transverse notch cut across the
weld metal laid on the specimen, as shown in Figure 12-21. The
basic philosophy of these tests is as follows. Since weld metal,
heat-affected base metal, and unaffected base metal are exposed at
the root of the notch, fracture can originate in the most sensitive
structure; thus, the cleavage-fracture sensitivity of a welded
joint can be evaluated. In the Lehigh test, duplicate test results
may be obtained from a single specimen, since two notches are used.

Investigators at Lehigh used various criteria, including bend
angle of maximum load, total energy absorbed, lateral contraction,
appearance, and mode of fracture, for the evaluation of test
results.[58,59]

Kommerell Test. The Kommerell-type longitudinal-bead-weld
bend specimen was developed in Germany after a number of failures

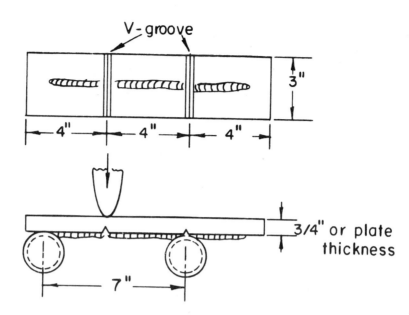

FIGURE 12-21 Lehigh Test

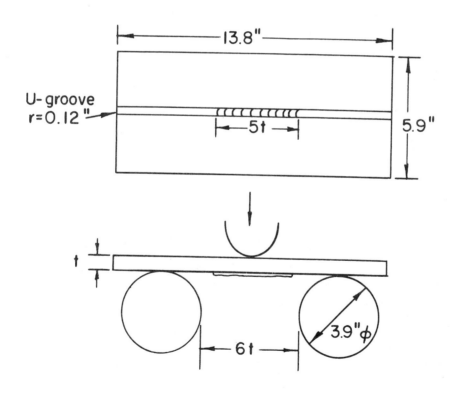

FIGURE 12-22 Kommerell Test

A-49516

Dimensions of the Kommerell test change depending upon the plate
thickness. Dimensions shown in this figure are for plates less
than 1 inch thick.

had occurred in welded bridges.[33] A single-pass weld bead is
made along a test plate, as shown in Figure 12-22. The test plate
is bent with the weld on the tension side, whereupon small cracks
appear in the weld metal or in the heat-affected zone. As bending
proceeds, these cracks extend into the base plate. Brittle steel
is unable to slow up the progress of the cracks, and the specimen
breaks suddenly at a small bending angle; ductile steel, on the
other hand, stops the advance of the cracks, and the specimen breaks
after considerable deformation.

The bead-bend test of Austrian Standard M3052 is a modification
of the Kommerell test.[37] Dimensions of the specimen and required
bend angles at room temperature are specified for different plate
thicknesses.

Fracture Tests of Weldments by Dynamic Loading

Dynamic loading such as those by explosives and projectiles
have been utilized as simple means of fracturing full-size weldments. Various tests have been proposes and used, as follows:

(1) Crack-starter explosion test developed by Pellini
    and associates[47,49]

(2) Explosion bulge test developed by Hartblower and Pellini[18.19]

(3) Direct explosion test proposed by Mikhalapov, et al.[38]

(4) H-plate shock test used by the Ordnance Corps, U. S. Army[5,62]

(5) Explosion tests of welded tubes conducted by Folkland,[8]
    Hauttmann,[20] and Kihara, et al.[30,31]

Explosion Crack Starter Test. The explosion crack starter
test was developed by Pellini, et al.[47] at the Naval Research
Laboratory as a method of determining the fracture-propagation

characteristics of the steel when fracture initiation is forced.
A plate is prepared for testing by the deposition of a short bead of
brittle, hard-surfacing weld metal which is then notched to half
thickness of the deposit by means of a disk abrasive wheel. The
test is performed by bulging a 14 x 14-inch plate over a die, as
shown in Figure 12-23. Static tests have shown that on loading to
the yield point of the plate the hard-surfacing weld bead cracks
in a completely brittle manner.

The explosive charge, standard 4-pound pentolite, is used at
a high standoff of 24 inches for 3/4- and 1-inch plate. A series
of tests, conducted over a range of temperatures, will indicate
that with increasing temperature, the fracture changes from (1)
"flat breaks," indicating little or no deformation prior to fracture
to (2) "bulge breaks," featuring forced initiation but easy propagation through the elastically loaded edge section, to (3) breaks
limited to the plastic load region at the center, and finally to
(4) ductile tears.

Some of the results of a series of tests of a mild-steel ship
plate of World war II are shown in Figure 12-13. Two criteria for
explosion-crack-starter test performance are represented: (1) fracture
transition, defined as the temperature of change from complete to
partial fracture shown in Figure 12-13 by the two vertical dotted
lines, and (2) ductility transition, defined as the highest temperature at which a brittle fracture is developed in the absence of
appreciable plastic deformation (a flat break).

Explosion Bulge Test. The explosion bulge test was developed
by Hartbower and Pellini[18,19] at the Naval Research Laboratory.

FIGURE 12-23 Explosion-crack-starter Test Setup.
(This figure supplied through the courtesy of P. P. Puzak, Naval Research Laboratory

Two plates, usually 10 x 20 inches, are butt welded to form a
square and are then placed over a circular die and explosion-loaded
by successive shots to the point of failure or to the development
of a full-hemispherical bulge. The temperature of the test is
controlled and equalized between successive shots. The source
point of fracture may be observed. The amount of deformation at
fracture is determined by measurements of plate thickness and is
expressed as per cent reduction of plate thickness. A "bulge
transition temperature" has been defined as the temperature range in
which the per cent reduction of thickness decreases from 10 to 1
per cent. These tests were aimed at investigation of the performance
of welded joints.

Tensile Tests of Wide-Plate Specimens

The primary objective of tensile tests of wide-plate specimens
is to reproduce in a laboratory brittle failures of actual structures.
The tests that have been proposed and used may be classified into
the following two groups:

(1) Fracture propagation tests

(2) Tests on welded-and-notched wide plates.

The major objective of the tests in the first group is to
study fracture propagation characteristics of steel plate. Tests
in this group include the Robertson test, ESSO (or SOD) test, and
the double tension test.

The tests in the second group have been used to study brittle
fracture characteristics of welded plates.

Robertson Test. Robertson[51,52] at the Naval Construction

Research Establishment, Scotland, developed a specimen shown in Figure 12-24. A cleavage crack is started at a notch on one edge of the plate by a bullet driven against a special nub. the amount of energy available from the explosion is insufficient to make the crack grow more than a very short distance but large enough to start it. The uniform tensile load tends to keep the crack growing. The specimen contains a temperature gradient in the width direction, with the notched side being cooler. A crack travels across the specimen until it reaches a zone where the temperature is high enough to permit the material to flow sufficiently to stop the crack. Thus, a temperature above which a crack does not propagate can be determined at each stress level. A typical test result is shown in Figure 12-17. The crack-propagation characterisitcs are expressed in terms of arresting temperature and critical stress.

ESSO (or SOD) Test. Feeley, et al.[6] at the Standard Oil Development Company developed a fracture-propagation test specimen as shown in Figure 12-25. The specimen is uniformly cooled to the desired temperature, and a selected load is applied. A cleavage crack is initiated by means of a hardened steel wedge driven into the previously prepared notch by a shot from an impact gun. If complete cleavage fracture occurs, the material is considered to be subject to cleavage fracture at the prevailing conditions. If failure does not occur, the tensile load is increased and the plate is subjected to impact at succeedingly high stress levels until cleavage fracture occurs. Thus, the critical stress for cleavage-fracture propagation at a certain temperature is obtained. Tests are conducted at various temperatures to determine the arresting

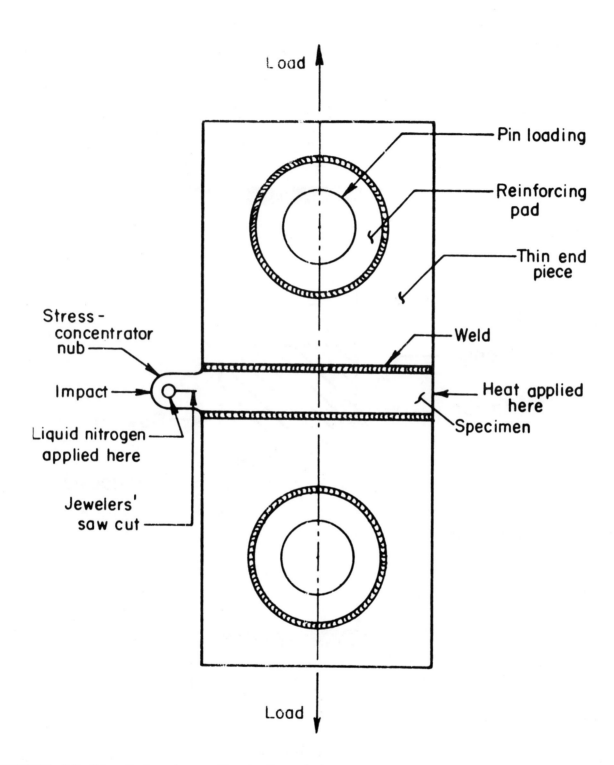

FIGURE 12-24   Robertson Test Specimen

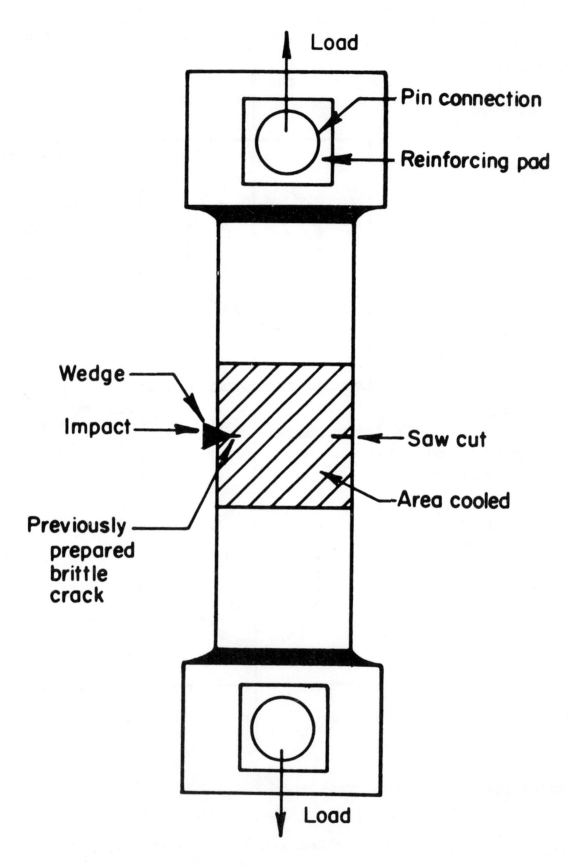

FIGURE 12-25 SOD Test Specimen

temperature. the results appear in a form very similar to those shown in Figure 12-17.

Double Tension Tests. Yoshiki and Kanazawa[69,70] at the University of Tokyo developed the double-tension test in an attempt to eliminate the influence of impact loading, used in the Robertson and the ESSO test, on test results. The double-tension test specimen, shown in Figure 12-26, is composed of two parts: the crack initiation part and the main part. These two parts are connected by a narrow passage and loaded independently under static tension by two sets of testing apparatus. A cleavage crack is initiated in the crack-initiation part and is driven into the main part through the passage. The critical stress at a specific temperature is determined as the lowest stress in the main part at which complete fracture of the specimen occurs.

Welded-and-Notched Wide-Plate Tensile Tests. A characteristic of brittle fracture is that actual failures usually occur at stresses well below the yield stress of the material. In many cases, fractures have occurred without any repeated or impact loading, and fractures have frequently started from a weld-joint flaw. However, low-stress fracture does not occur in most laboratory tests. Even in a specimen which contains very sharp notches and fractures with low energy absorption and with brittle-fracture appearance, the fracture stress is as high as the yield stress. In fracture-propagation tests the fracture propagates at a low stress; however, such expedients as the impact loading in the Robertson and the SOD tests or high-tensile stress applied at the auxiliary part of a specimen in the double-tension test are necessary to initiate a

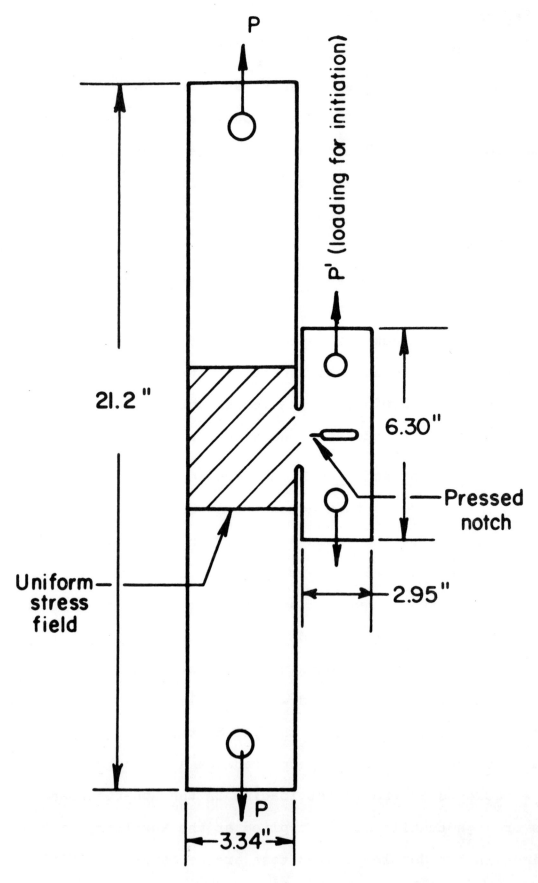

FIGURE 12-26   Double-tension Test Specimen

brittle crack.

Following the work done by Greene[14] and Wells[64,65], extensive research on low-applied-stress fracture of weldments has been conducted during the last several years. It has been found that a low-applied-stress fracture can be obtained experimentally from a notch located in an area containing high residual tensile stress. A book published in 1967 by Hall, Kihara, Soete, and Wells summarizes results of studies conducted in the United States (primarily at the University of Illinois)[16], Japan, Belgium, and Great Britain.[17] The following pages describe typical results obtained by Kihara and Masubuchi.[28]

Figure 12- 27 shows the specimens used by Kihara and Masubuchi.[28] Figure 12-28 shows general tendencies of their experimental results. The figure shows the effects of a sharp notch and residual stress on fracture strength of welded carbon-steel specimens on fracture strength.

When a specimen does not contain a sharp notch, fracture will occur at the ultimate strength of the material at the test temperatures, as shown by Curve PQR. When a specimen contains a sharp notch (but no residual stress), fracture will occur at the stresses shown by Curve PQST. When the temperature is higher than the fracture transition temperature, T_f, a shear-type fracture occurs at high stress. When the temperature is below T_f, the fracture appearance changes to a cleavage type, and the stress at fracture decreases to near the yield stress. When a notch is located in areas where high residual tensile stresses exist, various types of fracture can occur:

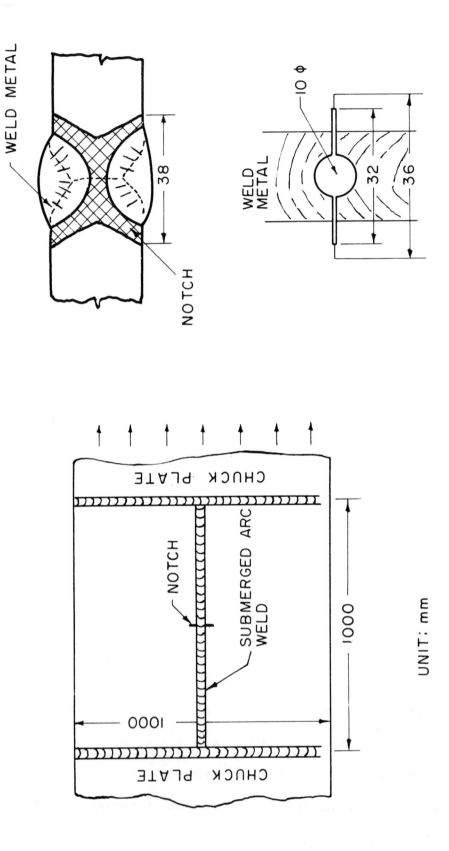

UNIT: mm

FIGURE 12-27 Welded-and-Notched Wide-Plate Tensile Test Specimen
Used by Kihara and Masubuchi (28)

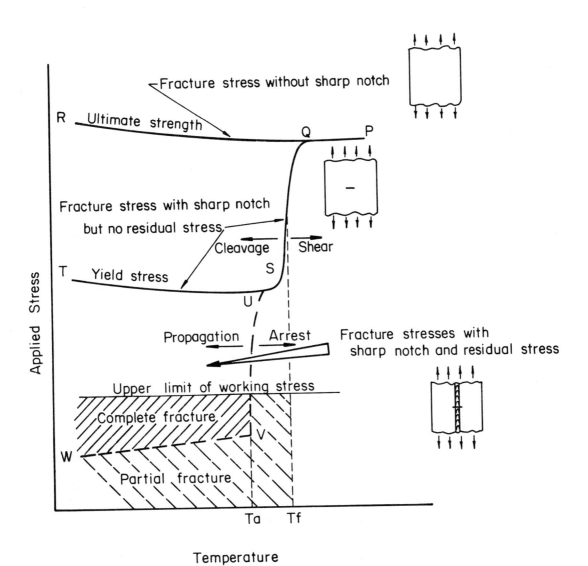

FIGURE 12-28   Effects of Sharp Notch and Residual
              Stress on Fracture Strength (Kihara
              and Masubuchi)[28]

(1) At temperatures higher than T_f, fracture stress is the
    ultimate strength (Curve PQ). Residual stress has no
    effect on fracture stress.

(2) At room temperatures lower than T_f but higher than the
    crack-arresting temperature, Ta, a crack may initiate
    at a low stress but it will be arrested.

(3) At temperatures lower than Ta, one of two phenomena can
    occur, depending upon the stress level at fracture
    initiation:

    (a) If the stress is below the critical stress VW,
        the crack will be arrested after running a short
        distance. Complete fracture will occur at the
        yield stress (ST).

    (b) If the stress is higher than VW, complete fracture
        will occur.

Effect of Mechanical and Thermal Stress Relieving. When a
load is applied to a weldment, residual stresses are redistributed
due to local plastic deformation. These stresses are reduced
when the load is removed. This is called mechanical stress
relieving. Figure 12-29 shows results obtained by Kihara, et al.[14]
They used welded-and-notched wide-plate specimens in carbon steel,
as shown in Figure 12-27. In a series of tests, they applied
external loads at 20° C, which was above the critical temperature
for crack initiation, Tc, to different stress levels, 5, 10, 15, 20,
and 23 kg/mm^2; then they reduced the load. After these mechanical
stress-relieving treatments, they cooled the specimens and applied
tensile loads again at temperatures below -30° C. As shown in

303

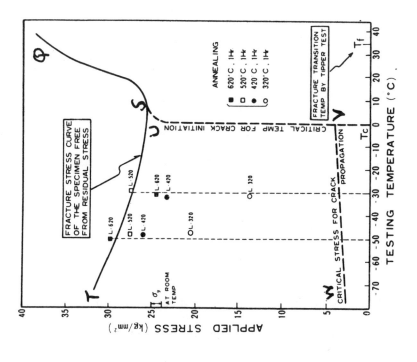

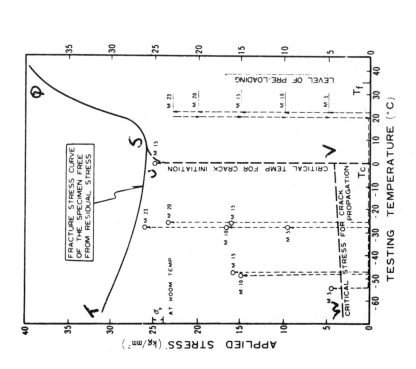

a. Effect of Mechanical Stress Relieving   b. Effect of Thermal Stress Relieving

FIGURE 12-29   Effects of Stress Relieving Treatments on Brittle-Fracture Characteristics
of Welded-and-Notched Wide-Plate Specimens (29)

Refer to Figure 12-28 for the explanations of Curves QST and UVW.

Figure 12-29a, fractures occurrred after the preloaded stresses were exceeded.

In another series of tests, specimens were thermally stress-relieved by placing them in a furnace for 1 hour at 320, 420, 520, and 620° C. Fracture test results of these specimens are shown in Figure 12-29b. Fracture stresses were higher when welds were heat treated at higher temperatures, indicating that more stresses were relieved by heating at higher temperatures.

Correlation of Results Obtained by Various Tests

Correlations of notch toughness data evaluated by means of various tests have been studied by many investigators. The book by Hall, et al.[17] covers correlations of transition temperatures obtained by various tests.

As discussed earlier, transition temperatures obtained by different tests can be classified into three types: ductility transition temperature T_d, fracture appearance transition temperature T_f, and crack arrest temperature, T_a. In general, a fairly good correlation exists between transition temperatures of the same type.

For example, Figure 12-30 shows a correlation between fracture appearance transition temperatures determined with the standard Charpy (machined) V-notch specimens, vTrs, and those determined with Charpy pressed V-notch specimens, pTrs.[17] A good correlation exists between these values which are similar.

Correlations are usually poor when comparisons are made between different type transition temperatures; for example between T_d determined by one test and T_f determined by another test.

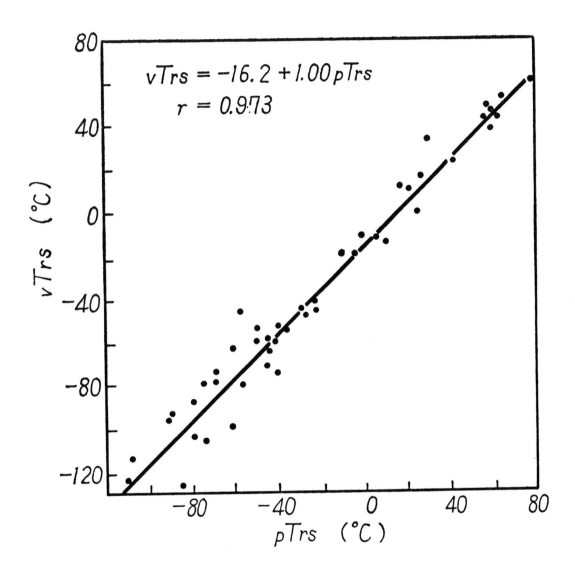

FIGURE 12-30   Correlation Between Fracture Appearance Transition
Temperature Determined With Standard Charpy (Machined)
V-Notch Specimens and Charpy Pressed V-Notch Specimens.

The figure has been prepared from data given in
Reference (17).

12.5 Effects of Chemical Composition and Manufacturing Processes [36,43]

Properties of steel are affected by chemical composition as well as manufacturing process variables such as grain size, roll finishing temperature, cold working, and prestraining. This section covers effects of these factors on:

(1) Tensile properties

(2) Notch toughness

(3) Weldability

Discussions in this section cover carbon steels and low-alloy high-strength steels used in as-rolled condition on which discussions are rather simple. This section does not cover quenched and tempered steels and maraging steels.

Effects of Chemical Composition

Effects of chemical composition on mechanical properties, notch toughness, and weldability of steel have been studies by many investigators. Tables 12-3 and 12-4 summarize some formulae for determining the effects of chemical composition of mechanical properties, notch toughness, and weldability. Included in Table 12-4 are formulae for determining the yield strength, ultimate tensile strength, elongation, and 15 foot-lb. transition temperature. Included in Table 12-4 are formulae for carbon equivalent, a parameter which expresses the effect of a certain alloying element as an equivalent amount of carbon.

Figure 12-31 shows effects of various alloying elements on the Charpy V-notch 15 foot-lb. transition temperatures. The figure is prepared by Kihara et al.[27] based upon experimental results obtained by Rinebolt and Harris.[50]

TABLE 12-3 TYPICAL FORMULAE TO DETERMINE EFFECTS OF CHEMICAL
             COMPOSITION ON MECHANICAL PROPERTIES AND NOTCH TOUGHNESS

---

1. Mechanical Properties

   a. Frazier-Boulger-Lorig's Formulae*[12]

   σ_{yu}, psi = (23,000 ± 1,500) + 29,200 x %C + 7,200 x %Mn

   σ_{uts}, psi = (30,800 ± 2,200) + 104,000 x %C + 13,000 x %Mn

   Elongation, % = (38.2 ± 2.4) - 32.6 x %C - 3.2 x %Mn

   b. Kihara-Suzuki-Tamura's Formulae**[27]

   σ_y, psi = (53,400 x %Cy + 23,900) ± 5,700

   σ_{uts}, psi = (87,000 x %Ct + 34,500) ± 5,000

   Elongation, % = (55.9 - 51.2 x %Ce) ± 6,400

2. Transition Temperature

   a. Williams-Allinger's Formula***[67]

   T_{r15}, °F = (70 ± 30) + 300 x %C + 1,000 x %P - 100 x %Mn

   + 300 x %Si - 5(ASTM Ferrite Grain Size Number)

---

*Results were obtained for carbon steel. σ_{yu} = Upper yield strength,
σ_{uts} = Ultimate tensile strength

**Results were obtained for Mn-Si high-strength steels, Values of Cy,
Ct, and Ce are given in Table 12-4.

***Formula based on experimental results on fractured ship plates.

TABLE 12-4 FORMULAE FOR CARBON EQUIVALENT FOR MECHANICAL PROPERTIES, NOTCH TOUGHNESS, AND WELDABILITY

---

1. Mechanical Properties

     a. Kihara-Suzuki-Tamura's Formula*[27]

$$Cy(Y.P.)\% = C + 1/5Mn + 1/7Si + 1/7Cu + 1/20Ni + \text{Zero} \times Cr + 1/2\,Mo + 1.1V$$

$$Ct(T.S.)\% = C + 1/5Mn + 1/7Si + 1/7Cu + 1/20Ni + 1/9Cr + 1/2Mo + 1/2V$$

$$Ce(\text{Elongation})\% = C + 1/9Mn + 1/12Si + 1/10Cu + 1/20Ni + 1/4Cr + 2/5Mo + 4/5V$$

2. Notch Toughness

     a. Kihara-Suzuki-Tamura's Formula**[27]

$$Ceq = C + 1/2.6\,Mn + 3.9P - 1/3.7\,Ni + 1.3Mo + \text{Zero}\,(Si + S + Cu + Cr)$$

3. Weldability

     a. Formula in Welding Handbook (for cold cracking)[63]

$$Ceq = C + 1/6Mn + 1/10Cr - 1/50Mo - 1/10V + 1/40Cu + 1/20Ni$$

     b. Formula Based on Sims-Banta's Work (for cold cracking)[43]

$$Ceq = C + 1/6Mn + 1/24Si + 1/29Mo + 1/29Mo + 1/14V + \text{Zero}\,Cr$$

     c. Kihara-Suzuki-Tamura's Formula (for maximum hardness)[††]

$$Ceq = C + 1/6Mn + 1/24Si + 1/15Ni \_ 1/5Cr + 1/4Mo$$

---

*Results obtained for Mn-Si high-strength steels.

**The formula was obtained by Kihara, Suzuki, and Tamura by analyzing results obtained by Rinebolt and Harris.

††The formula was obtained to express the effects of alloying elements on the maximum hardness of the weld heat-affected zone.

309

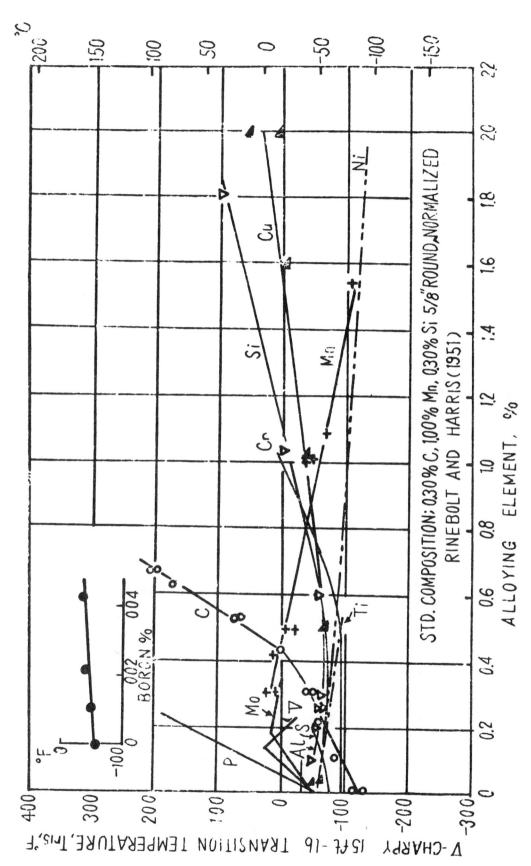

FIGURE 12-31   Effect of Various Alloying Elements on V-Charpy 15 ft-lb Transition
Temperatures. Summarized by Kihara, eta al., from the data by Rinebolt
and Harris (27,50)

In using formulae given in Tables 12- 3 and 12- 4, it is important not to extrapolate the formulae beyond the stage of alloy contents investigated in the experiments on which the formulae were based. Figure 12-31 may be used as a guide for practical limits of various alloying elements contained in structural steels.

Carbon.[43] Carbon added to steel causes an increase in strength (yield and ultimate), a reduction in elongation, a reduction in notch toughness (an increases in transition temperature), and a reduction in weldability (an increase in HAZ hardness and an increases in susceptibility for cracking). As carbon is added, a greater proportion of carbide is apparent in the structures observed under a microscope.

Managanese. Manganese is added to steel because it is a fairly good deoxidizer and it improves notch toughness (lowers transition temperature).[43] An increase in manganese up to around 1.2 per cent results in an improvement in notch toughness, but an addition of manganese over about 1.6 per cent causes a rapid deterioration in notch toughness.

Managanese also is known to retard hot cracking of weld metal and in the heat-affected zone near the fusion line by reacting with sulphur which causes the hot cracking.

Silicon. Evaluation of the effect of silicon is complicated by the fact that silicon acts as a deoxidizer and also as an alloying element.[43] Consequently, the influence of silicon on notch toughness will depend upon the concentration of other deoxidizing elements such as manganese and aluminum.

For example, in the experimental study conducted by Rinebolt and Harris, of which results are summarized in Figure 12-31, silicon up to 0.6 per cent had little effect on notch toughness, but an addition of silicon over 0.6 per cent caused an increase in transition temperature.

Phosphorus. An increase in phosphorus causes a drastic increase in transition temperature (or loss in notch toughness) at a rate that equals that of carbon.[43] Phosphorus also is known to cause hot cracking.

Sulphur. Sulphur in the form of sulphide inclusions often produce laminations in rolled steel plate.[43] Laminations, whether of the sulphide or oxide types, are oblate-shaped inclusions that have little strength under stresses in the thickness direction. Laminated plates tend to act like several thinner plates stacked together to form a thicker one. When the notch toughness of laminated plate is tested, the results tend to scatter, making accurate analysis difficult. Without laminations, the effect of sulphur on notch tougness is rather minor. Sulphur is known to cause hot cracking, since FeS has a low melting point.

Nitrogen and Oxygen. The effects of nitrogen and oxygen on notch toughness are complex because they react with steel and other elements present.[43] However, nitrogen and oxygen are known to have detrimental effects on notch toughness.

Aluminum. Aluminum modifies the embrittling effect of oxygen; it alters the structure of sulphide inclusions, and it tends to combine with nitrogen.[43] Aluminum is frequently used

as a deoxidizer during steel making. Aluminum tends to improve
notch toughness.

Nickel. Nickel is recognized as being beneficial for notch
toughness.[43] Nickel is frequently used for improving notch
toughness of low-alloy steels.

Effects of Mill Practices[36,43]

The items of major concern are those affecting the notch
toughness, such as the deoxidation practice, the rolling operation,
and the sizing operation.

Effect of Deoxidation. Steels can be classified into rimmed,
semi-killed, and killed steels, depending upon the degree of
deoxidation the steel receives during steelmaking. Killing or
deoxidation is ordinarily achieved by adding silicon or silicon
and aluminum.

In killed steel, the deoxidation is sufficiently complete
that during freezing there is essentially no evolution of carbon
monoxide gas. This minimizes segregation during solidification,
with the result that the rolled skelp is more uniform in composition and properties. The absence of gas evolution also causes
"piping," or shrinkage voids, so a larger amount of steel must be
cropped and discarded from the top of the ingot. Killed steels
can be made to various carbon, manganese, or alloy contents and
are generally more expensive than semi-killed or rimmed steels.

Semi-killed steels are only partially deoxidized and, as
a result, are not as sound as killed steels. They usually exhibit
more segregation than killed steels, and the center and surface

may differ noticeably in composition.

Effect of Finishing Temperature. The hot mechanical-working range for mild steel, in practice, extends from approximately 1600 to 2100° F. During the hot rolling of the plate, the temperature at which mechanical working is finished has an influence on mechanical properties and notch toughness.

It is generally conceded that a high finishing temperature results in a coarse grain structure, slightly lower strength, and lower notch toughness. With a low finishing temperature, a finer grain size and better toughness is more likely to be obtained.

It is also well known that the yield and tensile strengths are increased by decreasing the grain size. Thus, as low a finishing temperature as is practical is desired from the standpoint of high yield and tensile strengths and also for good notch toughness.

Effect of Cold Work and Prestraining. Cold working or prestraining increases yield strength and transition temperature. A rule of thumb that is frequently used is that a permanent strain of 1 per cent will raise the transition temperature 15 to 20° F.[36]

Gensamer, et al.[13] found that cold work when combined with aging is more effective is raising the transition temperature than cold work or aging alone. Figure 12-32 shows typical results of their investigation. The figure shows than when an as-rolled semi-killed steel is subjected to various amounts of tensile strain, e.g. ε = 0 per cent to 19 per cent and aged for 1 month, the transition temperature is raised considerably.

314

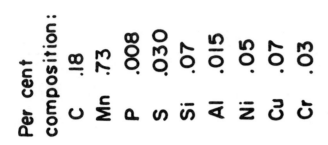

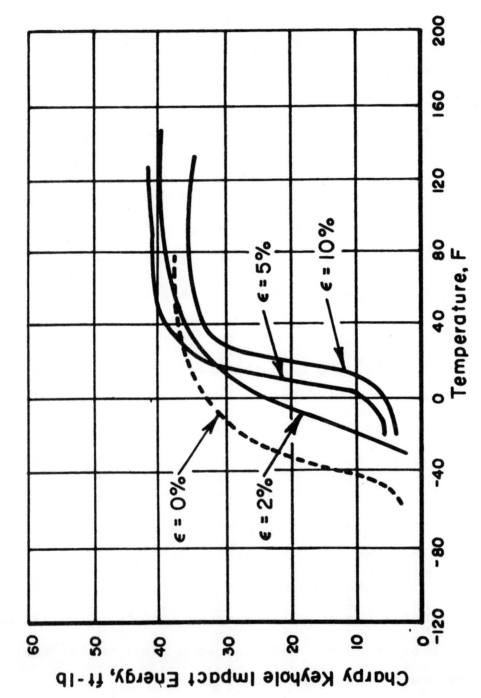

FIGURE 12-32 Effect of Cold Work and Aging on Charpy Properties. (13,36)

12.6 Theories of Brittle Fracture

This section covers briefly theories of brittle fracture.
Since a number of books have been written recently on theories
of fracture, the coverage in this section is limited to the
important information which would help readers to understand the
contents included in this book.

It is recognized thatmuch research has been done recently
on microscopic and atomic mechanisms of brittle fracture. However,
the discussion in this section covers primarily macroscopic mechanisms of fracture, because they have been developed to improve
the understanding of macroscopic fractures of engineering structures.

Theoretical Strength and Actual Strength[56]

The common demoninator of all thinking on the fracture of
solids is the ideal elastic solid, or one which would exhibit elastic
response to a load until atomic separation took place on a plane by
overcoming the interatomic forces. Such behavior is expected in
the case of an ideal crystalline solid which contains no defects
of any kind.

The calculation of the theoretical strength of an ideal elastic
solid is based on the proposition that all the energy of separation
is available for the creation of two new surfaces; the only energy
expenditure in creating these two new surfaces is assumed to be the
surface energy. Then the theoretical strength, σ_{th} is:

$$\sigma_{th} = \sqrt{\frac{ES}{a}} \qquad (1)$$

where,

        E = Young's modulus
        S = Surface energy per unit area
        a = lattice parameter of crystal

In the case of steel, for example, in which a = 3 x 10^{-8} cm,
E = 2 x 10^6 kg/cm^2, S = 10^{-3} kg.cm/cm^2, σ_{th} is approximately
2 x 10^5 kg/cm^2 (2000 kg/mm^2, or 3 x 10^6 psi), which is about E/10.

However, the strength of a real material is far less than the
theoretical strength, σ_{th}. For example, the ultimate tensile
strength of low carbon steel is only about 6 x 10^4 psi. Most
modern high-strength steels have ultimate strengths less than
3 x 10^5 psi, which is only 1/10 of the theoretical strength.

The average stress at brittle fracture is far below the above
figures. The average stress at the deck when the tanker "Schenectady"
fractured was only about 10^4 psi, which is only 1/300 of the
theoretical strength.

The major reason for this great difference between the theoretical strength and the actual strength is that real materials
contain many flaws which include atomic irregularities, cracks on
microscopic or even macroscopic scales, and metallic and non-metallic
inclusions.

Structure-Sensitive and Structure-Insensitive Properties.
Material properties, including mechanical, thermal, electrical, etc.
can be classified into two groups: the structure-sensitive and the
structure insensitive properties as follows:[3]

	Structure Insensitive	Structure Sensitive
Mechanical	Density, elastic moduli	Fracture strength, plasticity
Thermal	Thermal expansion, melting point, thermal conductivity, specific heat	
Electrical	Resistivity(metallic)	Resistivity (semi-conductor and at very low temperatures)

An important distinction between the two classes is that, whereas the structure-insensitive properties are well-defined properties of a phase of the material, the structure-sensitive properties are dependent not only on the composition and crystal structure of the material but also on structural details that depend upon the previous history of the sample.[3]

In the study of notch toughness, it is important to understand that such properties as fracture strength and plastic characteristics are structure sensitive. Such properties as density and modulus of elasticity are structure-insensitive and they are little affected by the imperfectness of a material.

The Griffith Theory and Its Modification

Griffith, in the 1920's, developed one of the earliest attempts to rationalize the observed strength of real brittle solids (elastic behavior up to fracture).[15] He proposed the existence of an array of pre-existing flaws in these materials. This model is the basis of many modern fracture theories, especially the fracture mechanics theory.

Griffith calculated the strain energy (per unit plate thickness) resulting from a crack of length $\ell = 2c$ in a thin plate under normal stress, σ, as follows:

$$\text{Strain energy} = \pi c^2 \sigma^2 / E \qquad (2)$$

where, E = Young's modulus.

The sign is negative because this is the energy which will be released in propagation of the flaw. The surface energy associated with the crack is given by :

$$\text{surface energy} = 4cS, \quad (3)$$

where S is the surface energy per unit area.

The equilibrium crack size can be calculated for the condition of net change of potential energy equal to zero, namely,

$$\frac{d}{dc} (4cS - \pi c^2 \cdot \sigma^2/E) = 0 \quad (4)$$

or

$$c = \frac{2SE}{\pi \sigma^2} \quad (5)$$

The critical stress (fracture stress) for a crack of size 2c is given by

$$\sigma_c = (\frac{2SE}{\pi c})^{1/2} \quad (6)$$

As the crack extends or as c increases, the stress necessary to propagate it decreases.

The theory predicts strengths reasonably well for bodies which behave in a brittle fashion, such as glass and ceramic materials. The fracture of metals is accompanied by dissipation of energy by plastic flow adjacent to the fracture surface, and this energy absorption is often at least ten times the energy corresponding to surface energy. Orowan[42] accordingly, proposed that the "S" term in Equations (5) and (6) should be replaced by p, a term which includes both the surface energy and the energy of plastic deform-ation, yielding.

$$C_c = \frac{2pE}{\pi \sigma^2} \quad (7)$$

$$\sigma_c = (\frac{2pE}{\pi C})^{1/2} \quad (8)$$

Fracture Mechanics Theory

Irwin and associates[23,24] have expanded the Griffith theory to what is commonly called the "fracture mechanics theory." The fracture mechanics theory has been used extensively to explain fracture of brittle materials, especially ultra-high strength materials for aerospace applications.

Important Technical Terms. When a crack of length $\ell = 2C$ exists in a plate with uniform thickness under uniform tensile stress, σ, as shown in Figure 12-33, the stress components (σ_x, σ_y) at a point, P (x,y), near the crack tip is expressed approximately by the following equation (see Figure 12-33):

$$\sigma_y = \frac{K}{\sqrt{2\pi r}} \cos \theta \left[1 + \sin \theta \cdot \sin \frac{3\theta}{2}\right]$$

(9)

$$\sigma_x = \frac{K}{\sqrt{2\pi r}} \cos \theta \left[1 - \sin \theta \cdot \sin \frac{3\theta}{2}\right]$$

where, K is called the "stress intensity factor," which is expressed as follows:

(10)

$$K = \sigma\sqrt{\pi c}$$

where a = constant.

Then the rate of elastic strain energy release due to the crack extension, $G = \frac{\partial U}{\partial c}$, is expressed:

$$\frac{dU}{dc} = G = \frac{K^2}{E}$$

(11)

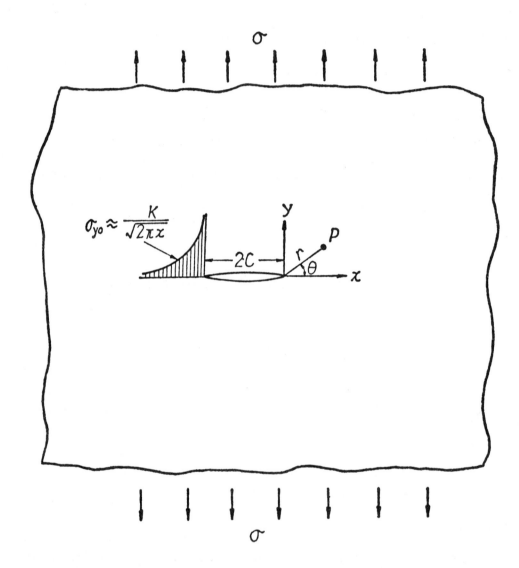

FIGURE 12-33   Infinite Plate Containing a Sharp Crack of
Lenght 2c Under Uniform Tensile Stress σ

Figure 12-34 shows schematically typical results when a series of fracture tests are made using specimens with different crack lengths in a certain high-strength material.[68] When the length of the initial crack is long enough, and the fracture occurs before general yielding, the relationship between the initial crack length, 2c, and the average fracture stress, σ, is expressed by:

$$K = \sigma\sqrt{\pi c} = \text{const.}$$

Consequently, fracture occurs when the K value reaches a certain critical value K_c for that material. K_c is called the "critical stress intensity fracture," or it is often simply called "fracture toughness."

When the fracture stress increases and exceeds approximately 80 per cent of the yield stress, experimental fracture stresses deviate from the K = constant curve, and fracture occurs in a ductile manner. As illustrated in Figure 12-34, the fracture mechanics theory should be used only when fracture occurs before general yielding.

Fracture toughness, K_c, has the following physical meanings:

(1) Relative to the locally elevated stress field at the leading edge of a crack K_c represents the intensity of local tensile stress necessary for unstable crack propagation.

(2) For a crack of length 2a in a large sheet, the K_c-value permits estimation of the membrane tensile stress σ necessary for unstable crack propagation through the relationship:

$$\sigma = \frac{K_c}{\sqrt{\pi a}}$$

(12)

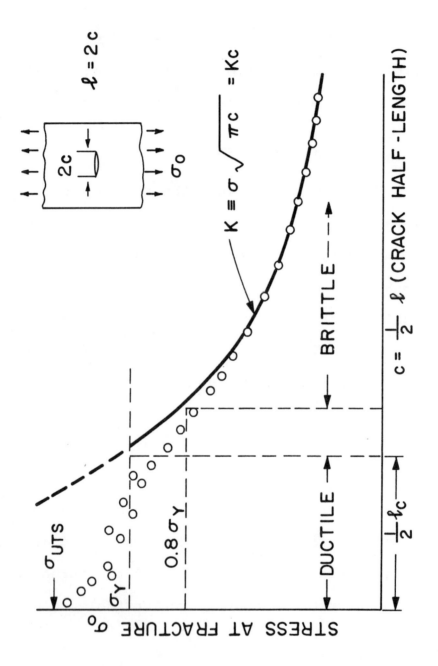

Figure 12.34 Effect of Crack Length on Fracture Strength

(3) Relative to the modified Griffith theory, the strain-energy-release rate G_c for unstable crack propagation may be directly expressed in terms of K_c by the relationship

$$K_c{}^2 = E \; G_c \qquad (13)$$

When further tests are conducted with specimens in different thicknesses, the K_c changes as shown schematically in Figure 12-35. As thickness increases, the K required to produce failure decreases and approaches a lower limit which is the critical stress intensity under plane-strain condition, K_{Ic}. From the engineering viewpoint, plane-strain fracture toughness, K_{Ic}, is considered as a material constant.

Fracture Toughness Tests. The ASTM Committee on Fracture Testing of High-Strength Sheet Materials[9] has described techniques of determining fracture toughness of high-strength sheet materials. Their interest has been limited to materials, both ferrous and non-ferrous, having a strength-to-density ratio of more than 700,000.

Figure 12-36 shows practical fracture toughness specimen types for general use, while Figure 12-37 shows specimens for determining K_{Ic}.[10] Also shown in these figures are formulas for determining K_c and K_{Ic}.

Fracture Toughness Values. Table 12-6 shows typical values of K_c for various sheet materials.

Figure 12-38 shows the general trend indicating how K_{Ic} values of different steels decrease with increasing yield strength.[45] For steels with over about 180 ksi yield strength, the K_{Ic} value is usually low enough that the fracture mechanics approach is applicable.

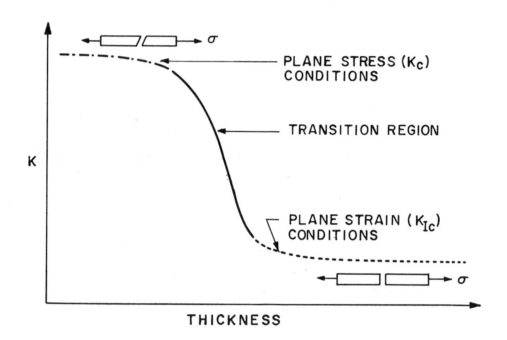

Figure 12.35 Behavior of Stress Intensity Required to Produce Failure as a Function of Material Thickness

325

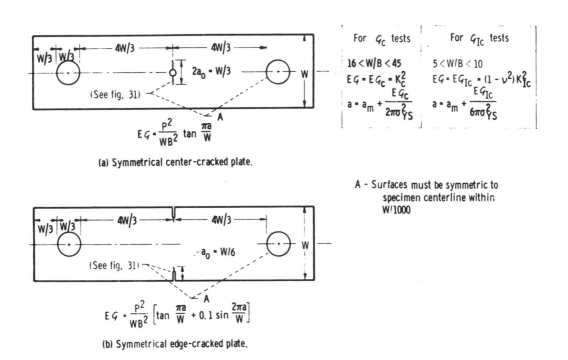

(a) Symmetrical center-cracked plate.

$$EG = \frac{P^2}{WB^2} \tan \frac{\pi a}{W}$$

For G_c tests

$16 < W/B < 45$

$EG = EG_c = K_c^2$

$a = a_m + \frac{EG_c}{2\pi \sigma_{YS}^2}$

For G_{Ic} tests

$5 < W/B < 10$

$EG = EG_{Ic} = (1 - \nu^2)K_{Ic}^2$

$a = a_m + \frac{EG_{Ic}}{6\pi \sigma_{YS}^2}$

A – Surfaces must be symmetric to specimen centerline within $W/1000$

$$EG = \frac{P^2}{WB^2} \left[\tan \frac{\pi a}{W} + 0.1 \sin \frac{2\pi a}{W} \right]$$

(b) Symmetrical edge-cracked plate.

FIGURE 12-36   Practical Fracture Toughness Specimen Types. Specimens for General Use. (The factor $(1 - \nu^2)$ is an Approximation.[10]

326

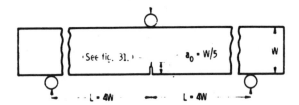

(a) Single-edge-notched plate (tension).

A - Surfaces must be true to specimen centerline within W/1000

$2 < W/B < 8$

$$E\mathcal{G}_{Ic} = \left(\frac{P}{B}\right)^2 \frac{L^2}{W^3}\left[31.7\frac{a}{W} - 64.8\left(\frac{a}{W}\right)^2 + 211\left(\frac{a}{W}\right)^3\right]$$

$$E\mathcal{G}_{Ic} = (1 - \nu^2)K_{Ic}^2$$

(b) Notch bend specimen (three-point loaded).

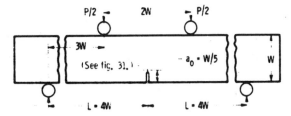

$2 < W/B < 8$

$$E\mathcal{G}_{Ic} = \left(\frac{P}{B}\right)^2 \frac{L^2}{W^3}\left[34.7\frac{a}{W} - 55.2\left(\frac{a}{W}\right)^2 + 196\left(\frac{a}{W}\right)^3\right]$$

$$E\mathcal{G}_{Ic} = (1 - \nu^2)K_{Ic}^2$$

(c) Notch bend specimen (four-point loaded).

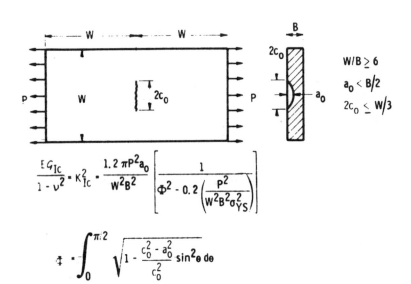

$W/B \geq 6$

$a_0 < B/2$

$2c_0 \leq W/3$

$$\frac{E\mathcal{G}_{Ic}}{1-\nu^2} = K_{Ic}^2 = \frac{1.2\pi P^2 a_0}{W^2 B^2}\left[\frac{1}{\Phi^2 - 0.2\left(\frac{P^2}{W^2 B^2 \sigma_{YS}^2}\right)}\right]$$

$$\Phi = \int_0^{\pi/2}\sqrt{1 - \frac{c_0^2 - a_0^2}{c_0^2}\sin^2\theta}\, d\theta$$

(d) Surface-cracked plate.

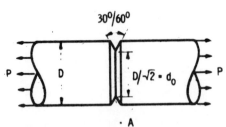

A - Surfaces must be concentric to within D/1000

$$\frac{E\mathcal{G}_{Ic}}{1-\nu^2} = K_{Ic}^2 = \frac{1.63 P^2 D}{d^4}\left[0.172 - 0.8(d/D - 0.65)^2\right]$$

$$d = d_0 - \left(\frac{E\mathcal{G}_{Ic}}{3\pi\sigma_{YS}^2}\right)$$

FIGURE 12-37 Practical Fracture Toughness Specimen Types. Plane-Strain Tests. (For all Specimens, $a = a_0 + E/6\pi\sigma_{YS}^2$; the factor $(1-\nu^2)$ is an Approximation (10).

TABLE 12-5 TYPICAL VALUES OF K_c FOR VARIOUS MATERIALS[9,26]

Material	Yield Strength, 1000 psi	K_c-Value, 1000 psi/$\sqrt{\text{in.}}$
AISI 4340 steel (air melt)		
Tempered at 350 F	208.3	192
Tempered at 425 F	203.9	204
Tempered at 500 F	197.9	174
Tempered at 700 F	181.6	204.5
6434 (vacuum melt)		
Longitudinal	190	205
Transverse	190	195.5
Tricent (vacuum melt)		
Longitudinal	190	197
Transverse	190	204.5
Ti-6Al-4V		
Longitudinal	159	114
Transverse	164	113

328

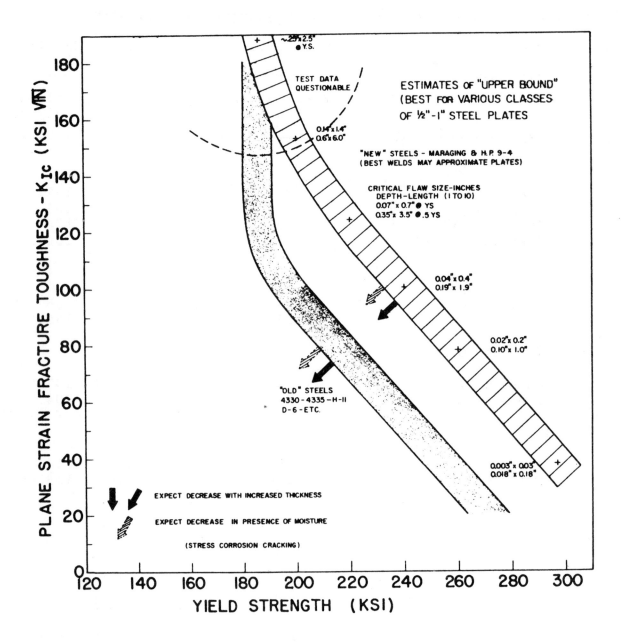

FIGURE 12-38   Upper bound, best value, limits of plane strain
fracture toughness (K_{IC}) for 1/2 to 1 in. plates
of high strength steels, as reported by various
qualified laboratories.[45]

However, for steels with less than 180 ksi yield strength, the K_{Ic} value increases drastically; consequently, the fracture mechanics approach is generally not applicable.

Unstable Fractures of Welded Structures and Effects of Various Factors

The above discussions show that:

(1) A crack will grow under stress when its size is larger than a critical size determined by the material properties for a given stress

(2) The critical crack length of steel decreases greatly as the yield strength increases.

However, there is evidence that fractures have occurred in actual structures from flaws smaller than the critical size. Many fractures in welded ships had their origin in flaws of various kinds. Most of the flaws were much smaller than critical crack lengths estimated from fracture tests. Additional factors are needed to cause the growth of a subcritical crack to a critical crack. Boyd[2] has discussed the effects of additional factors on initiation of brittle fracture. The effect of various factors, mechanical or metallurgical, may be classified according to the following three types.

Supply of Additional Energy. If additional energy is supplied either by residual stress, by stress concentration due to structural discontinuities, or by impact loading, a crack of subcritical size may grow to a critical size. According to the Griffith Irwin theory of brittle fracture, the relationship between crack length and the rate of release of strain energy G is as shown by

Line OA in Figure 12-39. When the G_c-value of the material is $2S_1$, an initial crack has to be longer than ℓ_1 in order that unstable fracture can occur. However, if a crack is located in an area of high tensile residual stresses the relationship between the crack length and the rate of release of strain energy may be as shown schematically by Curve ORB of Figure 12-39. Then for a G_c-value of $2S_1$, the crack can grow from a length ℓ_{1r}, which is much shorter than ℓ_1 for a uniform stress. The influence of residual stress on low-applied-stress fracture has been demonstrated experimentally (Figure 12-28).

Embrittlement of Material. Embrittlement by such mechanisms as strain aging, precipitation hardening, hydrogen embrittlement, etc., can change the properties of a material so that a subcritical crack may grow to a critical crack or may become of critical size without growth. Figure 12-39 shows that the critical crack length for uniform stress σ decreases from ℓ_1 to $\frac{1}{2}\cdot\ell_1$, when the G_c-value of a material changes from $2S_1$ to S_1. Mylonas[39] proposed that loss of ductility due to plastic straining is an important factor which causes low-applied-stress fracture of a steel weldment.

Materials may be locally embrittled by hydrogen or stress corrosion cracking, which are covered in Chapter 14. Masubuchi and Martin[35] applied the fracture mechanics theory to analyze crack patterns produced in weldments in SAE 4340 steel.

Fracture-Safe Design Concept

Many theoretical studies have been made of initiation and propagation of brittle fractures. Some of the studies have been covered in preceding pages. As investigators try to describe the

331

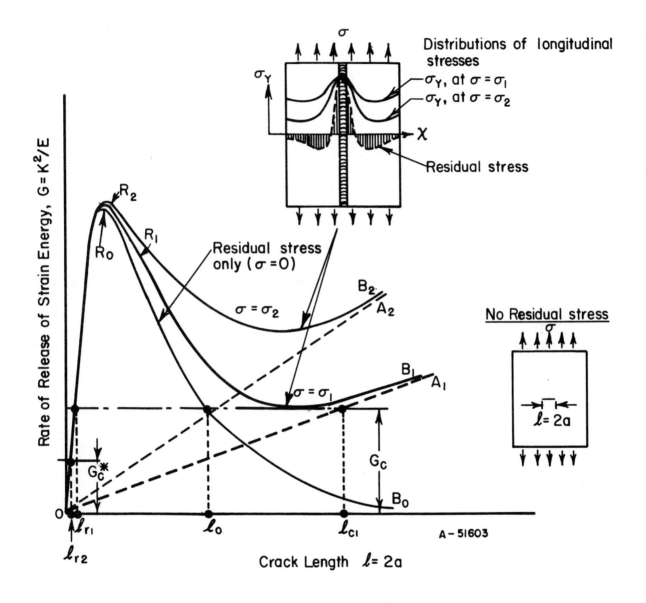

FIGURE 12-39 SCHEMATIC DIAGRAM SHOWING MECHANISMS OF THE
LOW-APPLIED-STRESS FRACTURE OF WELDMENTS
[KAMMER ET AL[40]]

phenomena more accurately, there is a tendency that theories become more complex, and they often become more difficult to apply to the design of practical structures. However, there is a strong need from practical engineers for a concept which could assist them in the design of structures. To meet this need, Pellini[44] has proposed a concept on "fracture-safe design," which is covered in the following pages.

Fracture Analysis Diagram. In Pellini's fracture-safe design concept, the "fracture analysis diagram (FAD)", as shown in Figure 12-40, plays a key role.[44]

The FAD provides a generalized definition of the flaw size, relative stress, temperature relationships by a "Δt" or "temperature increment" reference to the NDT temperature. The location of the generalized diagram to specific positions in the temperature scale requires the experimental determination of a single parameter-- the NDT temperature. The NDT temperature of a material can be determined by the drop-weight test or the dynamic tear test.

Definitions of important terms used in the FAD are as follows:

CAT: Robertson Crack Arrest Temperature transition curve

FTE: Fracture Transition Elastic, or the highest possible temperature for unstable fracture propagation through elastic stress fields

FTP: Fracture Transition Plastic, or the temperature point of fully ductile tearing

Pellini has used the FAD to explain many fracture problems, including analyses of failures in actual structures. NRL Report 6957 covers these analyses.

Ratio Analysis Diagram (RAD). Recently, Pellini[44] has

333

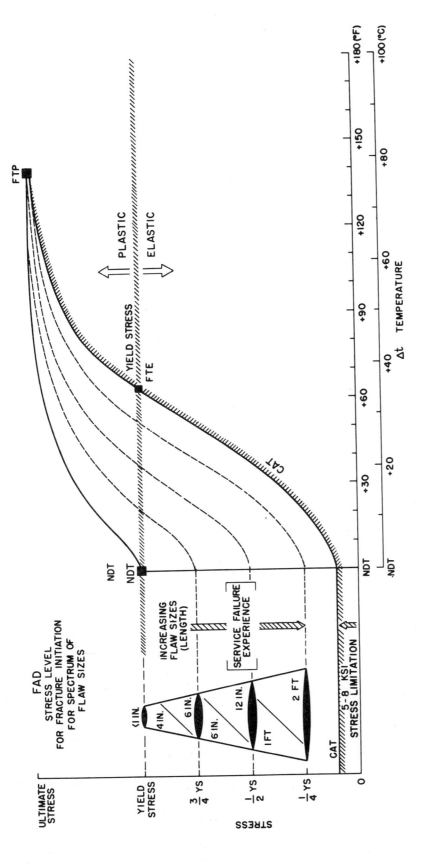

FIGURE 12-40   Fracture Analysis Diagram (FAD). Note that the stress level for plastic (over yield) fracture is not indexed because of the lack of analytical procedures for its definition. Ultimate stress signifies only that maximum load and strain tolerance is attained at FTP for the specific flaw size cited. It obviously does not indicate the equivalent of the tensile test specimen maximum load or maximum strain limits.

developed the Ratio Analysis Diagram (RAD), as shown in Figure 12-41. The RAD has been developed to obtain relationships between the fracture mechanics theory and the fracture-safe design concept, or, stated differently, relationships between the plane strain critical stress intensity factor, K_{Ic} and results obtained by the Charpy impact test or the dynamic tear test. As shown in the figure, the ratio between K_{Ic} and the yield stress, σ_{ys}, plays an important role in defining fracture toughness of a material.

The RAD may be separated into three general regions. The top region, above the ∞ ratio line, relates to ductile fracture. The bottom region, below the 0.5 ratio line relates to low levels of plane strain fracture toughness, involving flaw sizes which are too small for reliable inspection. These are generally in the order of tenths of inches for stubby flaws and decrease to hundredths of inches for long thin flaws subjected to high elastic stress levels. While fracture mechanics applies well in principle for this region, it applies poorly in practice due to the flaw detection problem. This region requires proof-test procedures for ascertaining pre-service structural integrity.

The width of the remaining "slice" of the RAD, the central region for which plane strain fracture mechanics calculations appear feasible, depends on the section size. For very thick sections (to over 10 inches) the applicable region available for possible plane strain calculations is between the 0.5 and ∞ ratio lines. For section sizes in the order of 1 to 3 inch thickness, this region is bounded by the 0.5 and roughly the 1,0 ratio lines. This range represents a very thin slice through the population of the steels represented by the diagram.

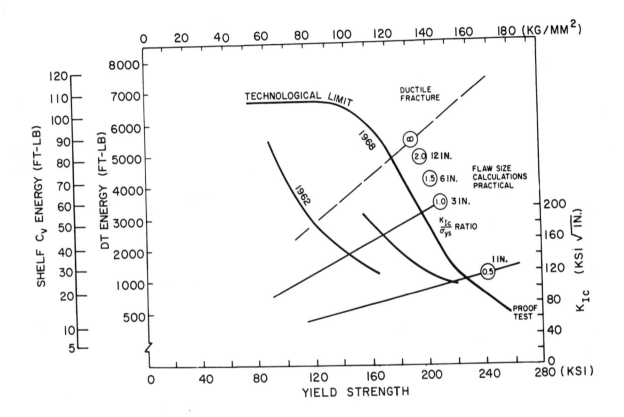

FIGURE 12-41   Ratio Analysis Diagram (RAD). The grid of K_{IC}/σ_{ys} ratio lines indexes the region of the diagram which relates to plane strain fracture.

REFERENCES

(1) Baker, J. F., and Tipper, C. F., "The Value of the Notch Tensile Test," Proceedings of the Institution of Mechanical Engineers, 170, 65-75, (1956).

(2) Boyd, G. M., "The Conditions for Unstable Rupturing of a Wide Plate," Trans. Inst. Naval Arch., 99 (3), Part II, 349-366 (July, 1957).

(3) Chalmers, B., Physical Metallurgy, John Wiley and Sons, Inc., 1959.

(4) de Graaf, J. E., and Van der Veen, J. H., "The Notched Slow-Bend Test as a Brittle Fracture Test," Journal of Iron and Steel Institute, 173, 19-30 (1-53).

(5) Babrykowski, Z. J., "Ferritic Welding of Steel Armor," The Welding Journal, 40 (4), 339-342 (1961).

(6) Feely, F. J., Jr., Hrkto, D., Kloppe, S. R., and Northrup, M. S., "Report on Brittle Fracture Studies," The Welding Journal, 33 (2), Research Supplement, 99s-111s (1954).

(7) Final Report of a Board of Investigation to Inquire into the Design and Methods of Construction of Welded Steel Merchant Vessels." Washington, D. C., Government Printing Office, 1947.

(8) Folkhard, E., "Le Comportment de Tuyaux pour Conduites Forcées, Soudés à l'arc, lors Déssais par Ecletement," Revue de la Soudre: Lastijdschrift (Brussels), 12 (1), 15-27 (1956.)

(9) "Fracture Testing of High Strength Sheet Materials: A Report of a Special ASTM Committee," ASTM Bulletin, No. 243, 29-40 (June, 1960); No. 244, 18-28 (February, 1960); and "Materials Research and Standards," 1 (11), 877-885 (November, 1961).

(10) Fracture Toughness Testing and Its Applications, ASTM Special Technical Publication No. 381, American Society for Testing and Materials, 1965.

(11) "Fractured Girders of the King's Bridge, Melbourne," The Engineering, 217, 520-522, (March 20, 1964).

(12) Frazier, R. H., Boulger, F. W., and Lorig, C. H., "An Investigation of the Influence of Deoxidation and Chemical Composition on Notched-Bar Properties of Semi-Killed Ship Steel," Ship Structures Committee Report SSC-53, (November, 1952).

(13) Gensamer, M., Klier, E. P., Prater, T. A., Wagner, F. C., Mack, J. D., and Fisher, J. L., "Correlation of Laboratory Tests With Full Scale Ship Plate Fracture Tests," Ship Structures Committee Progress Report SSC-9 (1947).

(14) Greene, T. W., "Evaluation of Effect of Residual Stresses," The Welding Journal, 28 (5), Research Supplement, 193s-204s (1949).

(15) Griffith, A. A., "The Phenomena of Rupture and Flow in Solids," Phil Trans. Roy. Soc., 221, 163-198 (1921), and "The Theory of Rupture," Proc. First Intern. Congr. Appl. Mech., 55-63 (1924).

(16) Hall, W. J., Nordell, W. J., and Munse, W. H., "Studies of Welding Procedures," The Welding Journal, 41 (11), Research Supplement, 505s-518s (1962).

(17) Hall, W. S., Kihara, H., Soete, W., and Wells, A. A., Brittle Fracture of Welded Plate, Prentice-Hall, Inc. Englewood Cliffs, New Jersey (1967).

(18) Hartbower, C. E., and Pellini, W. S., "Explosion Bulge Test of the Deformation of Weldments," The Welding Journal, 30 (6), Research Supplement, 307s-318s (1951).

(19) Hartbower, C. E., "Mechanics of the Explosion Bulge Test," The Welding Journal, 32 (7), Research Supplement, 333s-341s (1953).

(20) Hauttmann, H., "Erprobung Trennbruchsicherer Baustähle in in Berstversuchen," Schweisstechnik (Austria), 13 (2), 13-19 (1959).

(21) Hayden, H. W., Moffat, W. G., and Wulff, J., The Structure and Properties of Materials, Volume III, Mechanical Behavior, John Wiley and Sons, Inc., New York, 1955.

(22) Hodgson, J., and Boyd, G. M., "Brittle Fracture in Welded Ships," Quarterly Transactions of the Institution of Naval Architects, 100 (3), 141-180 (July, 1958).

(23) Irwin, G. R., "Fracture," Encyclopedia of Physics, Vol. VI Elasticity and Plasticity, Springer-Verlag, Berlin, 551-590 (1958).

(24) Irwin, G. R., "Structural Aspects of Brittle Fracture," Applied Materials Research, 3 (2), 65-81 (April 1964).

(25) Kahn, N. A., and Imbembo, E. A., "A Method of Evaluating Transition From Shear to Cleavage Failure in Ship Plate and Its Correlation With Large-Scale Plate Tests," The Welding Journal, 27 (4) Research Supplement

(26) Kammer, P. A., Masubuchi, K., and Monroe, R. E., "Cracking in High-Strength Steel Weldments--A Critical Review," DMIC Report 197, Defense Metals Information Center, Battelle Memorial Institute, Columbus, Ohio, February 1964.

(27) Kihara, H., Suzuki, H., and Tamura, H., Researches on Weldable High Strength Steels, Anniversary Series, Volume I, The Society of Naval Architects of Japan, Tokyo, 1957.

(28) Kihara, H., and Masubuchi, K., "Effect of Residual Stress on Brittle Fracture," The Welding Journal, 38 (4), Research Supplement, 159s-168s (1959).

(30) Kihara, H., Ichikawa, S., Masubuchi, K., Ogura, Y., Iida, K., Yoshida, T., and Oba, H., "Explosion Tests on Arc-Welded Tubes With and Without Stress Annealing," Journal of the Society of Naval Architects of Japan, 100, 179-187 (1965).

(31) Kihara, H., Ichikawa, S., Masubuchi, K., Iida, K., Yoshida, T., Oba, H., and Ogura, Y., "Explosion Tests on Arc Welded Tubes With Various Welding Procedures," Journal of the Society of Naval Architects of Japan, 104, 119-129 (1958).

(32) Kinzel, A. B., "Ductility of Steel for Welded Structures," The Welding Journal, 27 (5) Research Supplement, 217s-234s (1948); and Transactions of ASM, 40, 27-82 (1948).

(33) Kommerell, O., Stahlbau-Technik, 2, 51-52 (1938).

(34) Masubuchi. K., Monroe, R. E., and Martin, D. C., "Interpretive Report on Weld-Metal Toughness," Welding Research Council Bulletin, No. 111, January 1966.

(35) Masubuchi, K., and Martin, D. C., "Investigation of Residual Stresses by Use of Hydrogen Cracking," Parts I and II, Welding Journal, 40 (12), Research Supplement, 553s-563s (1961), and 45 (9), Research Supplement, 401s-418s (1966).

(36) McClure, G. M., Eiber, R. J., Hahn, G. T., Boulger, F. W., and Masubuchi, K., "Research in the Properties of Line Pipe," American Gas Association, Catalogue No. 401PR, May 1962.

(37) Melhardt, H., "New Austrian Welding Standards," The Welding Journal, 31 (7), 592-595 (1952).

(38) Mikhalapov, G. S., "Direct Explosion Tests of Welded Joints," The Welding Journal, 29 (3), Research Supplement, 109s-122s (1950) and "Evaluation of Welded Ship Plate by Direct Explosion Testing," The Welding Journal, 30 (4), Research Supplement, 195s-201s (1951).

(39) Mylonas, C., "Prestrain Size and Residual Stresses in Static Brittle-Fracture Initiation," Welding Journal, 38 (10), Research Supplement, 414a-424s (1959).

(40) Noren, T. M., "The Nominal Cleavage Strength of Steel and Its Importance for Welded Structures," Transactions of North East Coast Institution of Engineers and Shipbuilders, 73, 87-112 (1956-57).

(41) Orner, G. M., and Hartbower, C. E., "The Low Blow Transition Temperature," ASTM Proceedings, Vol. 58, pp. 623-634 (1958).

(42) Orowan, E., "Fracture and Strength of Solids," Report on Progress in Physics, Physical Society, London, Vol. 12, 185 (1949).

(43) Parker, E. R., Brittle Behavior of Engineering Structures, John Wiley and Sons, Inc., New York, 1957.

(44) Pellini, W. S., "Evolution of Engineering Principles for Fracture-Safe Design of Steel Structures," NRL Report 6957, Naval Research Laboratory, September, 1969.

(45) Pellini, W. S., Goode, R. J., Puzak, P. P., Lange, E. A., and Huber, R. W., "Review of Concepts and Status of Procedures for Fracture-Safe Design of Complex Welded Structures Involving Metals of Low to Ulta-High Strength Levels," NRL Report 6300, U. S. Naval Research Laboratory, June 1965.

(46) "Pipe Failures," Pipe Line Industry, 34-47, June 1963.

(47) Puzak, P. P., Eschbacher, E. W., and Pellini, W. S., "Initiation and Propagation of Brittle Fracture in Structural Steels," The Welding Journal, 31 (12), Research Supplement, 561s-581s (1952).

(48) Puzak, P. P., and Pellini, W. S., "Effect of Temperature on the Ductility of High-Strength Structural Steels Loaded in the Presence of Sharp Cracks," Naval Research Laboratory Report No. 4545, (June 22, 1955).

(49) Puzak, P. P., Schuster, M. E., and Pellini, W. S., "Crack-Starter Tests of Ship Fracture and Project Steels," The Welding Journal, 33 (10), Research Supplement, 481s-495s (1954), and Puzak, P. P., and Pellini, W. S., "Evaluation of Significance of Charpy Tests for Quenched and Tempered Steels," The Welding Journal, 35 (6) Research Supplement (1956).

(50) Reinbolt, J. A., and Harris, W. J., Jr., "Effect of Alloying Elements on Notch Toughness of Pearlitic Steels," Trans. ASM, Vol. 43, pp. 1175-1214 (1951).

(51) Robertson, T. S., "Propagation of Brittle Fracture in Steel," Journal of the Iron and Steel Institute, 175, 361-374 (December 1953).

(52) Robertson, T. S., "Brittle Fracture of Mild Steel," Engineering, Vol. 172, p. 444 (October 5, 1951).

(53) Schnadt, H. M., "A New Approach to the Solution of Brittle Fracture Problems in Modern Steel Construction," Document IX-292-61, Commission X of the International Institute of Welding, (1961).

(54) Shank, M. E., "A Critical Survey of Brittle Failure in Carbon-Steel Structures Other Than Ships," Welding research Council Bulletin, No. 17, 1954.

(55) Shank, M. E., editor, Control of Steel Construction to Avoid Brittle Fracture, Welding Research Council, 1957.

(56) Spretnak, J. W., "A Summary of the Theory of Fracture in Metals," DMIC Report No. 157, Defense Metals Information Center, Battelle Memorial Institute, Columbus, Ohio, August 1961.

(57) Srawley, J. E., and Esgar, J. B., "Investigation of Hydro-test Failure of SL-1 Motor Case," NASA TMX-1194, January 1966.

(58) Stout, R. D., McGeady, L. J., Sund, C. P., Libsch, J. F., and Doan, G. E., "Effect of Welding of Ductility and Notch Sensitivity of Some Ship Steel," The Welding Journal, Research Supplement, Vol. 26, No. 6, pp. 335s-357s (1947).

(59) Stout, R. D., and Doty, W. D., Weldability of Steel, Welding research Council (1953).

(60) Tetelman, A. S., and McEvilly, A. I., Jr., Fracture of Structural Materials, John Wiley and Sons, Inc., 1967.

(61) Van der Veen, J. H., "Influence of Steel-Making Variables on Notch Toughness," Ship Structure Committee Report, Serial No. SSC-128 (June 1960).

(62) Warner, W. L., "The Toughness of Weldability," The Welding Journal, 34 (1), research Supplement, 9s-22s (1955).

(63) Welding Handbook, Section 1, Fundamentals of Welding, Second Edition, 1963.

(64) Wells, A. A., "The Brittle Fracture Strength of Welded Steel Plate," Quarterly Trans. Inst. Naval Arch., 48 (3), 296-326, (July 1956).

(65) Wells, A. A., "Influence of Residual Stresses and Metallurgical Changes on Low-Stress Brittle Fracture in Welded Steel Plates," The Welding Journal, 40 (4), Research Supplement, 182s-192s (April 1961).

(66) Williams, M. L., "Correlation of Metallurgical Properties and Service Performance of Steel Plates from Fractured Ships," The Welding Journal, Research Supplement, Vol. 37, pp. 445s-454s (October, 1958).

(67) Williams, M. L., and Ellinger, G. A., "Investigations of Structural Failures of Welded Ships," Welding Journal, 32 (10), Research Supplement, 498s-527s (1953).

(68) Workshop in Fracture Mechanics, Text, August 16-28, 1964 in Denver, Colorado, arranged by Universal Technology Corporation, Dayton, Ohio.

(69) Yoshiki, M., Kanazawa, T., and Itagaki, H., "Double Tension Test With Falt Temperature Gradient," Proceedings Third Japan Congress on Testing Materials--Metallic Materials, pp. 103-106 (1960).

(70) Yoshiki, M., and Kanazawa, T., Studies on the Brittle Fracture Problems in Japan, Vol. 13 of the 60th Anniversary Series, published by The Society of Naval Architects of Japan, 1967.

CHAPTER 13   FATIGUE FRACTURE

13.1   Introduction to Fatigue Fracture

This chapter discusses fatigue fracture, i.e. fracture under repeated loading. When a material is subjected to repeated loading, fracture takes place after a certain number of cycles. The lower the applied stress, the larger the number of cycles before fracture takes place.

As far as fatigue fractures of engineering structures are concerned, the following two types of fractures are important:

1. High-cycle, low-stress fatigue

2. Low-cycle, high-stress fatigue.

In high-cycle fatigue, the endurance limit of a material after several million cycles or more is usually considered. In low-cycle fatigue, on the other hand, fracture after repeated loading less than 10^5 cycles is usually discussed.

High-cycle fatigue is a problem for portions of a structure subjected to fast, repeated loads. Such portions would include areas close to propellers, machinery and areas under fairly consistent vibrations. In such areas, stress cycles may reach several millions in a relatively short period. For example, when the frequency is 100 cycles per minute, the following cycles will be applied after:

1 hour:	6,000 cycles
1 day:	144,000 cycles
10 days:	144×10^6 cycles

However, for ordinary members of ships and ocean engineering structures in which major loads are induced by waves, low-cycle rather than high-cycle fatigue is more of a problem. When structural members are exposed to water, corrosion fatigue often becomes a serious problem; corrosion fatigue will be discussed in detail in Chapter 15.

Fatigue vs. Brittle Fracture

There are essentially two different types of fractures in engineering structures, namely fatigue fractures discussed in this chapter and brittle fractures discussed in Chapter 12. These two types are similar in certain respects but also have essential differences.

An initial fatigue fracture can grow fairly slowly. Years may elapse between the time a tiny fatigue crack initiates, and a catastrophic failure takes place. When the material is sufficiently ductile, final fracture does not occur until the crack has reduced the cross-sectional area so much that the nominal stress approaches the ultimate strength of the material. When the material is not ductile, however, brittle fracture takes place from the fatigue crack as the crack length reaches the critical crack length of the material. In fact, catastrophic brittle fractures in a number of structures started from fatigue cracks.

On the other hand, brittle fractures occur suddenly and have a rapid propagation. The structure's age is not important, but the temperature may be critical, as discussed in Chapter 12. Brittle fracture usually initiates from a sharp notch, which may be a weld defect or a fatigue crack.

Characteristics of Fatigue Fracture

A fatigue fracture can be grouped into the following three stages:

1. Initiation of a crack

2. Slow growth of the crack

3. Onset of unstable fracture.

A fatigue fracture has several characteristics, especially during the first stages, which enable an engineer to identify the fatigue fracture from other types of fracture.

Initiation of Fatigue Crack. In most cases, fatigue cracks originate at the surface. Smoothness of surface is very important to obtain high fatigue strength as discussed later.

On the other hand, brittle fractures often initiate at a subsurface defect where the triaxiality of stress is great.

Growth of a Fatigue Crack. A small crack initiated grows slowly as stresses are repeated. A fatigue crack is transgranular; it propagates within grains rather than along grain boundaries. A fatigue crack propagates in the direction perpendicular to the maximum tensile stress. More discussions on crack growth will be given in a later part of this chapter.

Onset of Unstable Fracture. As the crack progresses, the stress on the residual cross-section increases so that there is a corresponding increase in the rate of crack propagation. Ultimately, a stage is reached when the remaining area is unable to support the applied load and final rupture occurs. The fracture surface   of the final rupture area may be either crystalline or fibrous depending upon whether the fracture is brittle or ductile.

Fracture Appearance.   The above characteristics of fatigue fracture can be observed on fracture surfaces.

Figure 13-1 is a diagramatic representation of a typical fatigue fracture surface.[4]   The region surrounding the origin of a fatigue fracture has a smooth, silky appearance which extends to the limit of the fatigue fracture proper.

In the immediate vicinity of the crack's origin the surface may appear extremely smooth, a feature which is probably accentuated by rubbing of the surface as the crack propagates.   There is often a tendency for this smooth region to grow slightly, but progressively, rougher in texture as the distance from the origin increases.   Careful examination of this smooth part frequently reveals the existence of concentric rings or beach markings around the fracture nucleus and radial lines emanating from it. The beach markings or striations are more evident under an electron-microscope as discussed in a later part of this chapter.

Figures 13-2 and 13-3 show typical specimens in which the initial cracking was due to fatigue, but in which the final fractures were brittle and ductile, respectively.[4]

13.2 High-Cycle Fatigue

This section discusses primarily high-cycle fatigue. However, general discussions on fatigue including definitions of stress cycle, the S-N curve and mechanisms of fatigue fractures

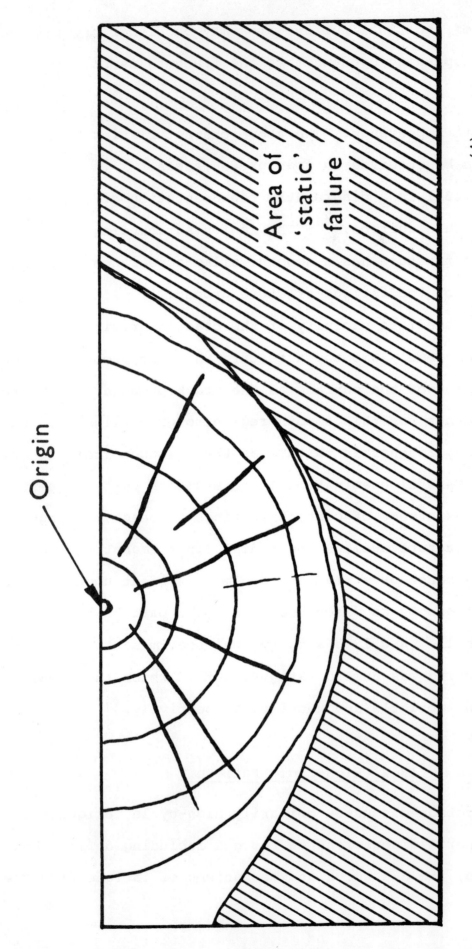

Origin

Area of 'static' failure

FIGURE 13-1   DIAGRAMATIC REPRESENTATION OF A TYPICAL FATIGUE FRACTURE SURFACE[4]

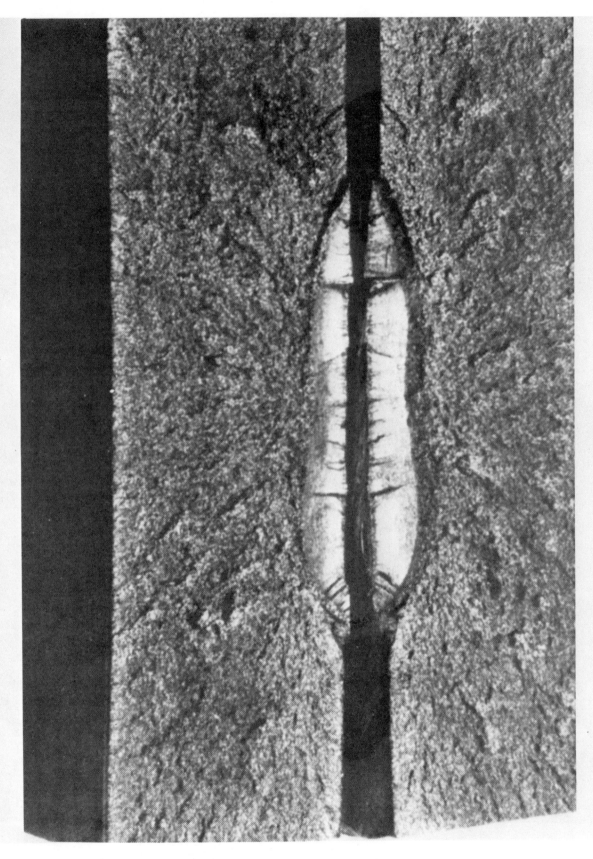

FIGURE 13-2   FRACTURE SURFACE OF A SPECIMEN IN WHICH FATIGUE
CRACKING WAS FOLLOWED BY BRITTLE FRACTURE

FIGURE 13-3   FRACTURE SURFACE OF A SPECIMEN IN WHICH FATIGUE
CRACKING WAS FOLLOWED BY A DUCTILE FRACTURE

apply to both high-cycle and low-cycle fatigue.

Fatigue Testing - Stress Cycles

The majority of our knowledge of the fatigue behvior of materials has developed from laboratory tests of relatively simple specimens. Some 200 laboratories in the United States are engaged in fatigue studies. The number outside the United States is estimated to be even larger.[3]

There are four possible basic parameters which can be used in the definition of the stress cycles to which a fatigue specimen is subjected. They are:[4]

$$\text{The minimum stress in cycle} \quad S_{min}$$

$$\text{The maximum stress in cycle} \quad S_{max}$$

$$\text{The mean stress} \quad S_m = \frac{1}{2}(S_{min} + S_{max})$$

$$\text{The stress range} \quad S_r = S_{max} - S_{min}$$

The stress cycle is fully defined when any two of the above four quantities are given.

Figure 13-4 shows typical stress cycles used in fatigue tests. The cycle shown in Figure 13-4a, in which the stress is varied between zero and tension, is often called a "pulsating tension" cycle. This type of cycle is most commonly used for fatigue testing of plates and welded joints. A fairly large number of results also exist for the case shown in Figure 13-4b, where $S_{min} = -S_{max}$. This type of cycle is usually refrred to as an "alternating " cycle.

In the cycle shown in Figure 13-4c, both the maximum and minimum stresses are tensile. The particular case in which

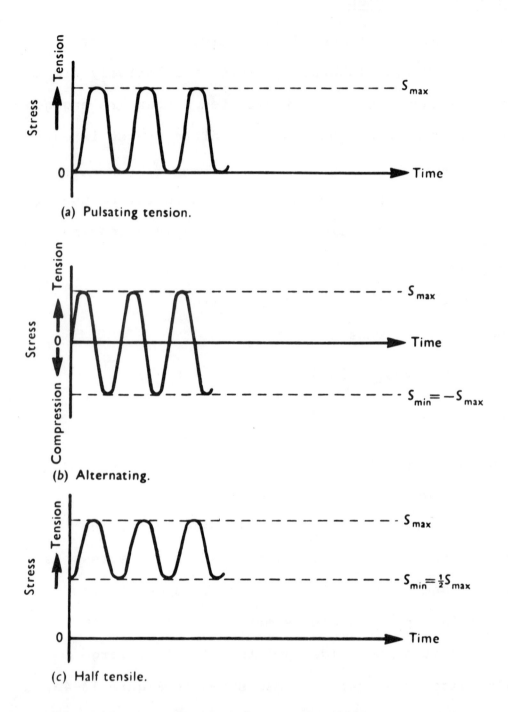

(a) Pulsating tension.

(b) Alternating.

(c) Half tensile.

Figure 13-4   TYPICAL STRESS CYCLES USED IN FATIGUE TESTS

$S_{min} = 0.5\ S_{max}$ is often called a "half tensile" cycle. Other cycles involving unequal tensile and compressive stresses, or wholly compressive stresses, also may be used.

The S-N Curve and Fatigue Strength

In order to determine the fatigue strength of a particular material under a given load condition, it is necessary to test several similar specimens. Each of three specimens is subjected to a given cyclic stress and the number of loading cycles required to produce failure in each specimen is recorded, and a relationship is obtained between the applied stress, S, and the number of cycles to failure, N.

Several different ways of reporting fatigue resistance have been used. In some cases, presentations are made on the basis of stress or strain while in other cases comparisons are based on such parameters as load, moment, or torque. Most commonly used are nominal stresses on the cross section rather than the local stress which, at notches, will be much higher than the nominal stress. In case of a bending test, it is necessary to specify to what point the stress refers, since the stress distribution is not uniform over the cross section. The stress at the extreme fiber is most commonly used.

Figure 13-5 shows the Wohler or S-N curve which represents the relationship between maximum stress and cycles to failure.[9]

Figure 13-6 shows a relationship between the logarithms of both stress and the number of cycles, log S - log N.[9] Because log S - log N relationships for many materials are approximately linear, most fatigue data are presented on the basis of log-log

352

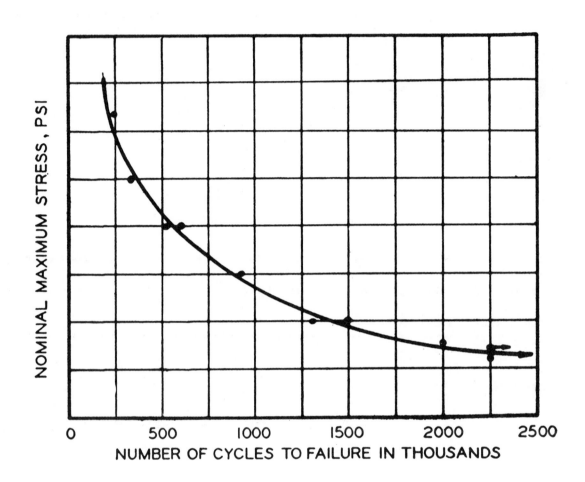

Figure 13-5   REPRESENTATION OF RELATIONSHIP BETWEEN MAXIMUM

STRESS AND CYCLES TO FAILURE (WOHLER OR S-N

CURVE) [9]

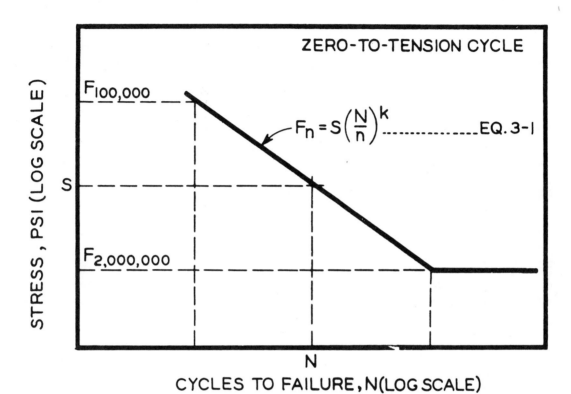

Figure 13-6 S-N CURVE PRESENTED ON A LOG-LOG SCALE[9]

relationships.

For plain specimens of ferrous metals it has been found that, after about 2 to 5 million cycles, the curve is almost parallel to the N-axis, indicating that at a very slightly smaller stress the specimen would have an infinite life. This limiting stress is called the endurance limit of the material. Most other materials do not exhibit an endurance limit, although the S-N curve becomes substantially horizontal as N increases.

The sloping line of Figure 13-6 can be expressed by the relationship

$$F_n = S(N/n)^k$$

where

F_n = the fatigue strength computed for failure at n cycles

S = the stress which produced failure in N cycles

K = the slope of the best-fit straight line representing the data.[9]

In such a relationship, the fatigue strength can be computed over the range covered by the sloping line for any selected number of cycles of the same types of stress cycle, if the slope of the line and one point on the line are known.[9]

Only one type of stress cycle is represented on an individual S-N curve. Therefore, for general understanding of the fatigue behavior of a material or a joint, it is necessary to have a number of S-N curves, one for each different type of stress cycle.

The data for various S-N curves can be summarized in another type of diagram called a Goodman diagram. Figure 13-7 shows such

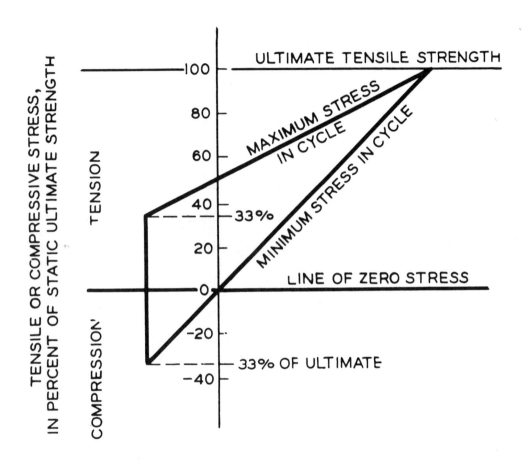

Figure 13-7   GOODMAN DIAGRAM -- A COMPOSITE REPRESENTATION

OF THE EFFECTS OF VARIOUS TYPES OF STRESS CYCLES

ON FATIGUE LIFE[9]

a diagram for an as-rolled flat plate. It provides a composite representation of the effects of various types of stress cycles, such as those ranging from static tension, through zero-to-tension, to complete reversal.

The range in stress is shown by the vertical distance between the two heavy sloping lines. For example, at the extreme left side of the diagram, the stress range is a complete reversal from a compressive stress to a numerically equal tensile stress. At the extreme right side of the diagram, the maximum stress line intersects the minimum stress line at the level of the ultimate tensile strength, and the range of stress is zero. At an intermediate point, the minimum stress line intersects the line of zero stress; the maximum stress represents the fatigue strength under a pulsating load (0 to tension). Unless a finite life is stated as the basis for the Goodman diagram, it usually pertains to the fatigue limit, as in the case of Figure 13-7.[9] There are several modifications of the Goodman diagram.

Factors Influencing Fatigue Limit

In this section, several factors pertinent to fatigue limits will be considered, including the material, notch, residual stress and corrosion.

Materials. Figure 13-8 shows the relationship between the endurance limit and the ultimate tensile strength of various steels. As Figure 13-8 illustrates, the fatigue strength of materials increases as the strength of the material itself increases at a ratio of about 50%. This relationship is true, however, only when the specimen is polished and the surface is very smooth. If

357

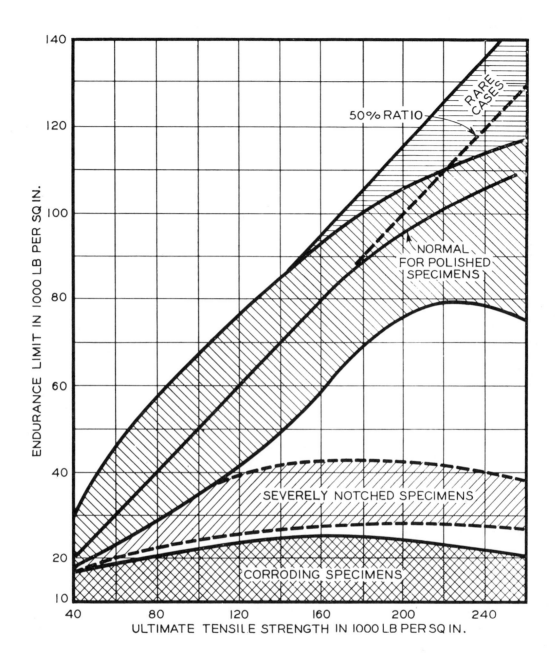

Figure 13-8   RELATIONSHIP BETWEEN THE ENDURANCE LIMIT AND
ULTIMATE TENSILE STRENGTH OF VARIOUS STEELS[9]

the material is severely notched, the endurance limit is reduced
drastically. The change is even greater when a corroding specimen
is involved. These latter effects--a reduction in the endurance
limit due to a severely notched surface or a corroding material--
are most severe when the material is a high-strength steel.[9]

Stress Concentration. The most deleterious factor affecting
the fatigue life of metals is the localized concentration of
stress by geometric discontinuities including cracks, notches,
fillets, holes, surface imperfections, etc. The shape and size
of the discontinuity determine its ability to concentrate stress.
The highest stress concentrating effects are caused by cracks
while generous fillets with a smooth, polished surface have the
lowest.[3]

One of the most important considerations of the notch effects
on the endurance limit is that a notch on the surface of a
material is much more detrimental than a notch in the base of the
same material.

The stress concentrating effects of discontinuities may
be calculated by using methods based on geometry, dimensions,
and assumed elastic behavior. The effect arrived at is called
the theoretical stress concentration factor (K_t). The actual
effect of a given stress concentration, however, may vary both
within and among materials. It is possible, through tests, to
establish the reduction in fatigue strength caused by a particular
stress concentration factor for a particular material at a
particular strength level. By comparing these data with unnotched
(smooth) test data, we can calculate the so-called fatigue

strength reduction factor (K_f). From K_t and K_f the notch sensitivity index (q) of the material can be calculated as follows:

$$q = \frac{K_f - 1}{K_t - 1}$$

In most materials, the notch sensitivity tends to increase with increasing strength. Therefore, in the presence of sharp notches, it is not unusual to find little or no advantage in fatigue strength for high-strength materials.[3]

Corrosion. As shown earlier in Figure 13-8, a material subjected to a corrosive environment is drastically affected in its fatigue strength. Corrosion, in fact, is one of the most detrimental factors resulting in low fatigue strength. Fatigue under corrosive environment is called "corrosion fatigue."

Many stressed structures in the sea fail because the sea water accelerates the fatigue process. A common solution to the problem is to design with a low strength alloy at a low working stress, say below 5,000 psi, and make a big bulky marine structure. Within recent years, with the need for better performance, attempts have been made to use higher strength materials in ocean environment. However, it has been found that if the designer also includes the effect of the marine environment the useful increase in strength is more than cancelled out by the material's reaction with its environment and an acceleration of the rate of crack propagation across the component. In fact, the advances made with AISI 4340 steels and other high-

strength materials in the aerospace structures are often lost
when these materials are submerged in sea water.

Figure 13-9 shows effects of notch and sea water on flexural
fatigue strengths of HY-130/150 steel and special grade Ti-6Al-4V.[2]
The figure shows the fatigue strength of HY-130/150 to be more
severely damaged by notch and sea water than that of Ti-6Al-4V.

Residual Stresses. If there are compressive stresses in
the surface of the plate, the fatigue limit may be increased.
This is true for both smooth and notched specimens which may show
increase in tensile strength beyond that 220,000 psi level. An
example of this is a driveshaft which was induction hardened with
a scanning coil that covered all of the deep grooves. No fatigue
problems with any shafts made this way have been recorded.
LaBelle[5] conducted a series of fatigue tests on actual shafts
which had been processes to introduce compressive stresses at
stress raisers.

"The original shaft, made of AISI 5046 steel, was hardened
and tempered to Bhn 226 to 269....rolling the fillets increased
the shaft's endurance limit from 40,000 to about 55,000 psi,
induction hardening raised it to nearly 80,000 psi, and
nitriding to a depth of 0.010 inches elevated it to 118,000 psi,
three times the load carrying capacity of the orginal shaft.
Although the nitrided shaft was made of a different steel (VCM,
containing 0.35 C, 1.00 Cr, 1.00 Mo, 0.65 Ni), it was hardened
and tempered to about the same hardness as the production shaft
and would have been expected to behave similarly. This large
increase in endurance was brought about by nitriding to a case

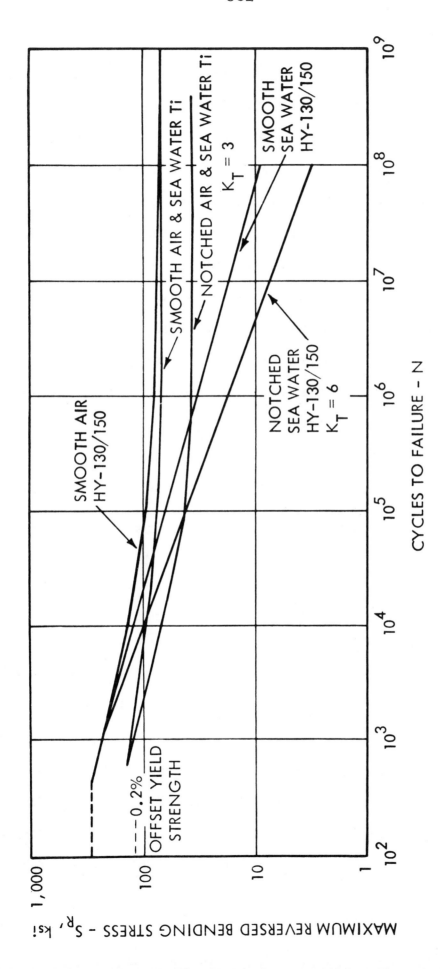

FIGURE 13-9 FLEXURAL FATIGUE CURVES FOR SPECIAL GRADE Ti-6Al-4V

depth of only 0.010 inches."[5]

The effects of residual stress, especially on steel structures, is still a matter of debate. A number of investigations have reported that the fatigue strength increased when specimens had compressive residual stresses, especially on the specimen surfaces. On the other hand, a number of investigators believe that the effects of residual stress on fatigue strength of weldments are negligible.

In most experimental programs conducted so far, fatigue tests were made on specimens under as-welded and stress-relieved conditions, and data were evaluated on the bases of the S-N curve. Little information has been obtained on:

(1) The effect of residual stresses on the growth of a fatigue crack

(2) Redistribution of residual stresses during repeated loading.

Scientific investigations on these problems are necessary to clarify the effect of residual stresses on fatigue strength.

Studies of Fatigue Crack Growth

In recent years many studies have been made of fatigue crack growth. An increasingly commonly used research technique is the electron-microscope fractography which is the observation under an electron microscope of fracture surfaces (in most cases plastic replicas of fracture surfaces). It has been established that a fatigue fracture surface is characterized by patterns called "striations," as shown in Figure 13-10 for Ti-8Al-1Mo-1V alloy. It is believed that a striation is formed on each cycle of loading.

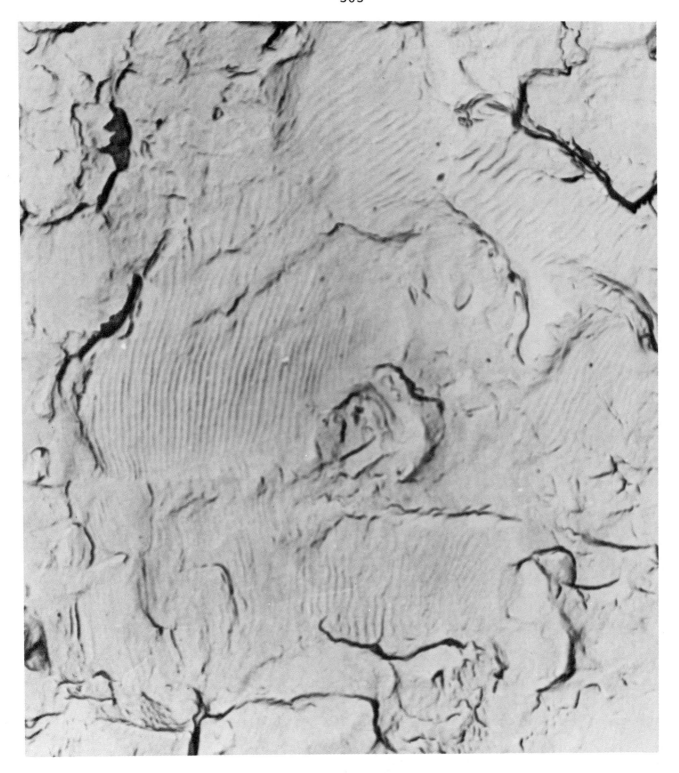

FIGURE 13-10   ELECTRON MICROSCOPE FRACTOGRAPH OF A
Ti-8Al-1Mo-1V ALLOY

Figure 13-11 shows that a reasonable agreement exists between:

(1) The crack growth rate dℓ/dn determined macroscopically from the number of cycles N and the length of crack ℓ observed on a specimen subjected to repeated loading.

(2) The striation spacings μ observed microscopically on the fracture surface.[11]

The results shown here were obtained with a specimen in 2024-T3 aluminum alloy.

A number of investigators have applied the fracture mechanics theory, which is introduced in Chapter 12, to fatigue fracture.[11,12] For example, the relationship between the crack-growth rate and the stress-intensity factor has been obtained. Figure 13-12 shows data on 7075-T6 aluminum-alloy. Experiments were made on four types of specimens subjected to different loadings.[11]

Information was obtained by periodically measuring the length of the crack which was artificially initiated from a sharp saw cut. After knowing the crack length, 2a, and maximum stress, σ g, the stress intensity factor for Case (a) was computed using

$$K_{max} = \sigma g \sqrt{a} \cdot \alpha_1$$

where α is the correction factor for finite panel width w = 2b

$$\alpha_1 = \sqrt{\frac{w}{\pi a} \tan \frac{\pi a}{w}}$$

Other equations must be used for computing K-values for other cases.

The crack growth rate, Δ2a/ΔN, was obtained by dividing

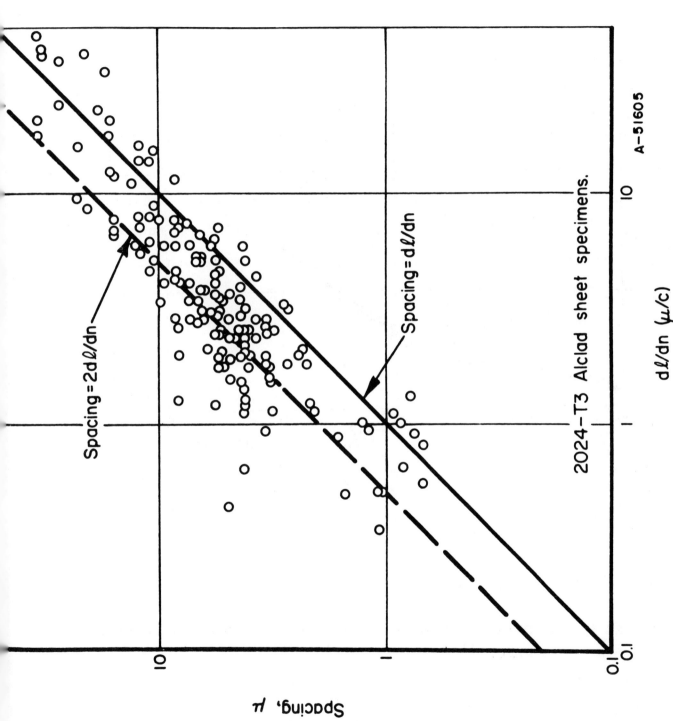

A–51605

2024–T3 Alclad sheet specimens.

FIGURE 13-11 CRACK–GROWTH–STRIATION SPACING VERSUS CRACK–GROWTH RATES IN 2024–T3 ALUMINUM ALLOY (11)

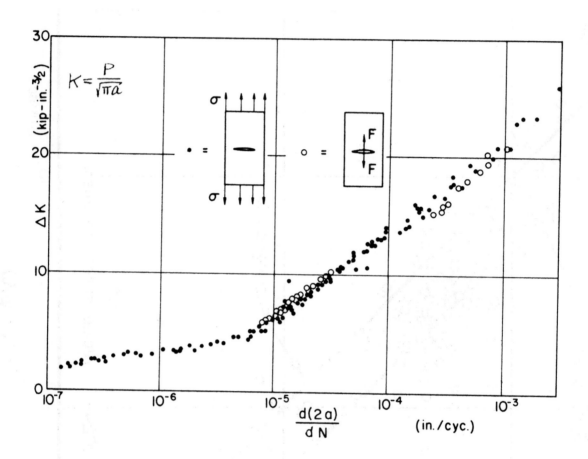

FIGURE 13-12   DATA ON FATIGUE CRACK GROWTH IN 7075 T6 ALUMINUM
ALLOY [11]

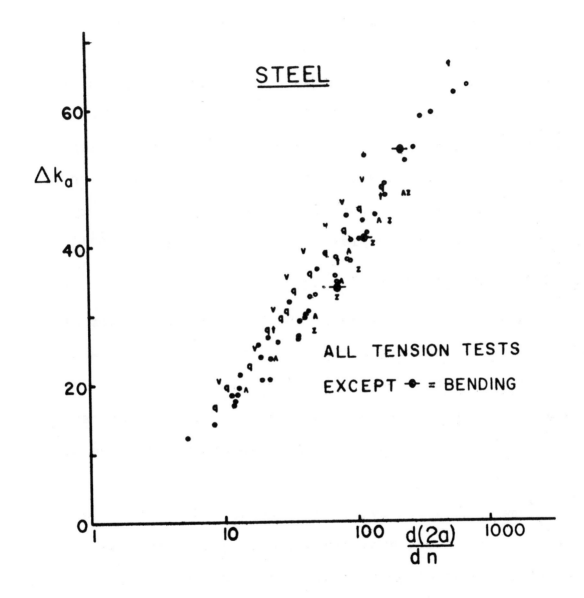

FIGURE 13-13   DATA ON FATIGUE CRACK GROWTH IN CARBON STEEL[11]

the difference in the crack length reading, $\Delta 2a$, by the cyclic number change, ΔN, for each measurement.

Figure 13-13 shows similar results on medium strength steels for both centrally cracked panels and notched beams in bending.[11]

Figures 13-12 and 13-13 show that crack growth rates depend upon stress-intensity factors. As a fatigue crack grows, ΔK increases as does $d(2a)/dN$. Diagrams of ΔK vs. $d(2a)/dN$ show the relative resistance to crack growth in various materials. The ability to correlate data from one configuration to another makes it possible to correlate data from simple laboratory test configurations to more complex structural situations.[11,12]

Fatigue Strength of Welded Joints

Table 13-1 gives fatigue strengths of various types of welded joints in ordering structural steels.[16] Data shown are fatigue strength at 100,000 cycles and 2,000,000 cycles for pulsating (0 to tension) and alternating (reversed) stresses. Figure 13-14 shows some of the welded specimens referred to in Table 13-1.

Specimens with transverse butt welds and reinforcements machined or ground off show fatigue strengths almost as high as those for plain unwelded plates. Therefore, when a joint with high fatigue strength is required, it is very important to remove the reinforcement of a butt weld. It is possible to improve the fatigue strength of a butt weld by adding reinforcement plates, unless the butt weld is grossly defective or deficient in the throat section.

TABLE 13-1 FATIGUE STRENGTHS OF VARIOUS TYPES OF WELDED JOINTS IN ASTM-A7 STRUCTURAL STEEL[15]

Type of Joint	0 to Tension		Reversed	
	F100,000	F2,000,000	F100,000	F2,000,000

Fatigue strengths of flat-plate specimens with and without butt welds, as-welded and with reinforcement removed flush

Plain plate (A-7 steel) *	47.8	31.7	26.8	17.5
Transverse butt-welded joint (single-Vee)—as-welded	34.9	23.2	22.4	14.9
Transverse butt-welded joint (a) single-Vee—(b) single-U—reinforcement off	37.5 (a)	28.7 (a)	27.4 (b)	16.0 (b)
Longitudinal butt-welded joint (single-Vee)—as-welded	39.6	26.0	....	....
Longitudinal butt-welded joint (a) single-Vee, (b) double-Vee—reinforcement off	47.1 (a)	30.5 (a)	21.0 (b)	15.6 (b)

Fatigue strengths of fillet welds of various arrangements in specimens designed for failure in the welds

(a) Fillet-welded lap joints—5/16-in. transverse welds. Failure in welds	30.3	18.5+	16.2	11.3
(b) Fillet-welded lap joints—5/16-in. longitudinal welds. Failure in welds	27.2	19.7	15.2	10.7
(c) Fillet-welded lap joints—5/16-in. transverse and longitudinal welds. Failure in welds	28.3	20.5	13.1	8.9
(d) Tee joints—5/16-in. fillet welds. Failure in welds	19.1	9.6	13.3	6.2
(e) Moment connectors—5/16-in. vertical fillet welds. Failure in welds	46.8	26.5	26.6	17.3

Fatigue strengths of rolled beams with lateral connection plates attached to the tension flanges

(f) Beams with welded attachments. Longitudinal welds	24.4	13.6	....	....
(g) Beams with welded attachments. Transverse welds	19.2	12.2	....	....

Fatigue strengths of welded beams and girders

(h) Rolled beams, 12-in., 31.8-lb I, with continuously welded, partial-length cover plates, 4 x 9/16-in., and 3/8-in. transverse fillet welds across square ends of cover plates	22.8	12.0	....	....
(i) Rolled beams, 12-in., 31.8-lb I, with continuously welded, full-length cover plates 4 x 9/16-in. top plate and 6 x 3/8-in. bottom plate, 3/16-in. fillet welds	41.1	22.8	....	....
(j) Built-up girders with continuous web-to-flange fillet welds and with middle stiffeners welded only to compression flange and upper half of web	46.7	17.6	....	....

Fatigue strengths of plug- and slot-welded joints (stresses given are those in welds)

(k) Plug welds—weld failure	23.1	12.6	11.3	6.5
(l) Slot welds—transverse (balanced design for weld and plate failure)	20.0	10.2	10.0	5.3
(m) Slot welds—longitudinal (balanced design for weld and plate failure)	27.9	11.1	16.1	6.1

Fatigue strength of plug-welded lap joint designed to fail in plate material (stresses given are those in plate material)

(n) Plug-welded lap joint—plate failure	26.3	10.1	14.2	5.3
	24.1	11.7	14.1	6.6

*Test specimens consisted of plates approximately 5 x 7 inches.

NOTE: Recent fatigue tests in Belgium of larger built-up girders subjected to ratios of minimum to maximum stress similar to those encountered in highway bridges have revealed no detrimental effect of welding ends of stiffeners to tension flanges. These tests involved one million cycles of loading. The University of Illinois tests showed a very substantial detrimental effect of welding the load-bearing stiffeners to the tension areas of the rolled beams, for approximately two million cycles of loading, zero to maximum tension.

370

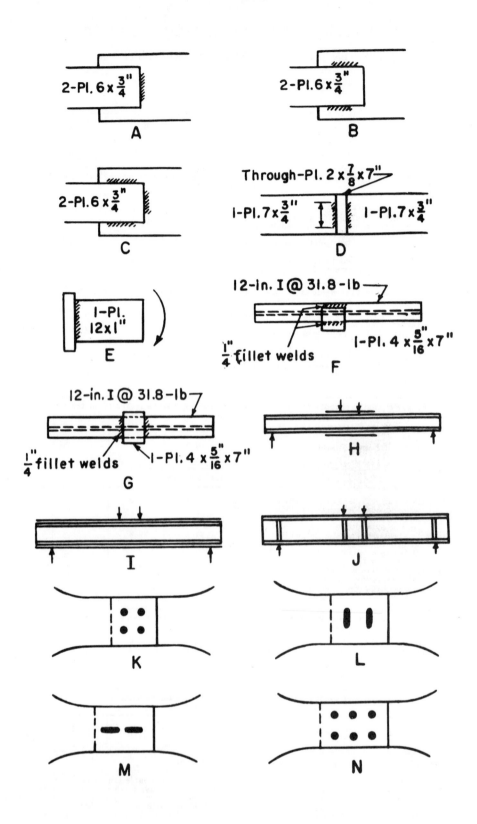

Figure 13-14 SOME OF THE WELDED SPECIMENS REFERRED TO

IN TABLE 13-1

Fatigue strengths of lap welds are considerably lower than those of butt welds. This is because lap welds disturb the flow of stresses more than do butt welds, as shown in Figure 13-15.

Data in Table 13-1 suggest that general rules for improving fatigue strength of a welded structure are as follows:

1. Use butt joints rather than lap joints

2. Avoid intermittent fillet welds

3. Avoid the use of joints that have a large variation in the ability to deform locally

4. Do not specify excessive sizes of fillet welds

5. Avoid making attachments at points where fatigue conditions are severe.

13.3 Low-Cycle Fatigue

To make the distinction between high-cycle and low-cycle fatigue more clear, we will use 10^5 cycles to failure to represent the dividing line between the two types of fatigue.[3]

Characteristics of Low-Cycle Fatigue

Many metals exhibit low-cycle fatigue failure during the loading and unloading cycles which involve plastic deformation. Under such conditions, true stresses are not calculable by elastic theory, and plastic theory is not developed enough to handle such problems. The control and analysis of low-cycle fatigue therefore is easier on the basis of strain rather than stress.

Cyclic Stress and Strain. Figure 13-16 illustrates the stress-strain relationships likely to develop under cyclic loading

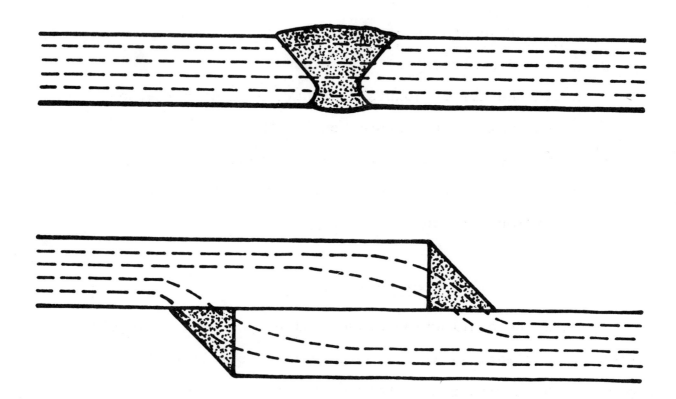

Figure 13-15   LINES OF STRESS FLOW IN A BUTT WELD AND LAP

WELDS

373

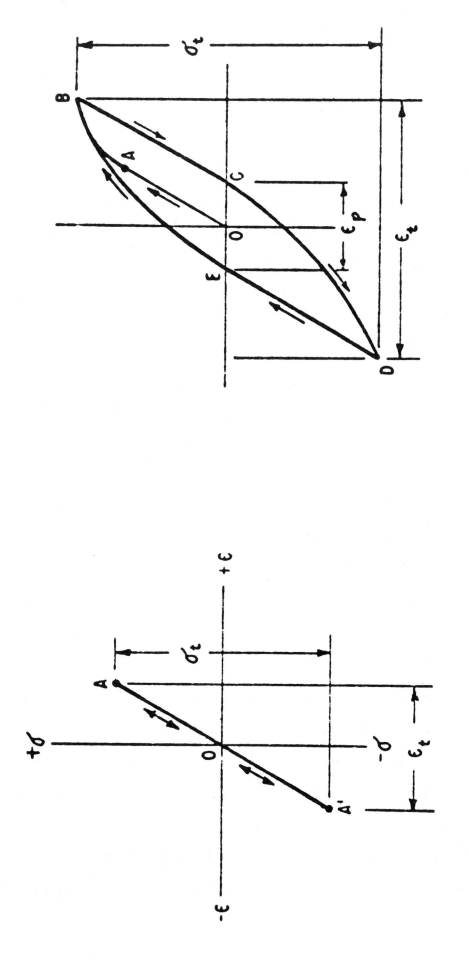

a. Elastic Range

b. Straining into Plastic Range

FIGURE 13-16   STRESS-STRAIN RELATIONSHIPS UNDER CYCLIC LOADING[3]

conditions. The relationship shown in Figure 13-16a happens when
the applied force or moment is completely reversed but within the
elastic region, σ_t is the total stress range and ε_t is the total
strain range. Figure 13-16b shows the relationship that develops
under reversed loading into the plastic region. The stress-strain
relationship is no longer linear but follows the hysteresis loop
BCDEB during each cycle. σ_t is the total stress range but in
this instance may be nominal or true. ε_t, the total stress range
involves two parts: (1) the elastic strain range ε_e, and (2) the
plastic strain range, ε_p, where $\varepsilon_t = \varepsilon_p + \varepsilon_e$.

S-N Relationships

High-cycle fatigue results usually are in the form of S-N
diagrams, such as Figure 13-5. Since the cyclic stresses needed
for failure in the high-cycle region are well within the elastic
region (see Figure 13-16a) it is not important whether we use
stress or strain as the independent variable.

With low-cycle fatigue, however, this is not the case. Due
to the great preponderance of evidence collected over the last
ten years, it has become clear that the strain becomes the
dominant factor with decreasing fatigue life.[3]

Figure 13.17 shows the low-cycle fatigue results obtained
for unnotched (smooth) specimens by the U.S. Navy Marine Engineering
Laboratory. The yield strength of the steels ranges from 41,000
to 230,000 psi and the plot is on the basis of total strain range
versus cycles to failure. The similarity of behavior irrespective
of yield strength is important to note. Perhaps even more significant is the fact that the behavior appears to be independent

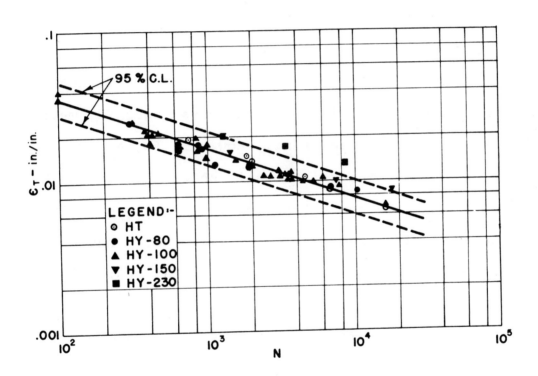

Figure 13-17   TOTAL STRAIN RANGE VS. CYCLES TO FAILURE FOR

STEELS IN AIR [3]

of material as illustrated in Figure 13-18. It seems that if strain is the controlling factor in low-cycle fatigue, that the strength or type of material used makes little difference. Recent tests on pressure vessels tend to support this view for steels with yield strengths from 30,000 to 90,000 psi.[3] However, high-cycle fatigue strength of materials varying in strength level are markedly different as shown in Figure 13-8. Consequently the relationships shown seem to be valid only in the restricted range of 100 to 10,000 cycles. Divergence from the relationship is apparent for the HY-230 steel shown in Figure 13-17.

In Figure 13-19, three steels with markedly different yield strengths are shown. Above 1000 cycles, advantages for materials at high yield strength are apparent. However, for a given cycle life greater than 1000, the relationship shows that the high yield strength materials can suffer low-cycle fatigue failure while apparently being cycled elastically. This actually has happened in tests of HY-230 steel and certain aluminum alloys.[2]

13.4 Fatigue Failures of Structures

In the design of airplanes, cars, and railroad bridges which are subjected to rapidly changing loads, the fatigue is considered very seriously. For example, the fail-safe design concept has been developed and applied to the design of modern aircraft.[11,12]

For a long time, naval architects have considered fatigue to be a problem of minor importance in shipbuilding on the grounds that ships meet slowly changing loads induced by waves.

377

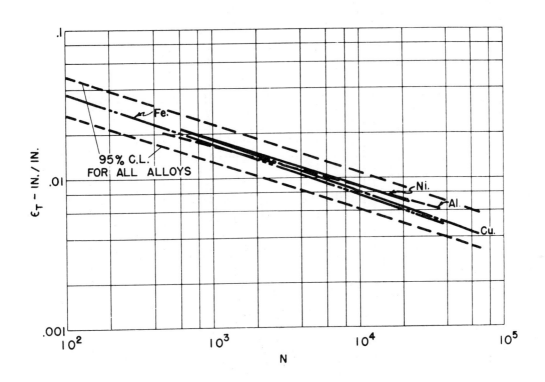

Figure 13-18 TOTAL STRAIN RANGE VS. CYCLES TO FAILURE FOR

ALLOYS IN AIR[3]

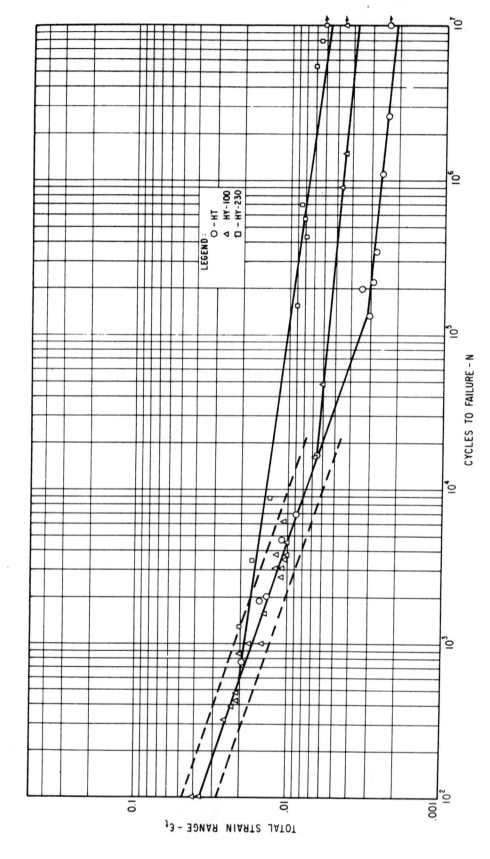

FIGURE 13-19 FATIGUE RELATIONSHIPS FOR STEELS (COMPLETELY REVERSED STRAIN) (3)

Consequently, except for some special areas such as those close to propellers and the machinery, high-cycle fatigue is not a problem. This statement is probably still valid in a general sense. However, there has been an increasing interest in the low-cycle fatigue of ship structures.

Information on fatigue failures of ocean engineering structure is scarce. Consequently, the following pages discuss fatigue failures, especially low-cycle failures, of ship structures.

Analysis of Stress Frequencies

Several investigators have studied frequencies of stresses which may occur in a ship structure.[10,14,15] Figure 13-20 shows a diagram presented by Yuille.[10,20] The straight lines for "one year" and "30 years" represent the cumulative frequency distribution of longitudinal bending stresses for the well-known ship "Ocean Vulcan," on which stresses were measured during service.

Figure 13-20 shows that only 100 cycles of stresses would surpass 800 kg/cm^2 (11,400 psi) during a service for 30 years. Stresses which would reach 10^5 cycles would be around 400 kg/cm^2 (5700 psi). Also shown in the figure is a line for the stresses at discontinuities with a stress concentration factor of 3. This curve nearly touches the fatigue-strength curve for notched specimens in corrosive environment. This is in conformity with the observation that during the tests with "Ocean Vulcan" new cracks did not appear and existing small cracks did not become larger.

Figure 13-21 shows stress frequencies for the deck plate of "Canada," a 9000 ton dead-weight cargo ship.[10] The figure

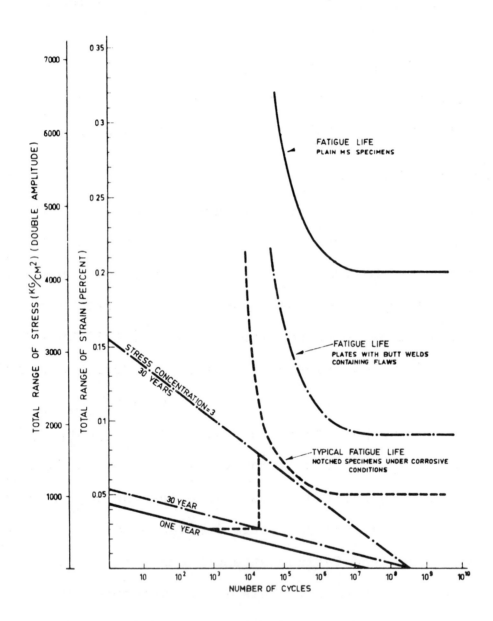

Figure 13-20   STRESS FREQUENCIES IN A SHIP (10,20)

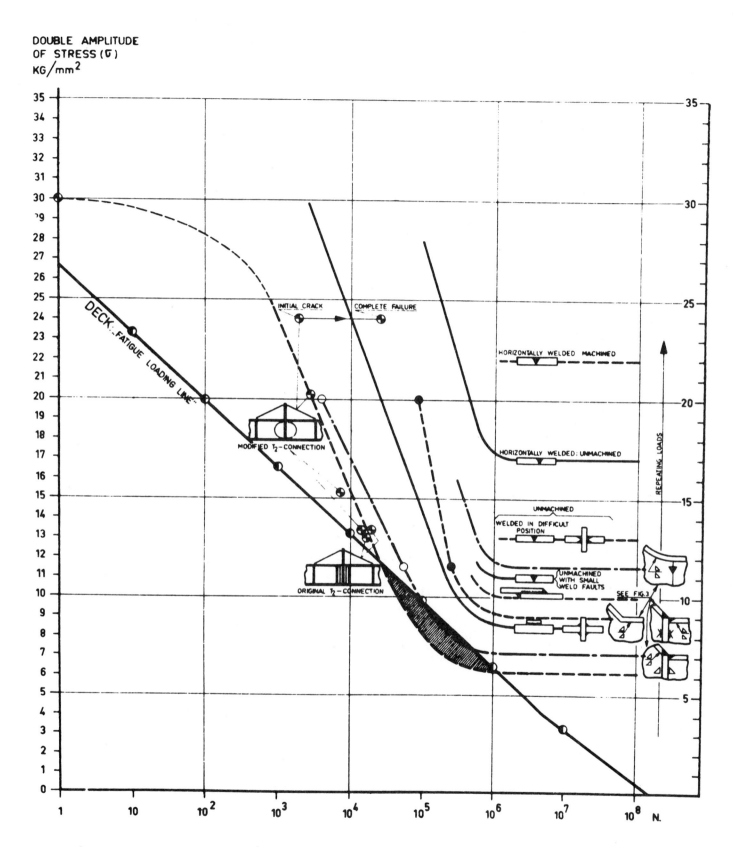

Figure 13-21   FATIGUE LOADING LINE FOR THE DECK OF

"CANADA" AND FATIGUE STRENGTH OF VARIOUS

WELDED JOINTS AND STRUCTURAL COMPONENTS

also shows fatigue strength curves for various welded joints
and structural components. The lowest S-N curve applies to inter-
connections of longitudinal frames of the types used in T-2 tankers
which were tested in the Ship Structure Laboratory of the Techno-
logical University Delft. Figure 13-21 indicates that in the
deck longitudinals of the "Canada," fatigue cracks will initiate
after approximately 10 years if these connections are of the
T-2 tanker type. The stress cycle at failure would be around
2×10^4.

Experience with Fatigue Failures

The preceding analysis on Figure 13-20 and 13-21 indicates
that low-cycle fatigue failures may occur in ship structural
members.

Several investigators have concluded that many cracks which
occurred in actual ships were indeed fatigue cracks. Vedeler,
for instance, stated that most of the cracks found in ships are
due to fatigue.[14,15] Figures 13-22 through 13-24 are examples
cited in a paper by Vedeler.[15]

Figure 13-22 shows the vertical transverse web frame in the
wing tanks of a tanker which had been in service for 9 years.
This is a true copy of the construction which the same shipyard
used when such ships were riveted. The sharp corners in the
bracket connections at top and bottom were found cracked when
surveyed 2 years ago. When repairs were made rounded knees were
fitted.

Figure 13-23 shows the wing tank frame of a tanker. Here
not only the floors were cracked, but also the girder under the deck

383

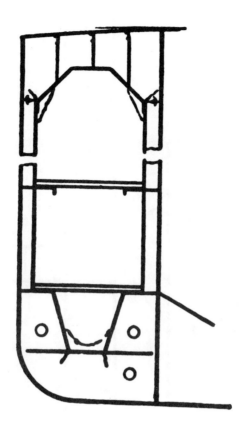

Figure 13-22 SECTION THROUGH WING TANK IN TANKER. CRACKS
            IN 4 SHARP CORNERS[15]

384

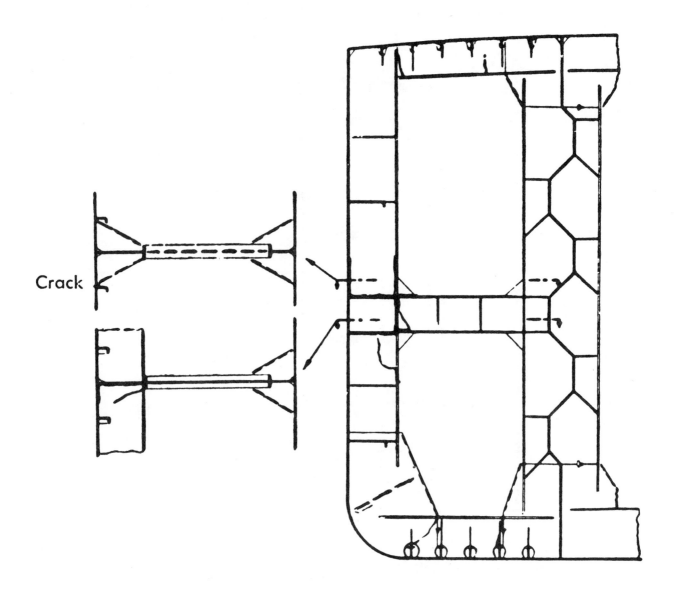

Figure 13-23   CRACKS AT TOP AND BOTTOM AND AT ALL CONNECTIONS

TO CROSS IE IN WING TANKS(15)

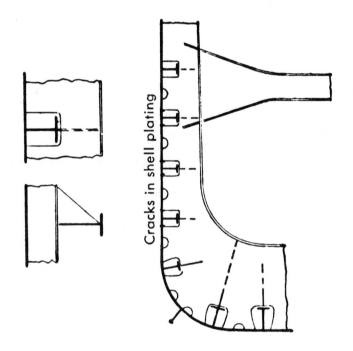

Figure 13-24   CRACKS IN SHELL PLATING AT INSUFFICIENT

CONNECTION BETWEEN WEB FRAME AND LONGITU-

DINALS[15]

and above all the connection of the center tie to the vertical web on the ship's side. This was virtually broken right across at two places, both at the top and bottom of the tie. Furthermore, the horizontal stringer at the bottom of the tie was broken right off. Finally there was a crack in the shell plate at the end of the horizontal bracket at the top of the tie. The vessel was 10 years old when the cracks were found.

Another element of danger when scalloping the frame, this time in a web frame, is shown in Figure 13-24. Here there are relatively large scallops where the longitudinal frames are carried through the web frame, and a relatively weak direct connectic between the longitudinal frames and the web frame. Consequently the horizontal and vertical stiffeners in the shell endeavor to follow a different form of deflection with the result that the shell plates are cracked at a number of points at the edges of the scallops.

13.5 Research and Development

Because of the prohibitive costs of experimentation with fatigue in field situations, most experimental work is done in the laboratory or with laboratory-type specimens. Therefore, comments on R&D efforts should take into account the practicalities of research and the problems inherent in the task. Laboratory efforts in the next several years should study the fatigue behavior of a wide variety of materials and strength levels. Such an expansion of the scope of materials would serve to more correctly determine the short- and long-range problems connected with material and design. More information is needed relative to the effects of notches, environment, crack propagation and stress history with respect to the evaluation of materials. And finally, long-range objectives should be directed toward incorporating laboratory test data into design criteria.[2]

References

(1) Coffin, L. F. Jr., "Low-Cycle Fatigue: A Review," Applied Materials Research, Vol. 1, No. 3, pp. 129-141, October 1962.

(2) Feige, N. G., "Welding of Titanium Alloys for Marine Application," Lecture presented at "Welding Fabrication in Shipbuilding and Ocean Engineering," Massachusetts Institute of Technology, August 20, 1969.

(3) Gross, M. R., "Low-Cycle Fatigue of Materials," NRL Report 6167, "Status and Projections of Developments in Hull Structural Materials for Deep Ocean Vehicles and Fixed Bottom Installations," U. S. Naval Research Laboratory, pp. 196-221, November 1964.

(4) Gurney, T. R., Fatigue of Welded Structures, Cambridge University Press, 1968.

(5) LaBelle, J. E., "Practical Aspects of Fatigue," Metal Progress, Vol. 87, No. 5, pp. 68-73, May 1965.

(6) Manson, S. S., "Interpretive Report on Cumulative Fatigue Damage in the Low-Cycle Range," Welding Journal, Vol. 43, No. 8, Research Supplement, pp. 344s-352s, 1964.

(7) Masubuchi, K., "Effect of Residual Stresses on the Behavior of Welded Structures," a Text Prepared for the Special Summer Session on "Welding Fabrication in Shipbuilding and Ocean Engineering," Massachusetts Institute of Technology, August 18-22, 1969.

(8) Masuda, Y., "Low-Cycle Fatigue Strength of Ship Structure," Journal of the Japan Welding Society, Vol. 37, No. 5, pp. 473-479, 1968.

(9) Munse, W. H., Fatigue of Welded Steel Structures, Welding Research Council, 1964.

(10) Nibbering, W., "Fatigue of Ship Structures," Report of the Netherlands' Research Center T. N. O. No. 555, 1963.

(11) Paris, P. C., "Fatigue Crack Growth," the Text Used for Workshop in Fracture Mechanics, August 16-28, 1964, held at Denver Research Institute, sponsored by Universal Technology Corporation, Dayton, Ohio.

(12) Tetelman, A. S., and McEvily, A. J. Jr., Fracture of Structural Materials, John Wiley & Sons, Inc., 1967.

(13) Vastu, J., "Fatigue Structural Models," NRL Report 6167, "Status and Projection of Developments in Hull Structural Materials for Deep Ocean Vehicles and Fixed Bottom Installations," U. S. Naval Research Laboratory, pp. 212-221, November 1966.

(14) Vedeler, G., "To What Extent do Brittle Fracture and Fatigue Interest Shipbuilders Today?" Houdremont Lecture at the Annual Assembly of the International Institute of Welding, Oslo, Norway, 1962.

(15) Vedeler, G., "One Learns from Bitter Experience," International Shipbuilding Progress, Vol.5, No. 42, pp. 67-77, 1958.

(16) Welding Handbook, Sixth Edition, Section 1, American Welding Society, 1968.

(17) Yamaguchi, Y., "Fatigue Failures in Ship Structures," Journal of the Japan Welding Society, Vol. 37, No. 10, pp. 1041-1046, 1968.

(18) Yamaguchi, Y., "Fatigue Strength of Welded Ship Structures," Journal of the Japan Welding Society, Vol. 34, No. 6, pp. 566-571, 1965.

(19) Yao, J. T. P., and Munse, W. H., "Low-Cycle Fatigue of Metals--Literature Review," Welding Journal, Vol. 41, No. 4, Research Supplement, pp. 182s-192s, 1962.

(20) Yuille, I. M., "Longitudinal Strength of Ships," R.I.N.A., 1962.

CHAPTER 14 <u>CORROSION, STRESS CORROSION CRACKING, AND HYDROGEN</u>
<u>EMBRITTLEMENT</u>

This chapter covers three subjects which are related to
the effects of environments on materials. These subjects include
corrosion, stress corrosion cracking, and hydrogen embrittlement.

Corrosion is often thought of only in terms of rusting and
tarnishing. However, corrosion damage also occurs in other ways,
for example in failure by cracking, or in loss of strength or
ductility. In general, each type with some few exceptions occurs
by electrochemical mechanisms. According to Uhlig[28], the five
main types of corrosion classified with respect to outward
appearance or altered physical properties are: (1) uniform attack
(or general wasting), (2) pitting, (3) dezincification and parting,
(4) intergranular corrosion, and (5) cracking.

The first section of this chapter (14.1) discusses corrosion
control problems in marine industry. Section 14.2 covers
corrosion in the marine environment, starting with general wasting
of steel and then discussing various types of corrosion in other
metals and alloys. Section 14.3 covers stress corrosion cracking,
and Section 14.4 covers hydrogen embrittlement.

14.1 Corrosion Control Problems in Marine Industry

The performance of metals in the marine environment has been
observed and studied for over 200 years. The results have been
published in a large number of technical journals. Manufacturers
and users of marine equipment have developed economical materials

that meet particular needs. Various control measures which have been developed over the years have offered impressive economics to the shipbuilding and ocean engineering industry.

Despite this wealth of information, corrosion is still one of the major problems which faces ocean engineers. There are so many aspects of the performance of materials in marine applications, that it is all too easy to overlook one, or more of the controlling factors. It is equally easy to underestimate both the extent and intensity of the corrosion processes--galvanic corrosion, pitting, crevice corrosion, velocity effects, etc.--in the marine environment where these processes are so greatly intensified.

Brown and Birnbaum[4] have summarized their views on corrosion control for structural metals in the marine environment. Their important conclusion is: "With new materials and new geometries of components and structures for various deep ocean projects, the primary need for corrosion control has shifted from one of maintenance economy (though of course this is as desirable as ever) to one of survival of the component or structure.

For example, important aspects of corrosion control in the past were to extend the period between drydockings and to reduce the number of plates needing replacement because of corrosion. By contrast, a typical corrosion control problem posed by deep ocean requirements is corrosion or corrosion fatigue of high-strength armor wire for communication cable.

Discussions given in the rest of this section come primarily from a report by Brown and Birnbaum.[4]

Figure 14-1 shows types of current or possible future corrosion control problems which face the Navy.[4] Engineers engaged in the design and fabrication of ocean engineering structures with high-strength materials will probably face similar problems.

The first case in Figure 14-1 is stress corrosion cracking of high manganese austenitic steels used in non-magnetic European submarines. Service failures occurred at welds. On the basis of experience to date it appears that an acidified salt solution containing H_2S will crack more steels, and in general crack them more quickly, than any other medium. However, it is not entirely certain that this is the case with all steels.

The second case of Figure 14-1 is a submersible of aluminum alloy 7079. This alloy has a high strength-to-density ratio, but it is susceptible to selective corrosive attack and stress corrosion. Even though the aluminum plates are protected by a coat of paint, cracking may occur at breaks in the coating, especi-when tensile residual stresses exist on the surface.

The third case in Figure 14-1 is the stress corrosion cracking of a proposed submersible of very high-strength steel (designated H-11). This steel is highly susceptible to stress corrosion cracking. If one attempts to counter stress corrosion in this material by the application of cathodic protection, a phenomenon intrudes to defeat this move. This phenomenon is hydrogen embrittlement, and all the high-strength steels (low alloy, hardenable stainless, precipitation hardening, and maraging) are susceptible to it when they are heat-treated to sufficiently

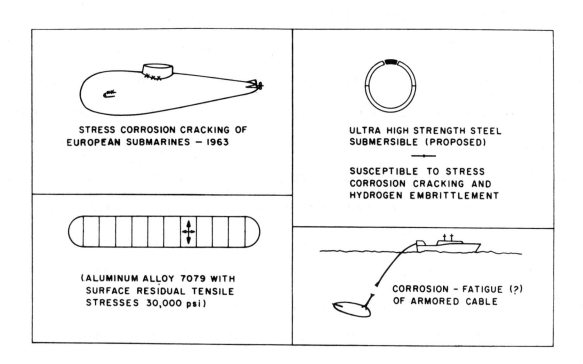

FIGURE 14·1   TYPES OF CURRENT OR POSSIBLE FUTURE
CORROSION CONTROL PROBLEMS [4]

high strengths.

The current estimate of the strength ranges for the three classes of steels at which they become susceptible to stress corrosion cracking is given in Figure 14-2.* (The precipitation hardening steels are lumped with the other hardenable stainless steels in this figure.) In preparing this figure, only data for sea water, salt solutions (used in the laboratory), distilled water (representing condensate), marine atmosphere, and humid air were used. Omitted were data for the acidified salt solutions containing H_2S, and nitrate solutions, on the grounds that these are unrealistic for the marine environment. Figure 14-2 also indicates the range at which hydrogen embrittlement under sustained load might be expected to become a practical problem. Also shown in Figure 14-2 are graphs for the expected cracking behavior of aluminum alloys, with the yield strength scale expanded. Titanium is also shown, although no incidence is known of stress corrosion cracking of any titanium alloy either in sea water or in other environments at room temperature.**

---

*The bar graphs of Figure 14-2 show the transition from "resistant" to "susceptible" strength levels as a stippled zone. Toward the left of these zones the stresses required for cracking may be lower, the times for cracking longer, some members of the class may be largely immune, or special heat treatments or special environments may be required to cause cracking. Toward the right-hand end of these zones the heavier shading is intended to indicate that most members of the alloy class are susceptible to cracking in short times at moderate stresses in common media.
**Two instances have been reported in which titianium alloy plate specimens containing circular welds were observed to be cracked after a period of exposure to laboratory atmospheres. But not enough information was given to be sure that these were not in fact brittle running fractures perhaps triggered by an aging reaction, and the appearance of the cracks supported this conjecture.

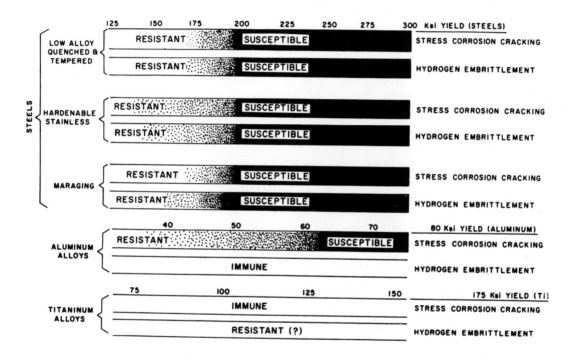

FIGURE 14-2   ESTIMATED STRENGTH RANGES FOR SUSCEPTIBILITY
TO CRACKING IN SEA WATER, FRESH WATER, OR
HUMID ATMOSPHERES[4]

Returning to Figure 14-1, the fourth illustrative case is one of corrosion fatigue or armored cable. It is obvious that if corrosion produces a pit, this can be expected to act as a stress raiser which could accelerate the fatigue process, and this is observed. Perhaps somewhat surprising, however, is the observation that although the titanium alloy B120VCA is essentially inert to salt water, when fatigued in salt water the life is sharply reduced compared with the life in air. It might be supposed that adequate cathodic protection should remove the "corrosion" component of cathodic protection, and this is observed for this alloy (Figure 14-3). But in the case of steels of almost any strength level, the application of cathodic protection beyond a certain level (defined for only a very few conditions for a very few steels to date), once again a form of hydrogen embrittlement appears to intrude to counter the effect of corrosion protection and place a limit on the available effectiveness of cathodic protection.* An example of data showing this is given in Figure 14-4. The study of corrosion fatigue is rather more formidable than that of stress corrosion cracking because of the requirement for machines in large quantities to produce the fatigue, whereas it is possible to use a bent specimen to produce its own stress field for stress corrosion cracking. Therefore it is imperative to design any program with the utmost care to ensure that the results have broad applicability.

---

*Such behavior was not observed in the few earlier studies of the application of cathodic protection to the corrosion fatigue problem because hydrogen embrittlement is strain-rate sensitive and does not appear at the high cycle rates used in the earlier studies, so that complete recovery was indicated even with indiscriminantly high levels of cathodic protection.

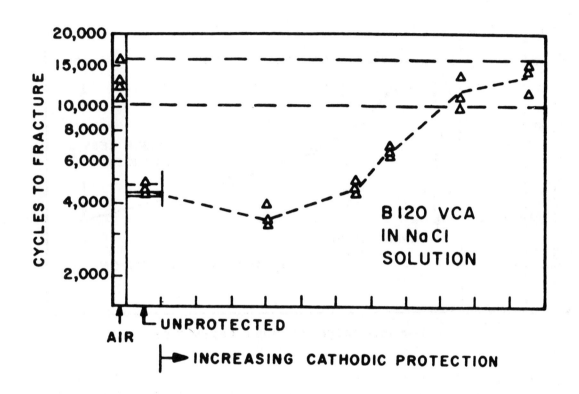

FIGURE 14-3   CATHODIC PROTECTION OF HIGH-STRENGTH
TITANIUM ALLOY AGAINST CORROSION FATIGUE[3]

398

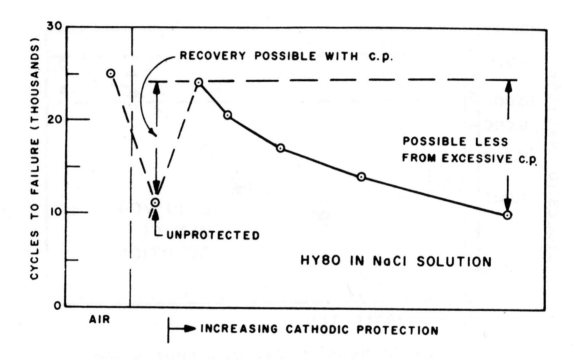

FIGURE 14-4   EFFECT OF CATHODIC PROTECTION ON CORROSION
FATIGUE OF STEEL AT LOW-CYCLE FREQUENCY
(100 c.p.m.)[4]

In addition to the major phenomena illustrated in Figure 14-1, there are a considerable number of relatively small corrosion problems usually not involving stress which will be needed to be solved on a continuing basis to answer the developing questions as deep ocean hardware continues to expand in types of materials and geometries. As an example of this is cited the problem of O-ring seals, which are widely used to exclude seawater from instrumented cases intended for prolonged service at great depths. A relatively minor amount of corrosion around the O-ring may break through the seal. It is possible that cathodic protection techniques can delay this for very long times--of the order of many years. But this must be examined experimentally at a marine site. A large number of similar new small problems requiring a sophisticated examination at a marine site could be cited based on experience with ocean surveillance and similar deep ocean projects.

14.2 Corrosion in the Marine Environment

This section covers corrosion in the marine environment, starting with general wasting of steel and then discussing various types of corrosion in other metals and alloys. Discussions in this section come primarily from an article entitled "Guidelines for Selection of Marine Materials," prepared by A. H. Tuthill and C. M. Schillmoller.[27]

It is presupposed that readers have a good comprehension of fundamental principles of corrosion--the basic corrosion cell--the barrier film--anodic and cathodic reactions--the effect of oxygen and the techniques used for measurement of corrosion rates. For those who may not have, or who may wish to review the matter, the following books are recommended:

(1) "Corrosion Theory Section" of the <u>Corrosion Handbook</u>[28]

(2) "Corrosion and Corrosion Control" by H. H. Uhlig[29]

Corrosion of Steel in Seawater*

Since steel is the basic materials of construction in the marine environment, its corrosion behavior must be understood in order to know when and where to upgrade to more durable materials.

It may be assumed that the rate of corrosion of steel in seawater, as in fresh water, is to a large extent governed by the oxygen content of the seawater. The pattern of corrosion on carbon steel in aerated seawater is shown in Figure 14-5. Average corrosion rates that have been found typical of steel piling in the atmosphere, in the splash zone, in the tidal zone, in clean seawater, and in the mud of the bottom are plotted. The differences shown are in part due to the availability of oxygen and in part due to other factors.

Atmosphere--Copson has provided us a very useful theory of the corrosion of steel in the atmosphere. He has proposed that the corrosion rate of steel in atmosphere is directly related to the rate at which a ferrous corrosion product is leached or washed from the barrier film of rust. When one of the products of corrosion is soluble, a fully protective barrier film is almost impossible to form.

-----

*This section is a copy of an article by Tuthill & Schillmoller[27].

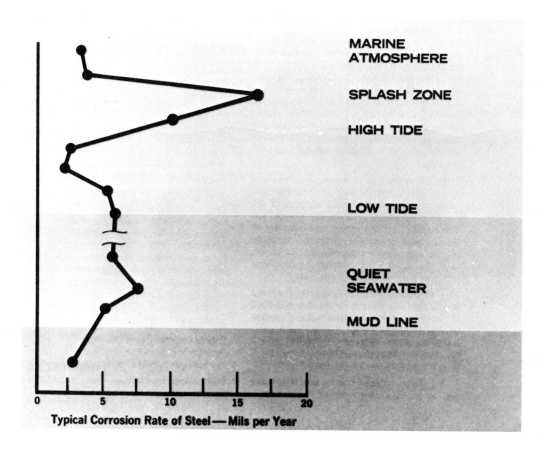

FIGURE 14-5   CORROSION OF STEEL PILING[27]

It has been postulated that small amounts of copper and
nickel in the low alloy steels enhance their corrosion resistance
in the marine atmosphere by altering the structure of the barrier
film of corrosion products so as to produce a tighter, denser
film with less tendency to be removed by leaching or spalling.
The films that form on these copper and nickel containing steels
are more impervious to moisture and more adherent. This results
in lower corrosion rates for the base metal and greater life of
protective paints than with plain carbon steel.

Splash Zone--Rust films that have little opportunity to
become dry do not develop protective properties. In the splash
zone above the high tide level such unfavorable conditions
exist, aggravated further by the high oxygen content of the
splashing seawater. It is not surprising, therefore, that
observations of old piling along the sea shore ususally reveal
holes just above the water line, where Figure 14-5 indicates
rates several times greater than for continuous immersion in
seawater.

Tidal Zone--Corrosion in the tidal zone reaches a minimum
as a result of the protective action of oxygen concentration
cell currents. Steel surfaces in the tidal zone in contact with
highly aerated seawater become cathodic to the adjacent submerged
surfaces where the oxygen content is less, especially when these
surfaces are covered with oxygen shielding organisms. The current
that flows from the anodic submerged surfaces to the cathodic
tidal zone areas is sufficient to provide substantial cathodic
protection.

Submerged--the attack of submerged surfaces is governed
principally by the rate of diffusion of oxygen through layers
of rust and marine organisms. It usually is in the range from
3 to 6 mils per year substantially independent of water temperature
and tidal velocity, except where industrial pollution leads to
higher rates.

Mud Line--The rate may go up in the vicinity of the mud
line because marine organisms can generate additional concentration
cell and sulfur compound effects. Reduction in the rate well below
the mud line is undoubtedly closely associated with two factors:

   (1) lower availability of dissolved oxygen

   (2) the fact that barrier films are relatively undisturbed.

This may be regarded as the typical pattern of corrosion
and the average corrosion rates of carbon steel in the marine
environment. It can be assumed that the pattern of corrosion
will remain the same but that the curve will be displaced to the
left or right as the oxygen content and temperature vary from
those of the test location--Harbor Island, N. C. At this
location, annual seawater temperature range from 40 to 80F
(4 to 26C), average monthly normal atmospheric temperatures range
from 40 to 80F (9 to 27C), oxygen content varies from 72 to 100%
of saturation, and maximum tidal flow is less than 1 foot per
second. See Figure 14-6.[16]

Normal seawater temperatures vary from the 36 to 40F
(2 to 4C) of the arctic and ocean depths to 85F (29C) of the tropics
and 110F (43C) outlet temperatures typical of condensers. The
present interest in saline water conversion plants has extended

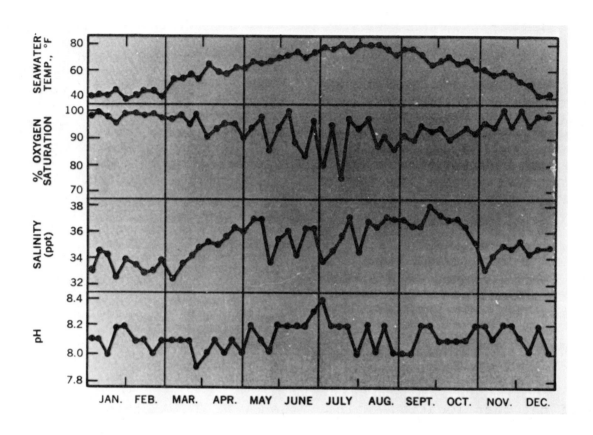

FIGURE 14-6   HYDROLOGY OF SEAWATER IN 1963 AT THE CORROSION
         LABORATORY IN HARBOR ISLAND, NORTH CAROLINA[16, 27]

this to 290F (143C) and even 400F (204C).

The effects of velocity may be summarized:

(1) Slight motion may tend to make the environment more uniform thereby tending to reduce local attack.

(2) More motion thins quiescent layers at the metal surfaces so that corrosives more readily reach the metal.

Figure 14-7 shows how the corrosion rate of carbon steel increases with velocity, tripling before 5 feet per second. Marine organisms can become attached during periods of low velocity and can serve as a barrier to reduce or eliminate the acceleration effect of higher velocities outside the protective barrier. Chemical treatment (chlorine) that prevents accumulation of protective layers of marine growth or slime may expose the metal to the expected velocity effects.

(3) Higher velocities may change the nature of the corrosion products and barrier film. Films on stainless type alloys become more protective, while those on steel and copper base alloys may be stripped away. Figure 14-7 also suggests that once the barrier film is fully stripped away carbon steel tends to corrode at several times the rate in quiet seawater where the barrier film is more protective.

Galvanized steel is often used in seawater, but as Field reports, adds but little to the life of steel. Figure 14-8 plots reported corrosion rates of zinc in seawater against velocities and suggests one reason why galvanizing is so limited in its effectiveness.

406

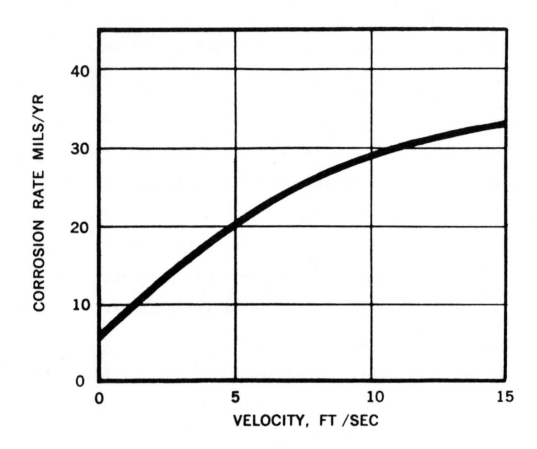

FIGURE 14-7   EFFECT OF SEAWATER VELOCITY ON CORROSION
OF STEEL AT AMBIENT TEMPERATURE, EXPOSED
38 DAYS[27]

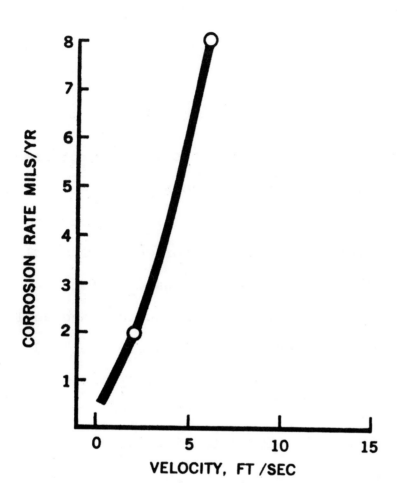

FIGURE 14-8   EFFECT OF SEAWATER VELOCITY ON CORROSION
RATE -- ZINC AMBIENT TEMPERATURE[27]

Zinc blocks and plates are widely used as sacrificial anodes to cathodically protect ship hull plate. The loss of a few more mils per year from velocity effects does not interfere with, and may even be beneficial to, the performance of block zinc as a sacrificial anode.

Protection of Steel

Steel may be painted, coated or cathodically protected with impressed current or sacrificial anodes in order to extend its usefulness in the marine environment. When evaluating these measures versus upgrading to inherently more durable metals, the following points should receive careful consideration:

(1) The cost of properly cleaning the surface and applying coatings is seldom less than 40 cents per square foot and commonly reaches $1 or more per square foot for *each* side that must be protected.

(2) The cost of cathodically protecting large areas of bare steel may be so high as to require the use of supplemental organic coatings. Such coatings reduce the current requirements to that needed for the reltively small area of bare steel exposed at the inevitable pinholes or other holidays in a coating.

(3) The effectiveness of cathodic protection falls off rapidly above the water line and is least effective where corrosion is most severe--in the splash zone.

(4) It is seldom practical to provide effective cathodic protection for the interior surface of piping even though the pipe may be full of seawater and might, in theory, be protected.

Other interior surfaces, and some external surfaces, present
similar practical limitations to protection.

(5) Protective coatings tend to fail first in the splash
zone and are difficult to renew effectively in this zone.

(6) Zinc and aluminum used as sacrificial anodes are
protective to steel. However, when these metals are employed
in structures protected by impressed current, they may suffer
accelerated attack from alkali generated in those areas which
receive above average current density.

In any event, both the first cost of coatings and cathodic
protection as well as their continuing annual costs are additive
to the cost of steel. When the full cost of protecting steel
is taken into account, the premium for more durable metals is
reduced. In many instances the more durable metals may actually
be more economical than fully protected steel.

Corrosion of Other Metals and Alloys in Seawater

Various metals used for marine applications are covered in
Chapter 6. This section covers corrosion in seawater of metals
and alloys other than steel.

Figure 14-9 shows typical general corrosion rates based on
weight loss immersed in *quiet* seawater for common marine materials.
Note that stainless steel, nickel copper and nickel chromium
alloys are rated inert, except for localized attack, i.e. pitting.
It is quite common, in quite seawater, in the presence of marine
organisms or other deposits, for 90 to 99% of the surface of these
relatively noble alloys to remain virtually unaffected, except

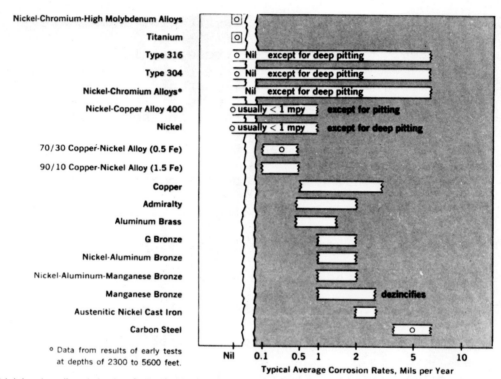

FIGURE 14-9   GENERAL WASTING OF VARIOUS METALS IMMERSED
IN QUIET SEAWATER

for a few deep pits. In some cases these pits penetrate the full
wall thickness.

The average corrosion rates for most of the materials listed
in Figure 14-9 are less than 3 mils per year in quiet seawater, a
low rate which would insure long service, *if the metal loss were
uniform and not localized*. The average corrosion rate data gives
no real idea of the durability of a material if all the weight
loss occurs in a few pits.

Pitting

Pitting is a localized type of attack, the rate of corrosion
being greater at some areas than at others. If appreciable
attack is confined to a relatively small area of metal, acting as
anode, the resultant pits are described as deep.

Figure 14-10 illustrates typical pitting behavior in quiet
seawater.[27] Very few metals exhibit the same degree of
resistance to pitting as to general wasting. In fact, many of
the more noble metals which suffer little or no overall wasting
are the most susceptible to deep, almost catastrophic pitting--
in *quiet* seawater.

Variations in the environment give rise to concentration
cells. These cells can be set up by local differences in oxygen,
temperature, agitation, liquid velocity and, in fact, by almost
any heterogeneity in exposure conditions. Oxygen concentration
cells are frequently encountered in marine service. Differences
in dissolved oxygen concentration lead to localized corrosion
of metal in hidden, secluded and shielded areas such as under

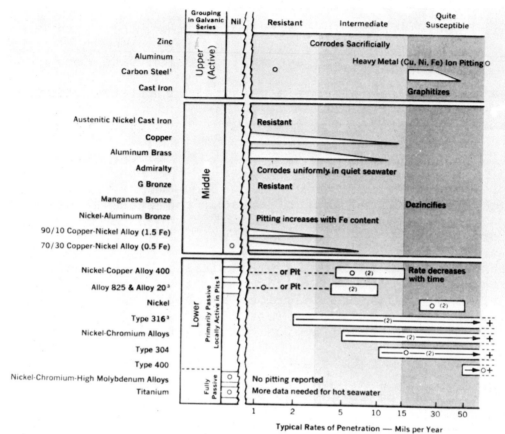

O Data from results of early tests at depths of 2300 to 5600 feet.
(1) Shallow round bottom pits.
(2) As velocity increases above 3 fps pitting decreases. When continuously exposed to 5 ft. per sec. and higher velocities these metals, except Type 400 series, tend to remain passive without any pitting over the full surface in the absence of crevices.
(3) These grades have an advantage over Type 304 stainless steel and related grades in that there is a substantial reduction in the number of pits, i.e., probability of pitting even though the depth of such pits as do occur is not greatly reduced.

FIGURE 14-10   PITTING OF VARIOUS METALS AND ALLOYS IMMERSED
IN QUIET SEAWATER (LESS THAN 2 ips)

deposits or crevices. Most of the more noble alloys in Figure 14-10 are particularly susceptible to such localized attack. Local breakdown of passivity leads to locally active areas, i.e. pits. Pits, once started, tend to be accelerated by galvanic currents flowing from the small active metal areas in the pit to the much larger, more noble passive areas surrounding the pit.

Eliminate Crevices. An important practive to avoid pitting is to eliminate, if possible, all shapes and joints which form crevices into which oxygen does not have ready access. In applications where crevices cannot be avoided, it may be useful to employ a zinc oxide compound in the assembly or such parts. This seals the crevice, and provides useful temporary, but not permanent, protection. Man-made crevices--under fastenings, washers, flanged joints, etc.--are favorite sites for pitting.

Fouling. Marine organisms (barnacles) which attach them-selves to any surface, grow and effectively seal off a small part of the surface from its environment. Fouling, so characteristic of the sea, also leads to concentration cells. Underneath the barnacle is a favorite site for pits to start and the deepest pits are quite often found there.

Velocity Effects

As velocities reach the 3 to 6 feet per second range, fouling diminishes, pitting of the more noble alloys alows down and even ceases. As velocities continue to increase, the corrosion barrier film is stripped away from one after another of the steel and copper base alloys, while the stainless type and many nickel

base materials remain passive and inert. The complete reversal
in the tolerance of metals for the marine environment as velocities
change is the source of much, seemingly conflicting information
on actual experience with marine metals. Velocity effects deserve
close study by all designers.

For example, turbulance creates a complicated problem in a
piping system. Some metals suffer extremely high rates of metal
loss where severe eddy turbulance is encountered. Another velocity
effect is cavitation errosion which often occurs in propeller
blades. This section does not go into further discussions of
velocity effects. Those who are interested in further details
are recommended to read the article by Tuthill and Schillmoller[27]
and other articles.

Galvanic Effects

One of the basic realities of seawater is the great intensi-
fication of galvanic effects as compared to fresh water. It is
not so much that galvanic effects are unfamiliar as that their
intensity in seawater is not fully appreciated. Galvanic corrosion
has been well known and widely studied since the H.M.S. Alarm
came back from a West Indies voyage in 1761 with iron nails
corroded away and most of the 12 oz. copper hull sheathing torn
loose.

The many components of well engineered modern equipment
require different metals to best accomplish the purpose of each
component. This "many metal" approach works quite well in fresh
water and many other applications. In seawater the result is,

all too often, a fine wet cell battery and a nightmare to the user and designer.

The following are useful rules, given in an article by Tuthill and Schillmoller,[27] to avoid galvanic corrosion:

Rule 1 - Where possible construct equipment for seawater service from *one metal*.

Example. *90/10 copper-nickel alloy pipe, 90/10 copper-nickel tube sheets and 90/10 copper-nickel tubes for seawater cooling systems.*

Rule 2 - Where not possible or desirable to construct from one alloy, make certain the *key components* are *more noble, i.e.* protected.

Example. *Use of 70/30 copper-nickel alloy for welds in 90.10 copper-nickel fabrication and use of 70/30 tubes with 90/10 copper-nickel water boxes would be appropriate galvanically.*

Rule 3 - *Expect* and allow for *increased corrosion* on the *less noble* metal by providing a large area or heavy wall to support the increased corrosion that will occur.

Example. *(1) Above, the large area of 90/10 copper-nickel alloy pipe, water box and tube sheet can support the increase in corrosion occasioned by the 70/30 copper-nickel weld metal and 70/30 copper-nickel tubes without significantly affecting overall serviceability.*

Example. *(2) Pump and valves - Provide a heavy wall casting with adequate corrosion allowance to protect the more noble internal trim - the key component.*

Rule 4 - Consider carefully the galvanic effect before painting or coating steel or cast iron. *It is frequently desirable to* paint *the* more noble (*stainless steel, nickel-copper alloy etc.*) material *and leave the steel bare*--A complete reversal of the more conventional practice.

Example. *Paint the nickel-copper alloy propulsion component containers to reduce galvanic currents flowing to steel hull of small submersibles.*

Many charts on the galvanic series in seawater have been published. Figure 14-11 is a chart prepared by LaQue.[12] The galvanic series are helpful in selecting materials for seawater--if they are properly used. Figure 14-12 has been prepared to assist the designer for speed. This figure applies to seawater and does not necessarily apply to either marine atmospheres or to fresh water where galvanic effects are more localized. In marine atmospheres and fresh water the galvanic effect is localized in the vicinity of the contact and is not spread over the whole area of the base metal as in seawater. The spreading of the galvanic effect in seawater is due to the high conductivity of the electrolyte.

Selective Corrosion

The internal structure of a metal or alloy sometimes influences corrosion and can lead to preferential or selective attack. There are many forms of selective corrosion including "graphitic corrosion" or "graphitization", "dezincification", "dealuminumification," even "denickelification." These related phenomena

417

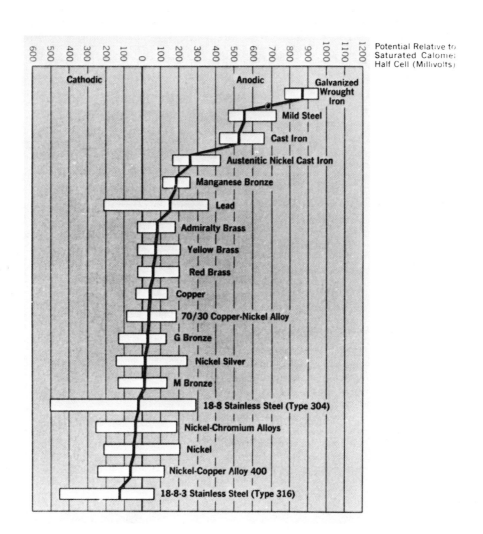

FIGURE 14-11   GALVANIC SERIES, SEAWATER (LaQue)[12]

BASE METAL ↓	FASTENER							
	Aluminum [1]	Carbon Steel	Silicon Bronze	Nickel	Nickel-Chromium Alloys	Type 304	Nickel-Copper Alloy 400	Type 316
Aluminum	Neutral	Comp. [2]	Unsatis-factory [2]	Comp. [2]	Comp.	Comp.	Comp. [2]	Comp.
Steel and Cast Iron	N.C.	Neutral	Comp.	Comp.	Comp.	Comp.	Comp.	Comp.
Austenitic Nickel Cast Iron	N.C.	N.C.	Comp.	Comp.	Comp.	Comp.	Comp.	Comp.
Copper	N.C.	N.C.	Comp.	Comp.	Comp.	Comp.	Comp.	Comp.
70/30 Copper-Nickel Alloy	N.C.	N.C.	N.C.	Comp.	Comp.	Comp.	Comp.	Comp.
Nickel	N.C.	N.C.	N.C.	Neutral	Comp. [3]	Comp. [3]	Comp.	Comp. [3]
Type 304	N.C.	N.C.	N.C.	N.C.	May Vary [4]	Neutral [3]	Comp.	Comp. [4]
Nickel-Copper Alloy 400	N.C.	N.C.	N.C.	N.C.	May Vary [4]	May Vary [4]	Neutral	May Vary [4]
Type 316	N.C.	N.C.	N.C.	N.C.	May Vary [4]	May Vary [4]	May Vary [4]	Neutral [4]

(1) Anodizing would change ratings as fastener.
(2) Fasteners are compatible and protected but may lead to enlargement of bolt hole in aluminum plate.
(3) Cathodic protection afforded fastener by the base metal may not be enough to prevent crevice corrosion of fastener particularly under head of bolt fasteners.
(4) May suffer crevice corrosion, under head of bolt fasteners.

NOTE: Comp. = Compatible, Protected. N.C. = Not Compatible, Preferentially Corroded.

FIGURE 14-12   GALVANIC COMPATIBILITY

are greatly intensified in seawater. Intergranular corrosion
of stainless is another, but quite different, form of selective
corrosion.

Table 14-1 lists alloys resistant to and subject to
selective corrosion. An article by Tuthill and Schillmoller[27]
discusses details of selective corrosion of various metals.

Corrosion of Metals in Deep Ocean

Corrosion of metals in the deep ocean has not been well
understood. The following discussion is presented in an article
by Tuthill and Schillmoller.[27]

Figure 14-13 shows that at 5500 feet temperature has dropped
to less than 4C (39F) from 24C (75F) near the surface. This alone
should result in a substantial decrease in corrosion rates of all
metals. Fouling is known to be most severe in the first several
hundred feet where both the highest temperatures and greatest
amount of light exist. Although there are reports of marine borer
activity below 2000 feet, metal surfaces can be assumed to remain
relatively free of fouling below 2000 feet.

Velocity in the great depths is a matter of local bottom
currents. The ocean is not still and quiescent at depth; it
is in motion. The currents are appreciable and measurable. So
far is as known, it appears reasonable to assume they are
generally in the range of 1 to 2 feet per second and comparable
to tidal currents at the surface, but could be higher in certain
locations.

There remains oxygen content. As Figure 14-13 (which is
only one measurement at one location) shows, oxygen does drop

TABLE 14-1 SELECTIVE CORROSION

	GRAPHITIZATION	DEZINCIFICATION	DEALUMINIFICATION	DENICKELIFICATION	INTERGRANULAR CORROSION OF AUSTENITIC STAINLESS STEEL
Susceptible	Cast Iron Ductile Iron	Copper alloys with more than 15% Zn. Examples: naval brass, admiralty, aluminum brass, muntz metal, manganese, bronze	Aluminum bronzes with less than 4% Ni	70/30 copper nickel refinery condenser tubes at high temperature and low flow	Heat of welding or slow cooling of castings lead to selective attack of stainless steel in seawater but has little effect in marine atmosphere
Solutions	Use austenitic nickel cast iron	1. Use inhibited grade 2. Use alloys with less than 15% Zn. Examples: red brass, silicon bronze, tin bronze, copper-nickels	Use 4% Ni grade	Don't run condenser dry. Keep 3 ft. per sec. min. flow	1. Anneal 2. Use low carbon grade- 304L, 316L, CF-4M 3. Use stabilized grade 347 or 321 4. Avoid welding after annealing of susceptible grades

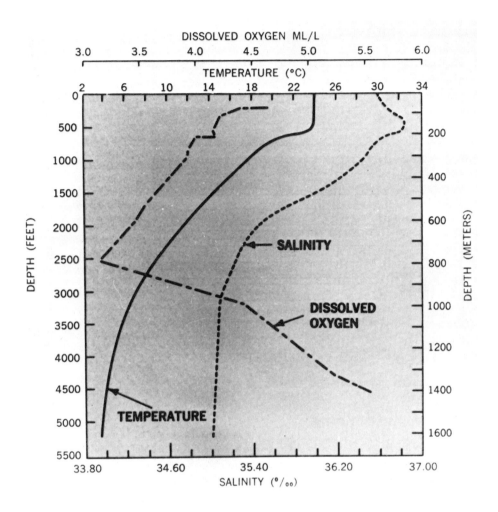

FIGURE 14-13   TEMPERATURE, SALINITY, AND OXYGEN
IN DEEP OCEAN[19,27]

with depth, but below 2500 feet, rises again well above the near surface concentration. The ocean appears layered rather than homogeneous and the various layers are distinguished by different oxygen and salinity contents.

Corrosion can be expected to decrease at depth as temperature decrease. Fouling, and pitting associate with fouling, but *not all pitting*, can be expected to decrease. Velocities may be high enough to sweep soluble corrosion products away and bring fresh reactants to metal surface but do not appear to be high enough to strip corrosion product barrier films from copper base alloys or high enough to maintain fully passive films on stainless and nickel-chromium alloys.

The variation in oxygen content with depth, in addition to its effect on corrosion at any given depth also introduces a "long line" effect on deep sea moorings. Galvanized steel mooring lines in deep mooring are reported to suffer accelerated corrosion at various depths with less corrosion in between. This is typical of oxygen cell concentration attack and is to be expected on steel wire passing through regions with distinctly different oxygen levels.

These then are the effects to be expected. Reinhart has published the preliminary results of corrosion tests at 5640 feet in the Pacific off the California coast and his conclusions are presented in Table 14-2. [20] Further data are needed but the general pattern of his findings is consistent with our knowledge of seawater corrosion.

TABLE 14-2 <u>Corrosion at 5640 ft.</u> (F. M. Reinhart, 1964)[20]

Temperature, 37F (2.8C)   Dissolved, O_2–0.83 ml or 1.2 ppm
pH, 7.25   Salinity, 34.6 parts per thousand
Velocity, 0.5 knots (0.9fps) Duration, 123 days

Preliminary results based solely upon visual observations
indicate:

1. The corrosion rates of aluminum alloys Clad 3003-H12 and Clad
    7178-T6 were greater than those normally found in surface sea-
    water exposures.

2. The pitting pattern was different from that of surface exposures
    in that there were usually fewer but deeper pits other than
    those above-mentioned on the aluminum alloys (30 to 60 mils).

3. The copper base alloys were corroded evenly over the surfaces
    of the specimens. They were covered with thin films of multi-
    colored corrosion products--blue, green, brown and black.

4. The magnesium base alloys were penetrated in scattered areas.

5. Only one nickel alloy, nickel-copper alloy 400, was covered
    with a thin tenacious film of mottled brown corrosion products.
    The other nickel alloys were untarnished.

6. Wrought iron, carbon steels, high strength low alloy steels and
    alloy steels all corroded in the same manner; uniform, relatively
    thin layers of red corrosion products (rust).

7. The chromium stainless steels (Types 405, 410, and 430), a
    chromium-nickel stainless steel (Type 301) and the precipitation
    hardening stainless steels were attacked in either a lacework
    pattern or an undersurface type of attack. Both of these types
    of attack are found to occur in stagnant surface seawater. The
    attack at the 5640 foot depth could be due to the absence of
    any appreciable water current (essentially stagnant conditions)
    or to the low oxygen concentration.

8. The other chromium-nickel stainless steels (Types 302, 316,
    316L, 321 and alloy 20) were not attacked by seawater.

9. The precipitation hardening stainless steels were also subject
    to severe crevice corrosion.

10. None of the titanium alloys, both unwelded and welded, were
    attacked.

11. Both parts of the aluminum-magnesium alloy galvanic couple were
    excessively corroded. There was slight galvanic corrosion of
    the aluminum portion of the 7075-T6 aluminum alloy--Type 321,
    stainless steel galvanic couple. There was no apparent galvanic
    corrosion of a precipitation hardening stainless steel--Type 301,
    chromium-nickel stainless steel couple.

TABLE 14-2   (Continued)

12. Nickel-copper alloy 400 fasteners were not attacked, stainless
    steel (Type 300 series) fasteners rusted, galvanized steel
    fasteners turned yellow, and cadmium plated steel fasteners
    rusted in the deep seawater.

13. Electroless nickel plating on steel cracked and bled red
    rust during the exposure.

14. There were no failures of organic coatings applied on low
    carbon steel or on fabrics.

15. All the plastic materials and elastomers became dull.

16. The permeability of conrete was not affected.

17. Bacterial slimes were found on some plastic materials and
    on electroless nickel plated shackles.

18. Cotton, manila and jute fibers showed degradation from
    microbial activity.

19. More definite conclusions will be forthcoming after the
    corrosion evaluations such as weight losses, pit depths,
    changes in mechanical properties and metallographic examinations
    have been completed.

14.3 Stress Corrosion Cracking

Stress corrosion cracking varies from other types of metal
attack in that it is a form of localized failure which is more
severe under the combined action of stress and corrosion than
would be expected from the sum of the individual effects of stress
and corrosion acting alone. Stress corrosion cracking involves
a brittle-type fracture in a material that is otherwise ductile.
The surface direction of the cracks is perpendicular to the
direction of the load.[1] Stress corrosion cracking should not
be confused with other types of localized attack such as pitting,
galvanic attack, intergranular corrosion, impingement or cavitation.

Causes of Stress Corrosion Cracking

There are many variables affecting the instigation of
stress corrosion cracking. Among these variables are alloy
composition, tensile stress (internal or applied) corrosive
environment, temperature and time.

Alloys. Pure metals, it is generally believed, do not
crack as a result of stress corrosion. Alloys prepared from
pure metals, may crack, however. In recent studies, a significant
improvement has been demonstrated in cracking resistance of some
alloys made from extremely pure metals. Some alloys in a particular base-metal system are more resistant to cracking than
others. Such metals include aluminum, copper, and magnesium-base
alloys. In these cases, cracking resistance improves as the
alloy content is reduced and the composition approaches that of

a pure metal.[1]

Stress. Since no cracking has been observed with metal surfaces in compression, only tensile stresses at the surface of the metal cause the cracking. These stresses may be due to strains within the material (residual) or they may arise from an external load (applied). Usual causes of internal or residual stresses include:

1. deformation of the metal near welds, rivets, bolts or in press or shrink fits

2. an unequal cooling of a section or structure from a relatively high temperature

3. the crystal structure within the metal going through a phase change or rearrangement which involves volume changes.[1]

Applied stresses are the result of operating conditions and include:

1. differential thermal expansion

2. dead loading

3. pressure differentials.[1]

Environment. The corrosivity of a chemical medium cannot be used as an indication of its capability of promoting stress-corrosion cracking. Many rather corrosive solutions do not cause cracking. In fact, the environments which are most conducive to stress-corrosion cracking are those which produce highly localized attack without a significant general surface corrosion. Examples of environments known to cause stress-

corrosion cracking in some of the more common metals and alloys
are shown in Table 14-3.

Figure 14-1, which shows four examples of corrosion control
problems in marine structures, covers stress corrosion cracking
by seawater of high-strength steels and a high-strength aluminum
alloy. Figure 14-2 also shows susceptibility to cracking in seawater, fresh water, or humid atmospheres of high-strength steels,
aluminum, and titanium alloys. Chapter 5 contains discussions of
stress corrosion cracking characteristics of titanium alloys
in salt water.

Temperature. As a general rule, the susceptibility of a
material to cracking increases with an increase in the temperature.
An exception, however, is the cracking of low alloy and carbon
steels in hydrogen sulfide. Under such conditions, an inverse
relationship exists and the cracking may be attributed to a
secondary effect of corrosion; embrittlement is due to the hydrogen released during corrosion. The embrittling effect of hydrogen decreases and the temperature increases and the tendency to
crack therefore decreases also.[1]

Detecting Stress Corrosion Cracking

Methods of discovering stress corrosion cracking vary from
cracks visible to the naked eye to those detected only with ultra-
violet-light-fluorescent penetrant. In this section, we will
discuss only a few of the possible detection processes.

428

TABLE 14-3 Materials and Environment Which Cause Stress Corrosion Cracking[1]

Material	Environment
Aluminum	Air, sea water, sodium chloride solutions
Copper-base alloys	Ammonia, steam
Steel	Alkalies, nitrates, hydrogen cyanide, hydrogen sulfate, anhydrous liquid ammonia, sodium chloride solutions, marine atmosphere
Stainless steels	Caustic, chloride solutions
PH stainless steels	Chloride solutions, marine atmosphere
Magnesium-base alloys	Chloride-chromate mixture, moisture
Nickel (commercial purity)	Aqueous or fused caustic at elevated temperature
Monel, Inconel	HF vapors
Titanium alloys	Red fuming nitric acid, HCl, dry molten chloride salts, salt water

Ultrasonic and Eddy-Current Methods. These techniques
are useful particularly where the cracks start on the interior
surface of a structure, such as the inside of a pipe. Three
types of ultrasonic devices are usually used:

1. pulse-echo

2. direct transmission

3. resonance.

The first involves a sound wave reflected from the back side of
the material being tested. A crack or other flaw gives a sound
r eflection from some intermediate distance, depending upon the
crack location. Direct transmission depends on the amount of
sound transmitted through the material. Transmission is reduced
by a crack or other flaw. The final method, resonance, determines
the thickness of a piece of material. The sound frequency is
varied constantly and periodically; a frequency is transmitted
with a wavelength which is an even multiple of the material thickness. The resulting resonant frequency is detected. Cracks or
flaws parallel to the surface being tested change the detection
signal.[1]

A magnetic field inducing eddy-currents within the material
being tested is the premise of the eddy-current method. Cracks
or other flaws affect the eddy-current pattern which then effects
the impedance in a pickup coil.[1]

Other Testing Methods. With the magnetic particle method,
the particles are attracted to localized leakage fields at discontinuities on the surface of a magnetized material. Another

method, the dye penetrant, involves a dye penetrant applied to the test material and then wiped off. A developing powder is then applied to the surface and reacts with the dye, giving a color indication where it seeps from the cracks. An ultraviolet light may be used in a similar manner to detect a fluorescent penetrant.[1]

Avoiding Stress Corrosion Cracking

Earlier in this section, we discussed the many variables contributing to stress corrosion cracking. Since many factors are involved, the reduction or elimination of one contributing factor may help the overall situation. In this section, we will touch on some of the methods for reducing the contributing factors.

Lower Tensile Stress. Stresses due to operating conditions often can be reduced by changing the design. Such changes may include lowering the operating pressures, avoiding misalignment of volted or welded connections, avoiding differential thermal expansion, and eliminating heavy loads on thin sections. Another possible method is increasing the metal thickness in loaded thin sections or in pressure vessels, when other design considerations will allow such actions. This will lower the applied stress by distributing it over a larger cross-sectional area. Residual stresses, however, cannot be reduced by increasing the metal thickness. These stresses may be eliminated by stress relief anneal after fabrication and installation. A final method for reducing residual tensile stress is to put the surface layers of the metal in compression.[1]

Altering the Corrosive Environment. This is perhaps the most obvious method of limiting stress corrosion cracking. For example, the elimination of chloride ions by ion exchange from aqueous solutions allows the use of stainless steel in high temperature waters. However, the addition of inhibitors to solutions to reduce stress corrosion cracking has not been promising in general. Design considerations can again aid in reducing cracking. For instance, placing internal heaters away from the side or bottom of vessels and near the bulk of the solution minimizes the tendency for concentration in relatively stagnant areas. A concentrated solution introduced into a dilute solution near the center of the vessel rather than near the edge or bottom permits the solution to become diluted before it reaches the vessel walls.[1]

Lower Temperatures. If the operating temperature of the bulk environment is fixed and cannot be lowered without loss in efficiency, this method is impossible. However, the raised local temperature associated with heat-transfer surfaces or the introduction of hot solutions into cooler ones can be avoided many times by the vehicle design.[1]

Cathodic Protection. Cathodic protection can be applied to the relief of stress corrosion cracking as well as other types of corrosion.

14.4 Hydrogen Embrittlement

Hydrogen embrittlement is one of the most common and serious types of time-dependent fracture.[25] In laboratory testing, the presence of hydrogen results in a decrease in the ductility of unnotched tensile specimens and hence a decrease in the tensile strength of notched specimens. In service, failure can occur without warning, minutes to years after a static load has been applied to a structure containing hydrogen.

The most spectacular and expensive failures have occurred in the petroleum industry, particularly in sour gas wells (containing H_2S) producing natural gas. Numerous failures of cadmium-plated high-strength steel parts, where hydrogen was introduced during electroplating, plagued the aircraft industry.

Characteristics of Hydrogen Embrittlement[25]

Besides steels having a body centered cubic (BCC) crystal structure, titanium and zirconium and their alloys also are susceptible to hydrogen embrittlement. The hydrogen can be introduced into these materials during processing, cleaning (acid pickling), and electroplating operations.

There are two characteristics of hydrogen embrittlement. First, it is not a form of stress corrosion cracking. In fact, fracture often occurs when the metal serves or has served as a cathode, during electroplating or in a cathodic "protection" operation. Second, embrittlement results from hydrogen contents that are greater than the equilibrium solubility limit (about 10^{-3} ppm by weight in iron and 20-35 ppm in Ti and Zr at room temperature at one atmosphere hydrogen pressure). The excess hydrogen which causes embrittlement can be as low as 1 ppm in high-strength steel, and 35 ppm in Ti and Zr.

Hydrogen Embrittlement of Steels

Figure 14-14 shows hydrogen induced delayed-fracture characteristics of various steels.[14] Curves in the figure were estimated by Martin and Masubuchi[14] from experimental results conducted by Simcoe and others[22] on SAE 4340 steel quenched and tempered at difficult strength levels.

Figure 14-14 shows a general tendency of steel to become more susceptible to hydrogen cracking as the strength level increases. For example, SAE 4340 steel oil quenched and tempered at 500F has the tensile strength of about 260,000 psi. When hydrogen is charged to this steel while it is subjected to tensile stress of 80,000 psi, it takes only 5 minutes before the steel fractures. When the stress is lowered to 40,000 psi, for example, it takes about 15 minutes before the steel fractures. When steel with a lower strength level is subjected to stress and hydrogen, it takes a longer time before the steel fractures. For example, when HY-80 steel is subjected to a tensile stress of 80,000 psi, it takes about 400 minutes before it fractures.

Masubuchi and Martin[14] investigated hydrogen induced cracking characteristics of weldments in various steels. Table 14-4 summarizes their experimental results as follows:

(1) Mild steel was immune to hydrogen embrittlement. Of
    six specimens which were charged with hydrogen up to
    379 hours, very small cracks were observed in two

434

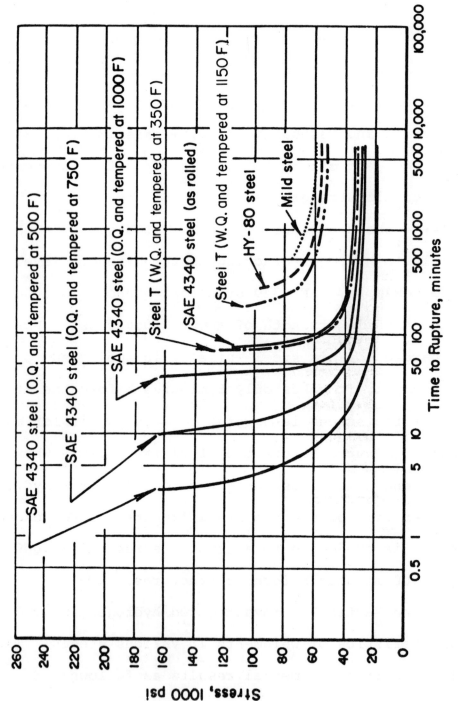

FIGURE 14-14 HYDROGEN-INDUCED DELAYED-FRACTURE CHARACTERISTICS OF
VARIOUS STEELS (Masubuchi and Martin) [14]

Note: Curves are estimated from results obtained by Simcoe, et al. [22]   The commercial
high-strength structural steel is identified above as Steel T. Table 14-4 contains
approximate tensile strengths of these steels.

TABLE 14-4 Summary of Results of Hydrogen-Induced-Cracking Tests on Welded Specimens Made From Various Materials

Base plate and heat treatment	Apprx. tensile strength, psi	Number of specimens tested	Plate thickness, in.	Hydrogen charging period, hr.	Summary of Test Results
1. Mild steel	75,000	6	1/2 to 2	Up to 379	Very small cracks in the heat-affected zone of 2 specimens, but no cracks in 4 other specimens hydrogen charged up to 126 1/2 hr.
2. HY-80 (quenched and tempered)	100,000	4	1/2 to 2	Up to 216	Small cracks in 2 specimens, but no cracks in 2 other specimens
3. Commercial high-strength structural steel water quenched and tempered at 1150°F	120,000	1	3/4	4 1/2	No cracking
4. Commercial high-strength structural steel water quenched and tempered at 350°F	150,000	5	1/2, 3/4	Up to 24	Several cracks were found in 3 specimens
5. SAE 4340 steel (as rolled)	150,000	1	3/4	14	No cracking
6. SAE 4340 steel (oil quenched and tempered at 1100°F)	175,000	1	1/2a	6 3/4	Several transverse cracks
7. SAE 4340 steel (oil quenched and tempered at 750°F)	220,000	1	1/2a	6	Fairly systematic cracks
8. SAE 4340 steel (oil quenched and tempered at 600°F)	240,000	1	3/4	1	Systematic cracks
9. SAE 4340 steel (oil quenched and tempered at 500°F)	260,000	25	1/4 to 3/4	Up to 16	Extensive cracks in all specimens except one which had been mechanically stress relieved; cracks were found after hydrogen charging for a few hours or less.

aGround from 5/8 to 1/2 inch thick.

specimens. No cracks were observed in the other four specimens hydrogen charged up to 126 1/2 hours.

(2) As the strength level of steel increased, weldments became more susceptible to hydrogen embrittlement.

(3) Weldments made in SAE 4340 steel quenched and tempered at a very high strength level were very susceptible to hydrogen embrittlement. Extensive cracks were obtained in all specimens except one which had been mechanically stress relieved. Cracks were found after hydrogen charging for a few hours or less.

In fact, Masubuchi and Martin conducted the study in an attempt to develop a technique for determining the distribution of residual stresses in a weldment by observing the pattern of hydrogen-induced cracks. Figure 14-15 shows a crack pattern in a simple butt weld made from SAE 4340 steel oil quenched and tempered at 500F after hydrogen charging test for 50 minutes. A system of transverse cracks was obtained; cracks were longer in the central region of the specimen. The crack pattern indicates that high tensile residual stresses in the direction parallel to the weld existed in areas near the weld.

Figure 14-16 shows the crack pattern in a complex weldment. In addition to a system of transverse cracks near the weld, long parabolic cracks occurred in the bottom plate near the end of the fillet weld. Parabolic cracks were found in the vicinity of every structural discontinuity indicating that high concentration of residual stresses existed in areas near such a discontinuity.

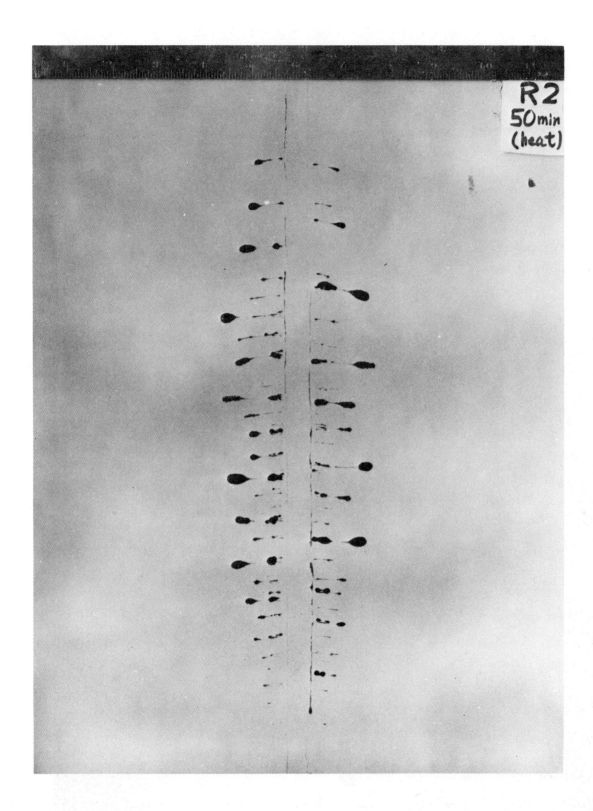

FIGURE 14-15   CRACK PATTERN IN A SIMPLE BUTT WELD MADE FROM SAE 4340
                STEEL OIL QUENCHED AND TEMPERED AT 500° F (SPECIMEN R2)
                AFTER HYDROGEN INDUCED CRACKING TEST FOR 50 MIN.

438

FIGURE 14-16 HYDROGEN-INDUCED CRACK PATTERN IN A COMPLEX WELDMENT IN 4340 STEEL

Masubuchi and Martin conducted a mathematical analysis
of crack pattern using the Griffith-Irwin fracture mechanics
theory.

REFERENCES

[1] Berry, W. E., "Stress-Corrosion Cracking - A nontechnical Introduction to the Problem," Battelle Memorial Institute, Defense Metals Information Center DMIC Report 144, January 6, 1961.

[2] Beck, T. R., and Blackburn, M. J., "Stress Corrosion Cracking of Titanium Alloys," AIAA Journal, Vol. 6, No. 2, pp. 326-332, 1968.

[3] Brooks, C. L., "Aluminum Alloys and Their Performance in Marine Environments," presented at Symposium on Materials Applications in Oceanography, Pacific Symposia, Hawaii District, Honolulu, Hawaii, July 2, 1968.

[4] Brown, B. F., and Birnbaum, L. S.,"Corrosion Control for Structural Metals in the Marine Environment," NRL Report 6167, "Status and Projections of Developments in Hull Structural Materials for Deep Ocean Vehicles and Fixed Bottom Installations," U. S. Naval Research Laboratory, November, 1964.

[5] Capson, H. R., "A Theory of the Mechanism of Rusting of Low Alloy Steels in the Atmosphere," Proceedings of the ASTM, Vol. 45, pp. 554-590, 1945.

[6] Carson, J. A. H., "Designing Cathodic Protection Systems for Ships' Hulls," Materials Protection, pp. 62-51, January 1966.

[7] Chalmers, B., Physical Metallurgy, John Wiley & Sons, Inc., 1959.

[8] Currer, G. W., "Cathodic Protection of Marine Structures," The Dock and Harbor Authority, pp. 75-78, July 1964.

[9] Dean, S. W., and Copson, H. R., "Stress Corrosion Behavior of Maraging Nickel Steels in Natural Environments," a paper presented at the 20th Annual Conference of the National Association of Corrosion Engineers, March 9-13, 1964, Chicago, Illinois.

[10] Groot, C., "Cathodic Protection of Aluminum," Materials Protection, pp. 10-13, November 1964.

[11] Judy, R. W., Jr., and Goode, R. J., "Stress-Corrosion Cracking Characteristics of Alloys of Titanium in Salt Water," NRL Report 6564, U. S. Naval Research Laboratory, July 1967.

[12] LaQue, F. L., "The Behavior of Nickel Copper Alloys in Sea Water," Journal of the American Society of Naval Engineers, Vol. 53, No. 1, pp. 29-64, 1941.

(13) LaQue, F. L., May, T. P., Uhlig, H. H., "Corrosion in Action," a publication of the International Nickel Company, Inc., 1966.

(14) Masubuchi, K., and Martin, D. C., "Investigation of Residual Stresses by Use of Hydrogen Cracking, Parts I and II," The Welding Journal, 40 (12), Research Supplement, pp. 553s-563s (1961), and 45 (9), Research Supplement, pp. 401s-418s (1966).

(15) Matsui, E. S., "Procedure Improved for Determining Corrosion Rate by Weight Loss," Materials Protection, pp. 31-32, July 1968.

(16) May, T. P., and Weldon, B. A., "Copper Nickel Alloys for Service in Sea Water," paper given before the International Congress on Fouling and Marine Corrosion, Canne, France, June 8-13, 1964.

(17) Melloy, G. F., "Galvanized High Strength, Low Alloy Steels," Metal Progress, pp. 123-130, August 1965.

(18) Metal Progress, Staff Report, "High Strength, Low Alloy Steels Hold Point Longer," pp. 141-144, August 1965.

(19) Milligan, S., "Effect of Deep Ocean Environment on Underwater Installations," paper given before the Marine Systems Conference of the American Institute of Aeronautics and Astonautics, San Diego, California, March 1955.

(20) REinhart, F. M., "Preliminary Examination of Materials Exposed on STU I-3 in the Deep Ocean (5640 feet of depth for 123 days). U. S. Department of the Navy, Naval Civil Engineering Laboratory, Technical Note N-605, June 1964.

(21) Routley, A. F., "The Compatibility of Paints with Cathodic Protection," Anti-Corrosion, pp. 8-10, June 1967.

(22) Simcoe, C. R., Slaughter, E. R., and Elsea, A. R., "The Hydrogen-Induced Delayed Brittle Fracture of High Strength Steels," unpublished manuscript.

(23) Slunder, C. J., and Boyd, W. K., "Environmental and Metallurgical Factors of Stress-Corrosion Cracking in High-Strength Steels," DMIC Report 151, Battelle Memorial Institute, DEfense Metals Information Center, August 1961.

(24) Taylor, B. M., "An Introduction to Cathodic Protection," BuShips Journal, pp. 30-33, November 1964.

(25) Tetelman, A. S., and McEvily, A. J. Jr., Fracture of Structural Materials, John Wiley and Sons, Inc., 1967.

(26) Troiano, A., Transactions ASM, Vol. 52, pp. 54-- (1960).

(27) Tuthill, A. H., and Schillmoller, C. M., "Guidelines for Selection of Marine Materials," paper presented before the Ocean Science and Ocean Engineering Conference, Marine Technology Society, Washington, D. C., during June 14-17, 1965.

(28) Uhlig, H. H., Corrosion and Corrosion Control, and Introduction to Corrosion Science and Engineering, John Wiley and Sons, Inc., New York, 1963.

(29) Uhlig, H. H., The Corrosion Handbook, sponsored by the Electrochemical Society, Inc., New York, John Wiley and Sons, Inc., 1948.

CHAPTER 15 WELDING*

15.1 Brief History of Welding Technology

Historians can trace joining techniques back to prehistoric days. Men were soldering with copper-gold and lead-tin alloys before 3000 B.C.[18] We have evidence that artisans were welding iron and steel into composite tools and weapons as early as 1000 B.C. However, the only sources of heat for welding in those days were wood and coal. The relatively low temperatures available severely limited the types and applications of welding processes.[15]

Development of modern welding technology began in the latter half of the 19th century when electrical energy became more easily available. Figure 15-1 shows important discoveries and break-throughs in joining techniques and their applications since the 1890's.[13]

Development of Welding Processes.[15] Most of the important discoveries leading to modern welding processes were made between 1880 and 1900. In 1881, a Frenchman, Moissan, first used the carbon arc for melting metals. Bernardos, a Russian, carried this step further and took a German patent for carbon-arc welding in 1885. Shortly thereafter, another Russian, Slavianoff, conducted experiments on metal-arc welding. The first metal-arc welding patents were issued in the United States to Coffin in 1889 as shown in Figure 15.1. Meanwhile, another American, Elihu Thomson, applied for the basic patents for resistance welding in 1886. Finally, in 1895, Le Chatelier,

---

*A Study of underwater welding processes and techniques is presented in Appendix.

444

First U. S. metal-arc welding patents issued to Coffin

1890--

1900--

1910--

Covered electrode introduced by Kjelborg in Sweden

First use of metal-arc welding in shipbuilding

1920--

Fullager, the first all-welded, sea-going ship built in England

1930--   First patent for inert-gas electric-arc welding issued to Hobart
           and Devers

Submerged arc welding introduced

1940--   War effort greatly advances use of metal-arc welding in shipbuilding

Development of gas tungsten-arc welding

Development of gas metal-arc welding

1950--   Electroslag welding introduced in Russia

CO_2 shielded metal-arc welding introduced

New processes, such as electron-beam, ultrasonic and laser welding
    developed

1960--   Battelle Narrow-Gap process introduced

1968--   Use of various processes in shipbuilding: Covered electrode-80-90%
                                                    Submerged arc-   10-15%
                                                    CO_2 shielded-   1%
                                                    Others-   .1%

1980--   Ships still built from steel but more use of high-strength steels
           and increasing use of nonferrous nickels such as aluminum and
           titanium. Welding processes: covered electrodes, 50%, Submerged
           arc, 15%, Gas metal-arc, 30%.

2000--   Ships will be built with various ferrous and nonferrous metals,
           possibly with some nonmetallic materials. Welding processes:
           both new and existing processes.

Figure 15-1. Development of welding processes and their present and future
use in shipbuilding.

a Frenchman, invented the oxyacetylene blowpipe.

This latter invention rocketed oxyacetylene welding to a leading position in the field of welding. Ferrous welds made with oxyacetylene blowpipes were so superior to carbon and bare-electrode arc welds that the latter processes were nearly forgotten for stressed welds. Then in 1907, a Swede named Kjellberg revolutionized arc welding when he introduced covered electrodes. It was soon found that their use greatly improved the properties of weld metals and allowed many new welding applications. Today, shielded metal-arc welding is the most commonly used welding process for fabricating steel structures. Efforts to mechanize the process have led to various automatic arc-welding processes, including submerged-arc welding which was developed in the United States during the 1930's.

Another significant advance was made in the early thirties. Two Americans, Hobart and Dever, explored the principle of welding by use of an electric arc operating in an inert-gas atmosphere. Their experiments led to the development of inert-gas tungsten-arc welding. In turn, inert-gas (consummable) metal-arc welding was developed several years later.

Many new welding processes have been developed since World War II. They include: CO_2-gas shielded arc welding, electroslag welding, ultrasonic welding, friction welding, electron-beam welding, and plasma-arc welding. The new laser technique is also expected to be used for metals joining.* As a result, almost

---

*A laser is a device which produces a concentrated coherent light beam through the manipulation and control of energy exchanges in solid-state transparent media. Welding Handbook, Section 3 (5th Edition), Applications of laser to welding and cutting.[31]

all metals used in present day applications can be welded.

Applications of Welding. In the early days of arc welding, application to metal fabrication was quite limited. But as processes improved, the fields of application gradually increased. During World War I, metal-arc welding was used for the first time in ship construction. However, even then it was used primarily in repair operations or for joining relatively unimportant parts.

Welded fabrication had been used for a 42-foot launch, Dorothea M. Geary, built at Ashtabula, Ohio, in 1915.[29] Then in 1921, the first all-welded ocean-going ship, Fullager, was built in a British shipyard.

From these beginnings, application of welding increased steadily until World War II. At that time, the great demands of war greatly accelerated the use of welding. The United States was able to build about 4,700 ships during World War II[5] only through the wide use of welding. Demand for reliable methods for welding light-metal alloys for aircraft accelerated the development of inert-gas metal-arc welding. This process has now almost completely replaced covered arc welding and gas welding for the welding of aluminum and magnesium alloys.

During the past twenty years, welding applications have grown at a fantastic rate. Various welding techniques are widely used for the fabrication of numerous metal products and structures, including space vehicles, airplanes, ships, cars, locomotives, bridges, buildings, pipelines, nuclear reactors, pressure vessels, machineries, household goods, as well as miniature electronic components.

Problems in Development. Although the progress in welding techniques has been substantial, the advancements in applications of new welding processes have not been as great as we might expect. For example, about 80 to 90 percent of the joints (in length) in a ship hull are still welded manually with covered electrodes, which were invented more than 50 years ago. The submerged-arc process is used for about 10 to 15% of these joints, but this process is also about 30 years old. The CO_2-gas shielded metal-arc process and electroslag processes, which are rather recent developments, are used to a very limited extent. The use of other processes is practically nil.

This situation has many causes. For example, over 1/2 of the joints in a ship hull are fillet-welded and it is difficult to apply automatic processes because many of these joints are interrupted by intercostal structural members. The quality of welding has improved significantly over the years, but it is true that primary welding processes used in shipbuilding today are over half a century old. This is not bad in itself, but only when we consider that there are many new and promising procedures available that are seemingly ignored.

Future Trends. Besides applying joining processes already commonly used, future ocean engineers will need to develop and apply welding techniques to meet the demands of advancing technology.

We can reasonably assume that most ocean engineering structures by about 1980 will still be built primarily with steels, although we undoubtedly will be using more high-strength steels. Non-ferrous metals, including aluminum and titanium alloys, will be used to some extent. There will be a shift in welding processes during this period. The use of covered electrodes will perhaps decrease to about 50% or less, with a corresponding increase in the use of gas metal-arc processes using a variety of shielding gases, including argon, CO_2, and mixtures of various gases. Electron beam welding may be used for some special applications.

By some future date, more radical changes will have taken place. Ocean engineering structures will then be built with a variety of ferrous and non-ferrous metals. Honeycomb structures might well be in use. Some non-metallic materials, such as plastics or fiberglass, may be used on a larger scale. Various joining processes will be used, including some that exist today and probably some that have yet to be developed. It is possible that some joining may be done by adhesive bonding. Most of these changes will probably occur first in military or research vehicles where economics is not such a confining factor. Later, after they prove to be economically competitive, these changes will find their places in commercial structures.

15.2 Welding Processes

Classification of Welding Processes.[15] More than 40 welding processes are used in present day metal fabrication. Figure 15-2 is a master chart of welding processes prepared by the American Welding Society.* There are eight basic processes

---

*There are several new welding processes which are not included in Figure 15-2. They include: (1) electroslag welding, (2) ultrasonic welding, (3) cold welding, (4) friction welding, (5) electron beam welding, (6) plasma-arc welding, and (7) laser welding.

449

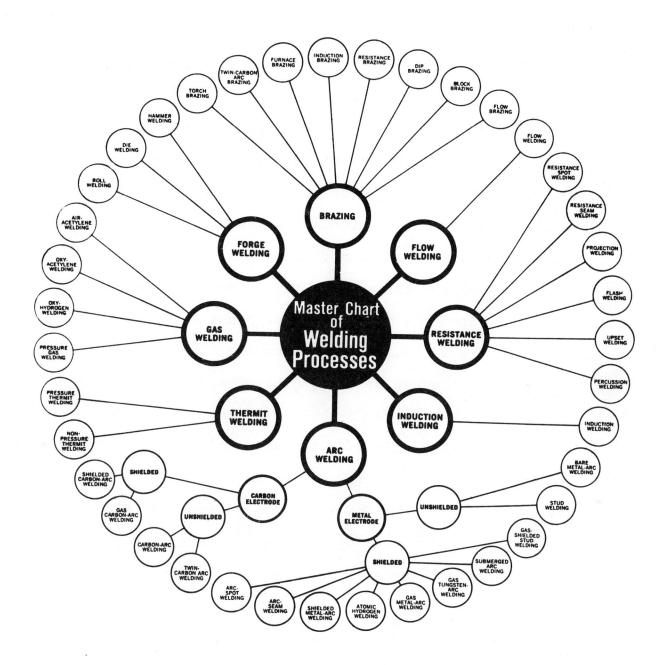

FIGURE 15-2 MASTER CHART OF WELDING PROCESSES

defined as follows:

1. *ARC WELDING:* A group of welding processes wherein
   coalescence is produced by heating with an electric
   arc or arcs with or without the application of
   pressure and with or without the use of filler metal.

2. *THERMIT WELDING:* A group of welding processes
   wherein coalescence is produced by heating with
   superheated liquid metal and slag resulting from a
   chemical reaction between a metal oxide and aluminum,
   with or without the application of pressure. Filler
   metal, when used, is obtained from the liquid metal.

3. *GAS WELDING:* A group of welding processes wherein
   coalescence is produced by heating with a gas flame
   or flames, with or without the application of
   pressure, and with or without the use of filler metal.

4. *FORGE WELDING:* A group of welding processes wherein
   coalescence is produced by heating in a forge or
   other furnace and by applying pressure or blows.

5. *BRAZING:* A group of welding processes wherein
   coalescence is produced by heating to suitable
   temperatures above 800° F., and by using a non-ferrous
   filler metal, having a melting point below that of
   the base metals. The filler metal is distributed
   between the closely fitted surfaces of the joint
   by capillary attraction.

6. *FLOW WELDING:* A welding process wherein coalescence
   is produced by heating with molten filler metal,

poured over the surfaces to be welded until the required filler metal has been added. The filler metal is not distributed in the joint by capillary attraction.

7. *RESISTANCE WELDING*: A group of welding processes wherein coalescence is produced by the heat obtained from resistance of the work to the flow of electric current in a circuit of which the work is a part, and by the application of pressure.

8. *INDUCTION WELDING*: A welding process wherein coalescence is produced by the heat obtained from resistance of the work to the flow of induced electric current, with or without the application of pressure.

The arc welding processes, with which this book is concerned, are by far the most commonly used. There are fourteen arc-welding processes. These are classified into four groups, depending upon the type of electrode used and whether the arc is shielded:

A. Metal-electrode arc-welding processes

    (1) Shielded metal-electrode arc-welding processes, including shielded metal-arc welding, submerged-arc welding, gas tungsten-arc welding, gas metal-arc welding, atomic hydrogen welding, gas-shielded stud welding, arc-seam welding, and arc-spot welding.

    (2) Unshielded metal-electrode arc-welding processes, including bare metal-arc welding, and stud welding.

B. Carbon-electrode arc welding processes

    (1) Shielded carbon-electrode arc welding processes, including gas carbon-arc welding, and shielded carbon-arc welding.

    (2) Unshielded carbon-electrode arc welding processes, including twin-carbon arc welding and carbon arc welding.

Metal electrodes may be either filler metal or tungsten. A filler metal electrode is deposited as weld metal and is called a *consumable electrode*. Tungsten and carbon electrodes are not deposited and are called *nonconsumable electrodes*. Filler metal also may be fed separately in arc-welding using a nonconsumable electrode.

In the shielded arc-welding processes, the welding arc is shielded from the surrounding air to protect the fused metal from oxidation and formation of nitrides, and to improve the quality of the weld metal.

The emphasis of the discussion in this book is placed on the following four shielded metal-electrode arc-welding processes:

    (1) Shielded metal-arc welding

    (2) Submerged-arc welding

    (3) Gas tungsten-arc welding

    (4) Gas metal-arc welding

These four processes are emphasized because they are by far the most widely used. Shielded metal-arc welding is the most commonly used welding process for fabricating metal structures,

especially steel structures. Submerged-arc welding is widely
used in the fabrication of heavy plate structures such as ships
and pressure vessels. Gas tungsten-arc welding and gas metal-
arc welding are two types of gas shielded-arc welding. They
are widely used for welding aluminum alloys, titanium alloys,
stainless steels, and many other alloys on which good-quality
welds are difficult to obtain with shielded metal-arc processes.

This book also discusses briefly the following two processes
which are not included in Figure 15-2:

(5) Electroslag process

(6) Electrogas process

The electroslag process was developed in Russia around
1950, while the electrogas process was developed in East Germany
around 1960. These processes have had a limited application in
the United States.

Further details of these six processes are given in the
Welding Handbook and other textbooks on welding.[3,31]

Shielded Metal-Arc Welding[15,20,21]

Shielded metal-arc welding is an arc-welding process wherein
coalescence is produced by heating with an electric arc between
a covered electrode and the work. Shielding is obtained from
decomposition of the electrode covering. Pressure is not used
and filler metal is obtained from the electrode.

In practice, the process is limited primarily to manual
covered electrodes in which the welding operator manipulates
the electrode.

Figure 15-3 is a schematic representation of the shielded metal-arc process. Equipment to operate the electrode usually consists of an electric power supply specifically designed for the process, insulated electrode holders of adequate electric and thermal capacity, cable, and grounding clamps. The process may use either alternating current or direct current with the electrode either positive or negative. Currents between 15 and 500 amperes with arc voltages between 14 and 40 volts, depending upon the covering characteristics are normal.

Coated electrodes for manual arc welding are classified by the American Welding Society (AWS) and the American Society for Testing Materials (ASTM) as shown in Table 15-1. The electrodes are classified according to the operating characteristics, the type of coating, and the strength level of the weld metal. For example, E-60XX electrodes produce deposited metal having a minimum specified ultimate tensile strength of 60,000 psi. The last two digits in the electrode designation refer to the type of coating and operating characteristics as shown. The yield strength and elongation of deposited metal are specified by the code; however, no notch-toughness requirement is included.

Electrodes as large as 5/16 inch diameter may be employed, depending upon the plate thickness and type of joint. The welds are built up in relatively thin layers, approximately 10 layers per inch of thickness. This will permit a partial progressive grain refinement of proceeding layers resulting in an improvement in ductility and impact resistance of the weld metal.

TABLE 15-1   ELECTRODES COMMONLY USED FOR WELDING LOW-CARBON STEEL (3)

AWS-ASTM Electrode Class	Coating	Current, Polarity*	Welding Position†
E6010	High cellulose, sodium	dcrp	F, V, OH, H
E6011	High cellulose, potassium	dcrp, ac	F, V, OH, H
E6012	High titania, sodium	dcsp, ac	F, V, OH, H
E6013	High titania, potassium	dcsp, ac	F, V, OH, H
E6014	Iron powder, titania	dcsp, ac	F, V, OH, H
E7016	Low-hydrogen, potassium	dcrp, ac	F, V, OH, H
E7018	Low-hydrogen, iron powder	dcrp, ac	F, H
E6020	High iron oxide	dcrp, dcsp, ac	F, H
E7024	Iron powder, titania	dcrp, dcsp, ac	F, H
E6027	Iron powder, iron oxide	dcrp, dcsp, ac	F, H

*dcrp--direct current reverse polarity, electrode positive; dcsp--direct current straight polarity, electrode negative; ac--alternating current

†F--flat; V--vertical; OH--overhead; H--horizontal

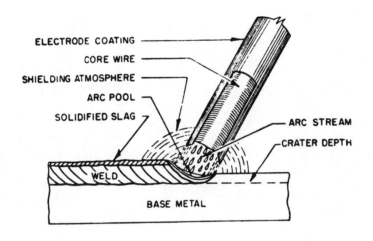

ELECTRODE COATING

CORE WIRE

SHIELDING ATMOSPHERE

ARC POOL

SOLIDIFIED SLAG

ARC STREAM

CRATER DEPTH

WELD

BASE METAL

Figure 15-3   SCHEMATIC REPRESENTATION OF SHIELDED

METAL-ARC PROCESS

Semi-automatic Processes. Some attempts have been made at mechanizing the shielded metal-arc process. The idea of making shielded metal-arc welding semi-automatic by laying an electrode along the joint has existed for more than 30 years. However, the electrodes were not advanced enough to be used for actual production in large scale. The development of contact electrodes using high iron oxide and iron-powder coating has shown that a weld metal with sound quality can be obtained without manipulating electrodes.[13]

A group of Japanese electrode manufacturers and shipbuilders has developed electrodes and welding devices which can be used for production in a semi-automatic operation. Electrodes are 25 to 40 inches long and one operator can handle four to six electrodes simultaneously.

Submerged Arc Welding[3, 31]

Submerged arc welding is an arc-welding process wherein coalescence is produced by heating with an electric arc or arcs between a bare metal electrode or electrodes and the work. The welding is shielded by a blanket of granular, fusible material on the work. Pressure is not used, and filler metal is obtained from the electrode and sometimes from a supplementary welding rod. Figure 15-4 shows how a submerged arc groove weld is made.

The fusible shielding material is known as "flux," "welding composition," or "melt." This is a finely crushed mineral composition and will be referred to as flux in this book. Flux is the basic feature of submerged arc welding and makes possible the special operating conditions which distinguish the process.

458

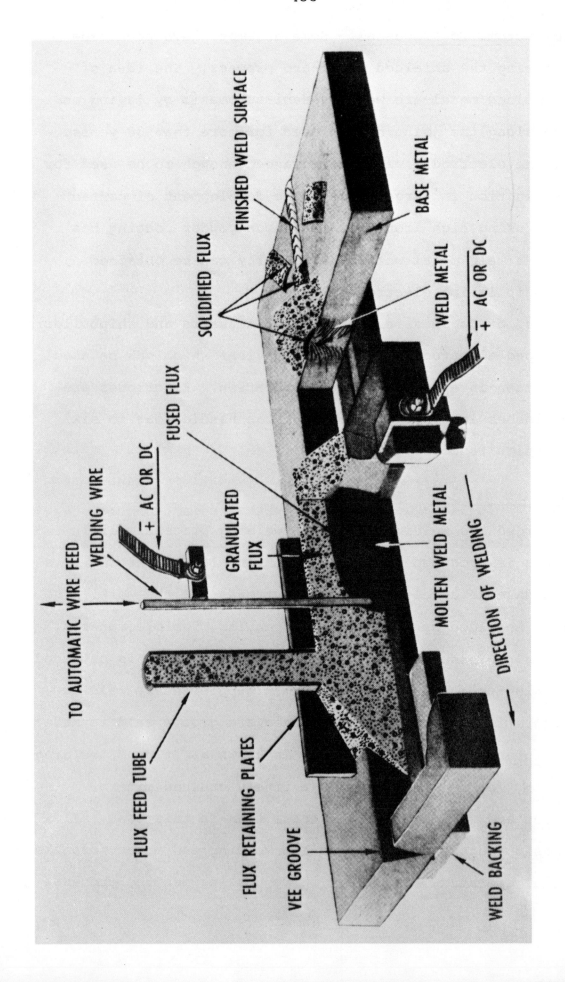

FIGURE 15-4 THE PROCESSES OF A SUBMERGED ARC GROOVE WELD (3)

Flux, when cold, is a non-conductor of electricity, but in the molten state it becomes highly conductive.

In submerged arc welding, there is no visible evidence of the passage of current between the welding electrode and the workpiece. The electrode does not actually contact the workpiece; instead, the current is carried across the gap through the flux.

The flux completely covers both the tip of the welding electrode and the weld pool. Thus, the actual welding operation takes place beneath the flux without sparks, spatter, smoke or flash. No protective shields, helmets, smoke collectors or ventilating systems are needed, with the exception of goggles which may be worn as routine protection for the eyes.

In its molten state, the flux provides exceptionally suitable conditions for unusually high current intensities. Thus, great quantities of heat may be generated. The insulating qualities of the flux enable the intense heat to be concentrated in a relatively small welding zone, where the welding electrode and base metal are rapidly fused. High welding speeds are possible under these conditions, and deep penetration can be obtained by this concentrated heat. Consequently, a relatively small welding groove can be used permitting the use of smaller amounts of filler metal.

The wide use of submerged arc welding stems from its ability to produce satisfactory welds at high rates of deposition. Currents used in submerged arc welding are much higher than those employed in manual shielded metal-arc welding. The maximum electrode usually employed is 5/16 inch diameter and the maximum current (for single electrode) is approximately 2000 amperes.

It should be pointed out that high amperage results in a very coarse columnar structure in the weld metal with an enlarged heat-affected zone in the base metal as compared to multipass welding wherein the energy input is relatively small. This is important in connection with vessels which will operate at low temperature. The coarse columnar structure of the high amperage weld usually results in lower impact resistance as compared to multipass welds. Therefore, particular consideration should be given to the welding procedure in terms of the service requirements involved.

During the last ten years, many studies have been made on: 1) How to improve notch toughness of two-pass submerged-arc deposited metals in heavy mild-steel ship plates and 2) how to improve notch toughness of submerged arc deposited metals in high-strength, notch-tough steels. Both of these areas will be considered later in this chapter.

Gas-Shielded-Arc Welding [3, 31]

In the gas-shielded-arc welding process coalescence is produced by fusion from an electric arc maintained between the end of a metal electrode, either consumable or nonconsumable, and the part to be welded with a shield of protective gas surrounding the arc and weld region. The shielding gas may or may not be inert; pressure may or may not be used; and filler metal may or may not be added.

At this time, there are two different types of gas-shielded arc welding. One is termed gas tungsten-arc welding. It employs a tungsten electrode or an electrode of some other refractory, high-melting-point material, such as graphite, which will not melt or be vaporized too rapidly in the intense heat of the arc. The other type is termed gas metal-arc welding. It employs a continuously fed electrode which melts in the intense arc heat and is deposited as weld metal.

Gas Tungsten-Arc Welding. Gas tungsten-arc welding is also known as GTA welding, while gas metal-arc welding is called GMA welding.*

Figure 15-5 shows basic features of gas tungsten-arc welding. The shielding gas is fed through the electrode holder which is generally referred to in this process as a torch. Argon, which is an inert gas, is commonly used for the shielding gas. The use of helium as the shielding gas is rather rare. A small percentage of oxygen is often added to the inert shielding gas. Carbon dioxide and a mixture of CO_2 and O_2 have been found to be effective as the shielding gas for use in welding carbon and low-alloy steels. More recently, fluxes have been added as a core within a tubular sheath or as a granular magnetic material which adhere to the filler wire surface.

Gas Metal-Arc Welding. Figure 15-6 shows basic features of gas metal-arc welding. The filler wire, which is manufactured in a coil form, is fed mechanically into the welding arc. The

---

*In the past, gas tungsten-arc welding was known as TIG (tungsten inert-gas) welding, and gas metal-arc welding was known as MIG (metal inert-gas) welding.

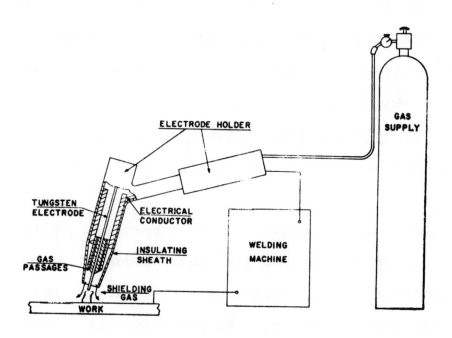

Figure 15-5   SCHEMATIC DIAGRAM OF GAS TUNGSTEN-ARC WELDING[31]

463

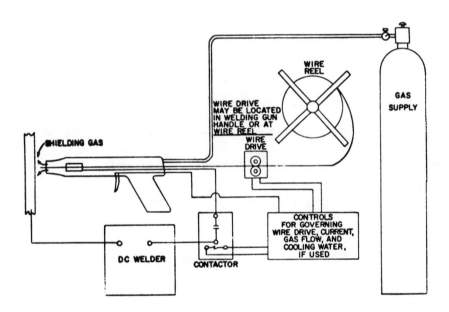

Figure 15-6 SCHEMATIC DIAGRAM OF GAS METAL-ARC WELDING (3)

arc travel is controlled manually in the semi-automatic process and mechanically in the automatic process.

A bare wire is commonly used for the electrode, but flux-covered wires are used also.

Various gases are used for shielding, including:

(1) Pure inert gases such as argon and helium

(2) Mixtures of argon, CO_2, O_2, and other gases

(3) CO_2.

Because of the high cost of inert gas, CO_2 and a mixture of CO_2 and O_2 are widely used for welding carbon steel and low-alloy high-strength steels. Argon and a mixture of argon and CO_2 are often used for welding quenched-and-tempered steels and ultra-high strength steels. Inert gases, which may contain small portions of other gases, are used almost exclusively for welding stainless steel, aluminum, titanium and other non-ferrous metals.

Compared to the gas tungsten-arc welding, the gas metal-arc welding is characterized by a high deposition rate. Therefore, the GMA process is chiefly used in welding heavy plates, while the GTA process is used mainly in welding thin sheets.

A new gas metal-arc welding process, which is called the Battelle Narrow-Gap Welding process, is being developed at Battelle Memorial Institute for the Bureau of Ships of the U.S. Navy.[17]* The new process, which has been developed primarily for automatically welding submarine hulls, can be widely used in the fabrication of pressure vessels made with quenched and tempered high-strength steels.

---

* Currently called "Naval Ship Systems Command."

As the name simplies, the narrow gap process differs from conventional welding procedures in the type of joint design used. The new procedure uses a square-butt joint with a narrow root opening (approximately 1/4 inch). The use of this type of joint results in a weld deposit with a small fusion zone. A typical weld deposited by this process in 2-inch-thick HY-80 steel is shown in Figure 15-7. The very high depth-to-width ratio and the very narrow, uniform heat-affected zone are apparent. A mixture of argon and CO_2 is used as the shielding gas. Welds are deposited from one side of the plate using a specifically designed guide tube that extends into the joint. Welds can be made in all positions--the weld shown in Figure 15-7 was made in the vertical position.

Electroslag Welding [3, 31]

Electroslag welding is a new process for welding thick sections that was developed and initially applied in the Soviet Union. The process is based on the generation of heat produced by passing an electrical current through molten slag. Figure 15-8 illustrates the basics of this process. In electroslag welding, the electrode is immersed in the molten slag pool between the components to be welded and the copper molding devices. The melt is heated to a high temperature by a current passing between the electrode and the base metal and electrical conductivity is increased. The temperature of the slag pool must exceed the melting point of the base and filler metal. Therefore, the slag melts the faces of the work and the electrode is immersed in the molten slag. The weld pool is formed when the molten base

466

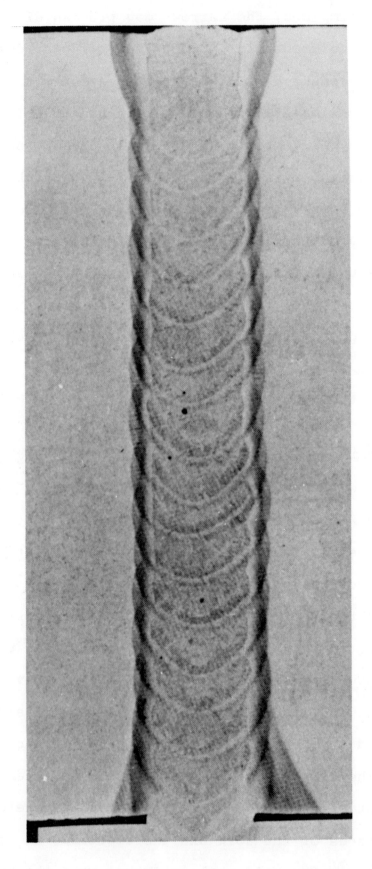

FIGURE 15-7 A WELD DEPOSITED BY THE NARROW GAP PROCESS[14]

467

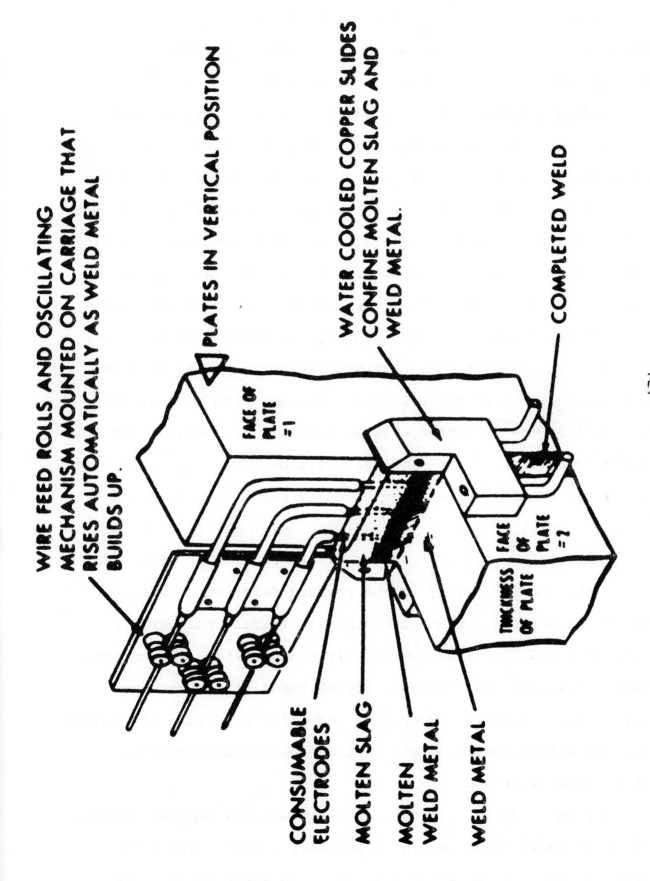

WIRE FEED ROLLS AND OSCILLATING MECHANISM MOUNTED ON CARRIAGE THAT RISES AUTOMATICALLY AS WELD METAL BUILDS UP.

PLATES IN VERTICAL POSITION

WATER COOLED COPPER SLIDES CONFINE MOLTEN SLAG AND WELD METAL.

COMPLETED WELD

FACE OF PLATE #1

FACE OF PLATE #2

THICKNESS OF PLATE

CONSUMABLE ELECTRODES

MOLTEN SLAG

MOLTEN WELD METAL

WELD METAL

Figure 15-8   THE PROCESS OF ELECTROSLAG WELDING (3)

and filler metal collect at the bottom of the slag pool. This weld pool solidifies and forms the weld and joins the faces of the components. The electrode is lowered as it melts.[3]

Welding Position. Welding with the electroslag process is done with the joint positioned vertically, as shown in Figure 15-8. This provides the best conditions for fusing the base metal and obtaining a deep slag pool. Consequently, electroslag welding is often provided with a molding device for the weld pool. Welding in the flat position is less convenient and generally not used. This process was used originally where the flat position suitable for arc welding could not be employed, for example in on-site welding of vertical points. The molding principle is based on mechanical cooling of the weld-pool surface. The pool becomes cup-shaped if the direction of heat transfer is changed, enabling welding to be done in the vertical position. Usually cooling is effected by means of shoes or stationary copper blocks which are water cooled.[3]

Electrical Conductivity. The transformation of electrical energy into thermal energy is the primary function of slag in electroslag welding. Therefore, the electrical conductivity of the slag and the variation of this conductivity with temperature is critical. The conductivity of fused slags increases sharply with a rise of temperature and below a certain temperature slags are almost non-conducting. This makes stabilizing the process more difficult.

Certain slags with conduction electrons (as opposed to ionic conduction possessed by slags in the molten state) are good conductors even in the solid state at room temperatures. The stability of electroslag welding depends on a constant slag pool temperature, presenting a balance between the heat received and the heat lost.

Applications. Electroslag welding originally was developed for joining thick sections. Theoretically, any thickness can be welded in a single pass, but the maximum reported thickness is 1000 mm. (40 inches). Single pass welds are made in any practical thickness of steel or alloy steel. Welding starts at the bottom of the joint and progresses upward, usually by moving the welding head and auxiliary equipment.

Although the electroslag process offers economic savings through reduced welding costs and lower capital-equipment inventories in heavy-machine-building factories, application of the process in the U.S. has been limited.

Advantages and Disadvantages. The primary advantages of the electroslag process are that no joint preparation is needed, a greater steel thickness can be welded, costs are reduced, and heavy equipment can be constructed from small wrought shapes, castings, or forgings. Disadvantages include hot-cracking problems, low notch toughness of the weld metal and the heat-affected zone and difficulty in closing cylindrical welds.

Electrogas Welding

A fully automatic process, electrogas welding provides a method of fusion welding of butt, corner and tee joints in the vertical position. Metal sections ranging from 1/2-inch in thickness can be joined in a single pass by adapting the operational concepts of electrogas welding equipment to spray-type arc welding.[3]

Basic Principles. Figure 15-9 illustrates the electrogas process. In electrogas welding, water-cooled copper shoes bridge the gap between the components being welded and form a rectangular pocket or cavity containing the welding operation. A wire guide feeds a flux-cored wire or solid wire into the pocket. An electric arc is established and maintained continuously between the electrode and the weld puddle. To provide a suitable atmosphere for shielding the arc and weld puddle, helium, argon, carbon dioxide or mixtures of these gases are continuously fed into the pocket. The weld metal is cleaned by deoxidizers and slagging materials provided by the flux core of the electrode.[3]

An ionized shielding gas preheats the work surfaces which are brought to the proper temperature for complete fusion when they are in contact with the molten slag. This molten slag flows toward the copper shoes and forms a protective coating between the shoes and the faces of the weld. This also prevents entrance of the atmosphere into the weld pocket by providing a seal between the shoes and surfaces of the workpieces.

As the weld metal is deposited, the wire feed, carrying the wire guide and copper shoes, moves steadily upward to provide a weld pocket of uniform depth in which the weld is deposited.[3]

Applications. Electrogas welding is used primarily for joining vessel seams in field storage tanks. These may include, for instance, oil storage tanks, water tanks, and blast furnace shells. It also is finding some application in structural welding of vertical joints.[3]

471

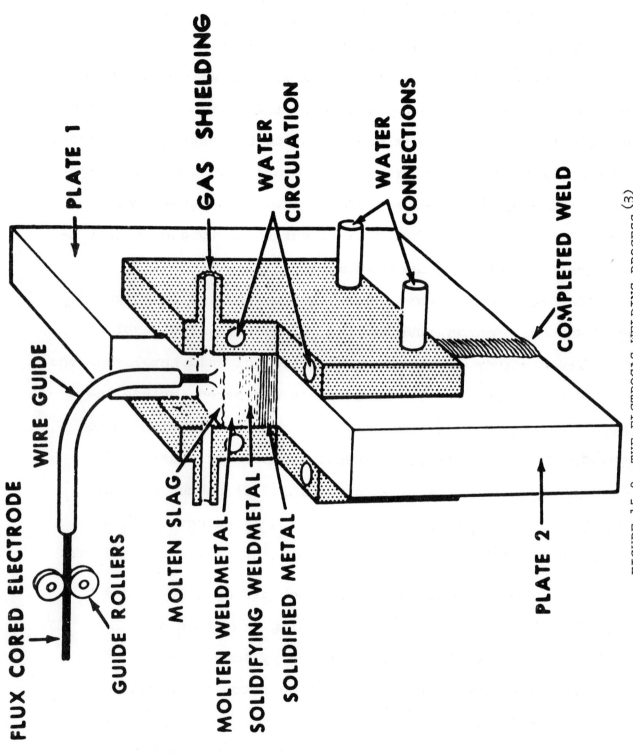

FLUX CORED ELECTRODE

WIRE GUIDE

GUIDE ROLLERS

PLATE 1

GAS SHIELDING

WATER CIRCULATION

WATER CONNECTIONS

MOLTEN SLAG

MOLTEN WELDMETAL

SOLIDIFYING WELDMETAL

SOLIDIFIED METAL

COMPLETED WELD

PLATE 2

FIGURE 15-9 THE ELCTROGAS WELDING PROCESS (3)

15.3 Notch Toughness of Weld Metal[14]

In order to avoid brittle fracture of welded structures,
it is important that both the base plate and weld metal have
adequate notch toughness. It is not difficult to obtain weld
metal with notch toughness equivalent to ordinary carbon steel.
However, it becomes a serious consideration when welding high-
strength, quenched-and-tempered steels. These steels have
both high yield strength and excellent notch toughness that is
difficult to match in the weld metal.

The Welding Research Council Bulletin No. 111, prepared
by Masubuchi, Monroe, and Martin[14] presents results of a
literature survey on the notch toughness of weld metals and the
heat-affected zones. The base metals discussed include low
carbon steel and low-alloy, high-strength steels with up to
120,000 psi yield strength. Welding processes considered
include (1) shielded metal-arc welding, (2) submerged-arc
welding, (3) gas metal-arc welding, and (4) electroslag and
electrogas welding. The report covers the following subjects:

(1) General trends of notch toughness of weld metals
    deposited with various welding processes as
    evaluated by the V-notch Charpy impact test.

(2) Effects of various factors on notch toughness
    including:

    a. Chemical composition and microstructure

    b. Factors related to welding procedures
       including heat input, multilayer welding
       techniques, welding position, preheating

and post-weld heat treatments.

(3) Notch toughness of the heat-affected zone evaluated
by the Charpy impact test.

(4) Evaluation of weld metal toughness with various tests
other than the Charpy V-notch.

This paper covers primarily the first subject as discussed
in the WRC Bulletin No. 111.

Requirements of Various Specifications for the Notch Toughness of Weld Metals. Table 15-2 summarizes the requirements
of various specifications for weld metal notch toughness in butt
joints. Also shown for comparison are the notch-toughness
requirements for some steel plates. Minimum values of Charpy
V-notch absorbed energy at certain temperatures are specified.*
Figure 15-10 shows notch toughness requirements for entries marked
in Table 2 with asterisks. No specification covers the notch
toughness of weld metals in fillet joints.

Base Metal. At an international conference held in London
in 1959, ship classification societies of major shipbuilding
countries approved an international specification for ship steels.[7]
In this specification, steels are classified into five grades,
A, B, C, D, and E. Charpy V-notch values of 35 ft-lb at 32 F.
for Grade D and 45 ft-lb at 14 F. for Grade E are specified.
These requirements are represented in 15-10 by points designated D and E.

---

*Notch toughness of the base metal is discussed in detail in
  Chapter 12.

TABLE 15-2   NOTCH TOUGHNESS REQUIREMENTS OF VARIOUS
SPECIFICATIONS

Base Metal or Weld Metal	Specifications and Classes	Charpy V-Notch	
		Minimum Energy Absorption, ft-lb	Temperature, F
	International Specification for Ship Steel		
Base metal	Grade D*	35	32
	Grade E*	45	14
	U. S. Navy HY-80 Steel		
	Thickness 2 inches or less*	60	-120
	Thickness over 2 inches*	30	-120
	AWS-ASTM (1964)		
Covered-electrode-deposited metal	E6012, E6013, E6020, E7014, E7024	Not required	
	E7028*	20	0
	E6010*, E6011, E6027, E7015, E7016, E7018	20	-20
	E8016-C3*, E8018-C3	20	-40
	E9015-D1, E9018-D1, E10015-D2, E10016-D2, E10018-D2	20(a)	-60
	E9018-M*, E10018-M, E11018-M, E12018-M	20	-60
	E8016-C1*, E8018-C1	20(a)	-75
	E8016-C2*, E8018-C2	20	-100
	International Specification for Ship Steel		
	Joining Grade A steel*	35	68
	Joining Grades B, C, D steels*	35	32
	Joining Grade E steel*	45	14
	IIW Commission II		
Tensile strength of weld metal 61,000 to 67,000 psi	Quality I*, flat position	35	63
	Quality I*, vertical position	29	68
	Quality II*, flat position	46	68
	Quality II*, vertical position	40	68
	Quality III*, flat position	57	68
	Quality III*, vertical position	52	68
Tensile strength of weld metal 74,000 to 88,000 psi	Quality II, flat position	46	68
	Quality II, vertical position	40	68
	Quality III*, flat position	57	68
	Quality III*, vertical position	52	68
	IIW Commission XII		
Submerged-arc-deposited metal	Minimum tensile strength of weld metal 60,000 psi 42C*	20	32
	42D*	20	-4
	Minimum tensile strength of weld metal 71,000 psi 50C	20	32
	50D	20	-4

(a) Stress-relieved condition.

* Notch-toughness values are shown in Figure 15-10.

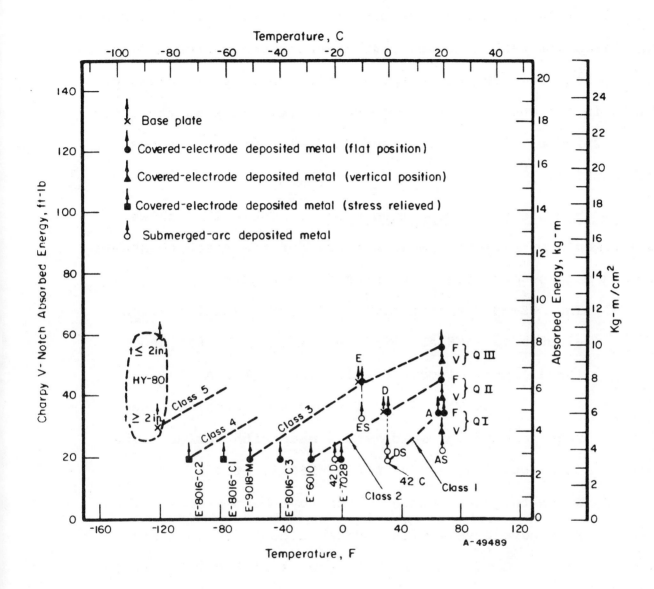

Figure 15-10 NOTCH TOUGHNESS REQUIRED BY VARIOUS SPECIFICATIONS

1. International specification for ship steel: A,D,E,(AS,DS,ES
    are 70% of A, D, and E, respectively; submerged-arc welding)
2. AWS: E-7028, E6010, E-8016-C3, E-9018-M, E-8016-C1, E-8016-C2
3. IIW, Commission II: QI, QII, QIII for flat and vertical positions
4. IIW, Commission XII: 42C, 42D (submerged arc-welding)

Many specifications cover the notch toughness of high-strength structural steels. The U.S. Navy Specification MIL-S-16216 covers notch-toughness requirements for HY-80 steel, which is a low-carbon nickel-chromium-molybdenum steel. The minimum requirements for Charpy V-notch impact test values of steels in the quenched and tempered condition are 60 ft-lb for plates 2 inches thick or less and 30 ft-lb for plate over 2 inches thick.

Covered-Electrode-Deposited Metals. Table 15-2 shows requirements of various specifications for the notch toughness of weld metals deposited with covered electrodes:

(1) The AWS-ASTM specifications for mild-steel covered electrodes (AWS A5 1-64 T and ASTM A-233-64) and for low-alloy-steel covered electrodes (AWS A5-64T and ASTM A316-64 T). Table 2 shows the notch-toughness requirements for weld metals made with various types of electrodes, E7028, E6010, E6011, etc.

(2) Notch-toughness values which have been proposed for weld metals made with covered electrodes used for joining Grades A through E steels of the international specification for ship steels.

(3) Notch-toughness values which have been recommended by Sub-Commission C, "Testing and Measuring of Weld Metal" of Commission II of the International Institute of Welding.[19] Notch-toughness values have been recommended depending on (1) the tensile strength of the weld metal, (2) the quality of the weld (QI, QII,

QIII), and (3) the welding position (flat or vertical).

Electrodes used for military applications are specified by MIL-E-22200/1. Electrodes MIL-9018, MIL-10018, MIL-11018 and MIL-12018 are similar to those of AWS specifications E-9018-M, E10018-M, E11018-M and E12018-M. The U. S. Navy specification NAVSHIP 250-637-3 specifies among other processes the use of covered electrodes in the fabrication of HY-80 submarine hulls. MIL-11018 and 9018 electrodes are used most commonly.

Submerged-Arc-Deposited Metals. Sub-Commission E, "Study of Weld Metal Deposited by All Processes," of Commission XII of IIW has proposed notch-toughness requirements for submerged-arc-deposited metals in carbon steel.[28] Values shown in Table 15-2 are recommended, depending on (1) the minimum tensile strength (60,000 or 71,000 psi) and (2) the class of weld (C or D).

Theoretically, requirements for the notch toughness of weld metal should be the same for all welding processes. However, several ship-classification societies currently accept submerged-arc-deposited weld metals with notch toughness lower than those of covered-electrode deposited weld metals. Apparently, this is because of the difficulty in obtaining submerged-arc-deposited metals which meet the requirements for covered-electrode-deposited metals. For example, a ship classification society requires the following values for submerged-arc-deposited metals:

   (1) 25 ft-lb at 68 F. for joining Grade A steel

   (2) 25 ft-lb at 32 F. for joining Grades B, C, and D steels

   (3) 35 ft-lb at 14 F. for joining Grade E steel.

These values are shown in Figure 15-10 by points designated
AS, DS and ES, respectively. These values are about 70% of
the values required for covered-electrode weld deposits
used in joining steels of corresponding grades.

Comparison of Various Specifications. To facilitate
discussions in later chapters of this report, the notch-toughness
requirements of various specifications are divided arbitrarily
into the following five classes, with Class 1 being the least
severe and Class 5 the most severe:

Class 1: QI, flat position; 42 C(35 ft-lb at 68 F: 20 ft-
lb at 32 F)

Class 2: QII, flat; Grade D; E6010 (46 ft-lb at 68 F; 35
ft-lb at 32 F; 20 ft-lb at -20 F)

Class 3: QIII, flat; Grade E; E9018-M (57 ft-lb at 68 F;
45 ft-lb at 14 F; 20 ft-lb at -60 F)

Class 4: E8016-C2 (20 ft-lb at -100 F)

Class 5: HY-80 steel base metal over 2 inches thick (30
ft-lb at -120 F).

Basis for Toughness Requirements. It is still debatable (1)
whether the Charpy V-notch impact test is an adequate test for
evaluating notch toughness and (2) what notch-toughness level is
really needed even for base metals. Nevertheless, attempts have
been made to establish realistic requirements for the notch toughness of the base metal by:

(1) Analyses of notch-toughness data obtained with
specimens taken from fractured ships.

(2) Comparison of notch-toughness data obtained with Charpy
specimens and the fracture behvior of large-size
specimens.

electrodes used by Pellini and Watkinson.

(2) Low-hydrogen electrodes: Modern E6015 and E6016
electrodes provide weld metals with notch toughness which
almost can meet the Grade E requirement.

(3) E6012 electrodes: E6012 electrodes do not meet even
the Grade A requirement. E6012 and E6013 electrodes are not
approved for use in joining main structural members because of
the low notch toughness of their weld metals.

Low-hydrogen, Low-Alloy Electrodes. Figure 15-12 shows
a band for notch toughness data of weld metals obtained with
modern low-hydrogen, low-alloy electrodes, E9018, E10016 and
18, and E11016 and 18. These data were obtained by Sagan-
Campbell[25] and Smith.[27]

Also shown in Figure 15-12 are some notch-toughness requirements as shown in Figure 15-10. The figure shows that the notch
toughness of weld metals deposited with modern low-hydrogen,
low-alloy electrodes is much superior to that required of
Grades D and E steels. However, no electrode specification
requires weld metals that meet the notch toughness requirements
for HY-80 steel base metal.

Submerged-Arc-Deposited Metals

Heavy Mild Steel Ship Plates. It has been known for some
time that the notch toughness of two-pass submerged-arc-
deposited metals decreases as the plate thickness increases. In
the fabrication of pressure vessels, which require high-quality

However, almost no information has been obtained which can be used to establish realistic requirements for weld-metal and heat-affected-zone notch toughness. Most current specifications are apparently based on the principle that the notch toughness required for the base metal should also be required for the weld metal deposited by a welding process used to join the steel. This basic principle, however, is not always obeyed. For example:

(1) The notch-toughness requirement for weld metals deposited with MIL-11018 electrodes, which have been used extensivley for the fabrication of submarine hulls from HY-80 steel, is considerably less severe than that for HY-80 base metal. The requirement is 20 ft-lb at -60 F for the weld metal and 60 or 30 ft-lb at -120 F for the base metal.

(2) Some ship-classification societies allow lower notch toughness for submerged-arc-deposited metals than for covered-electrode-deposited metal.

The low toughness requirements have been set in some cases because weld metals with higher notch toughness are not available at the present time, not because notch toughness is less important in the weld metal than in the base metal.

Covered-Electrode-Deposited Metals

There are abundant data on the notch toughness of weld metals deposited with different types of electrodes from various manufacturers. This paper discusses the general trends observed

in the notch toughness of weld metals deposited by a variety of
electrode types.

Mild Steel Electrodes E60XX. Figure 15-11 contains V-notch
Charpy transition data for various weld metals:

(1) A "band" which shows variation of transition curves
for weld metals deposited with mild-steel electrodes.
The band was obtained from data reported by Pellini[23]
on E6010 electrodes and by Watkinson[30] on three types
of British electrodes. A band for the notch toughness
of ship steels used in World War II production is
shown also.

(2) Ranges of notch toughness values expected in the
as-deposited condition at 70 F and -40 F, given in the
AWS Welding Handbook for weld metal deposited by
E6010, E6012, E6015 and E6016 electrodes.[21]

Also shown in Figure 15-11 are some notch toughness requirements. Figure 15-11 shows the general trends in the notch
toughness of weld metals deposited with mild steel covered
electrodes:

(1) E6010 electrodes: The band for Pellini's data for
American E6010 electrodes and Watkinson's data for British mild
steel electrodes is higher at all temperatures than the average
band for ship steels for World War II production, indicating
superior notch toughness. Data given in the latest AWS Welding
Handbook and the recent toughness requirement for the E6010
electrode indicate that modern E6010 electrodes provide weld
metals with better notch toughness than provided by the older

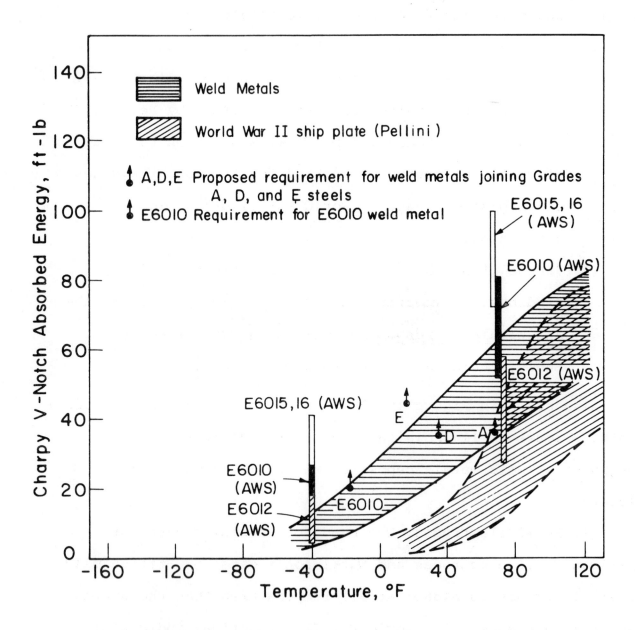

Figure 15-11 NOTCH-TOUGHNESS DATA OF WELD METALS OF MILD-STEEL
ELECTRODES

483

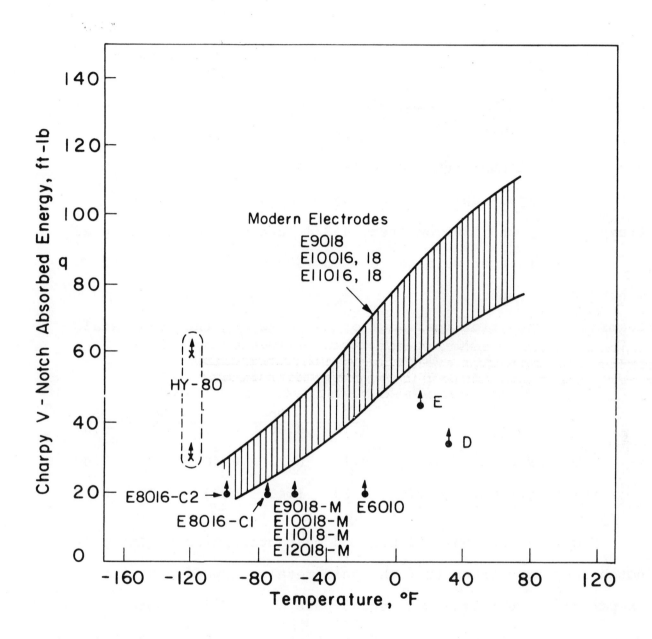

Figure 15-12   NOTCH-TOUGHNESS OF WELD METALS OF LOW-HYDROGEN
HIGH STRENGTH STEEL ELECTRODES

welds of very heavy plates, multilayer procedures are used commonly. In the fabrication of merchant-ship hulls, there has been strong interest in reducing the number of passes, primarily for economic reasons.

Augland and Christensen[1] investigated the notch toughness of submerged-arc weld metals made with two-pass and multilayer procedures in 1-9/16-inch-thick-carbon-steel plates. Welding conditions were:

(1) Two-pass welding: 1000 amperes, 4-1/2 inch/minute travel speed; and 1200 amperes, 4-3/4 inch/minute travel speed.

(2) Multilayer welding: 600 to 750 amperes, and 10 inch/ minute travel speed.

Chemical composition (percent) of the base and the weld metals was:

	C	Mn	Si	P	S
Base metal	0.15	0.66	0.19	0.02	0.03
Two-pass weld metal	0.10	0.73	0.46	0.04	0.02
Multilayer weld metal	0.09	1.43	0.33	0.05	0.02

Figure 15-13 shows transition characteristics of Charpy V-notch specimens taken from the weld metals. Specimens were prepared in the first- and second-pass weld metal. Notches were made in the thickness direction and parallel to the plate surface. The location and orientation of the specimen had little effect on the transition characteristics; therefore, Figure 15-13 shows bands only for two-pass and multilayer welding. The notch toughness of the multilayer-weld metal was excellent for mild steel, passing Grade D requirements. The notch tough-

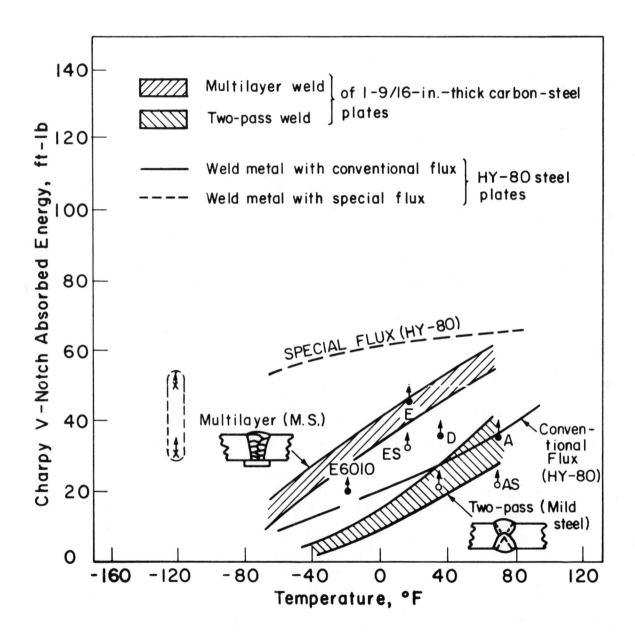

Figure 15-13 NOTCH-TOUGHNESS OF SUBMERGED-ARC DEPOSITED

METALS IN MILD STEEL AND HIGH STRENGTH STEELS

ness of the two-pass weld metal was poor; the weld did not meet Grade A requirements.

High-strength Notch Tough Steel. Many research programs have been and are being conducted for developing submerged-arc welding processes which provide weld metals with high strength and good notch toughness.

A research program was conducted for the Bureau of Ships at Battelle Memorial Institute[10,11] for developing fluxes and filler wires for submerged-arc welding of HY-80 steel. Figure 4 shows the Charpy V-notch transition curves of (a) the weld metal which had the best notch toughness and (b) a weld metal made with conventional wire and flux. The welds were made in 1/2 inch thick HY-80 steel plate by the multilayer technique with heat inputs of 45,000 joules per inch of weld bead. The improvement of the experimental weld metal was over 90,000 psi.

The oxygen content of the experimental weld metal was less than one half that of the conventional weld metal. Microscopic investigations revealed that the experimental weld metal was significantly cleaner (with fewer inclusions) than the conventional weld metal. The investigators believed that the improvement in notch toughness was a result of the lower oxygen content and fewer inclusions in the experimental weld metal.

Gas Metal-Arc Deposited Metals

Figure 15-14 shows two sets of data as follows:

(1) Notch toughness data obtained by Sekiguchi, et al.[26] on weld metals using low-carbon steel wires shielded by 100 percent CO_2 gas. The base plate was low-carbon

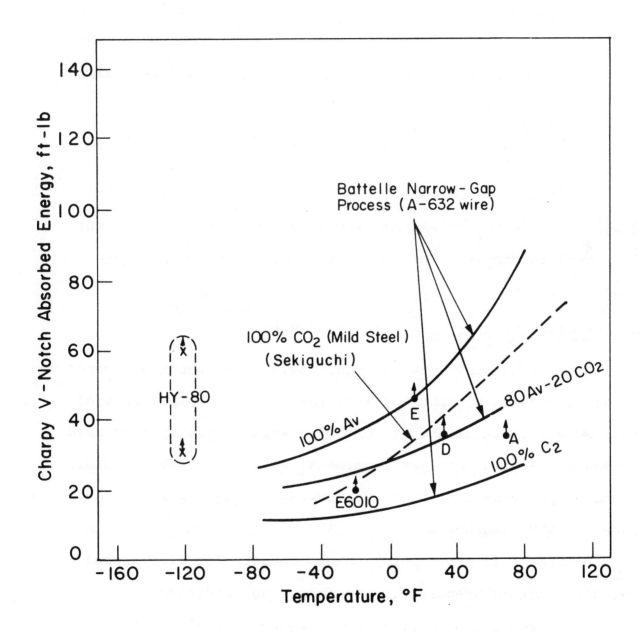

Figure 15-14   NOTCH-TOUGHNESS OF GAS METAL-ARC DEPOSITED

WELD METALS

steel.

(2) Notch toughness of weld metals made with the Battelle
Narrow-gap process using A-632 filler wire and shielding
gas mixtures of CO_2 and argon (100, 20, and 0 percent
argon).[17,22] The base plate was HY-80 steel. The
chemical composition (in percent) of the filler wire was:

C	Si	Mn	Ni	Cr	Mo	V	Zr	Al
0.04	0.57	1.36	1.22	0.13	0.45	0.15	0.005	0.013

The figure shows that the notch toughness of weld metal
deposited with CO_2 shielding is fairly good but not excellent.
The notch toughness of weld metals improves as the percentage of
argon increases.

Electroslag and Electrogas-Deposited Metals

Notch toughness data for electroslag and electrogas deposited
metals appear in a number of articles. Direct comparisons of
test data are different, since most Russian investigators used
2-mm-deep U-notch Mesnager specimens, while Charpy V-notch or
keyhole specimens were used by American, Western European and
Japanese investigators.

Figure 15-15 shows notch toughness data for electroslag
deposited metals prepared in the following conditions:

(1) Bands for data obtained by Burden, et al.[2] in
5/8-inch-thick carbon-steel welds:

(a) As-welded

(b) Stress relieved for 1 hour at 1200 F.

(c) Normalized from 1740 F., then stress relieved
for 1 hour at 1200 F.

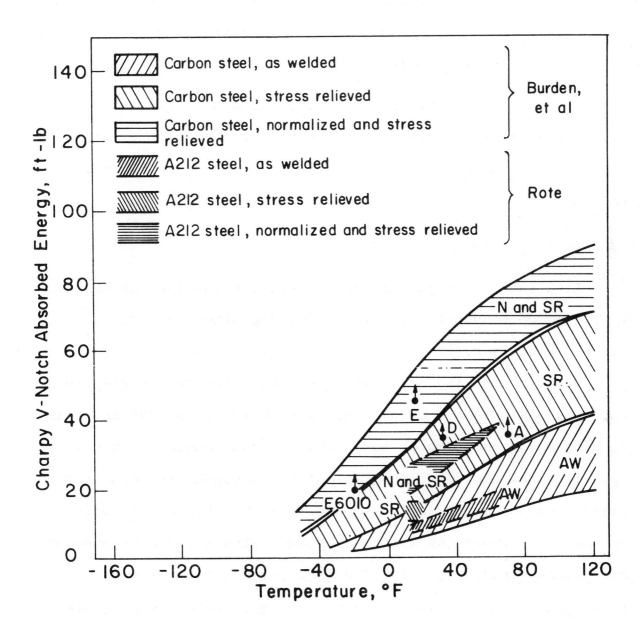

Figure 15-15   IMPROVEMENT THROUGH HEAT TREATMENT OF NOTCH-
TOUGHNESS OF ELECTROSLAG-DEPOSITED METALS

(2) Bands for data obtained by Rote[24] in carbon-steel
    A212 welds 3-3/4 inches thick:

(a) As-welded

(b) Stress relieved at 1150 F. and then furnace cooled

(c) Normalized from 1675 F., stress relieved at 1225 F.,
    and air cooled.

As shown in the figure, the notch toughness of electro-slag-deposited weld metal in the as-welded condition is very poor, but it can be improved by stress relieving and normalizing treatments.

Summary

As a summary, Figure 15-16 illustrates the general trends in the notch toughness of weld metals deposited by various welding processes.[13]

Here, notch toughness level is classified arbitrarily in five classes, Class 0 being the lowest notch toughness and Class 5 being the highest notch toughness. Notch toughness is not specified for Class 0. The typical notch toughness value for Class 1 is 35 ft-lb V-notch impact of 68°F; this corresponds to the requirement for weld metals made with covered electrodes for joining Grade A steel. The requirements for Class 2 correspond to requirements for Grade D and E6010. The requirements for Class 3 correspond to Grades E and E9018-M. Values for Classes 4 and 5 correspond to E8016-C2 and HY-80 base plate, respectively.

This figure is prepared to provide general guidance for

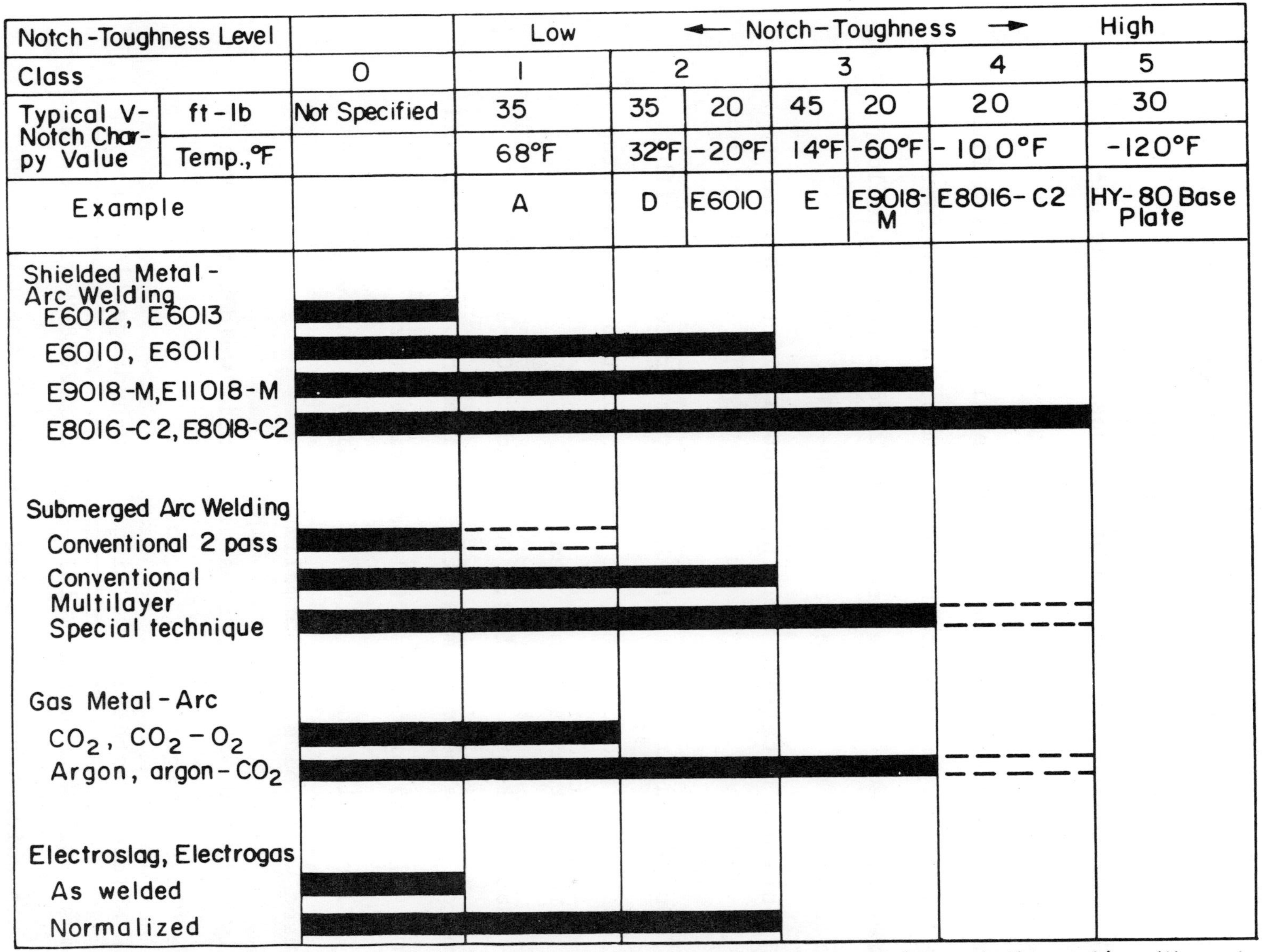

Notch-Toughness Level	Low			← Notch-Toughness →				High
Class	0	1	2		3		4	5
Typical V-Notch Charpy Value — ft-lb	Not Specified	35	35	20	45	20	20	30
Temp., °F		68°F	32°F	-20°F	14°F	-60°F	-100°F	-120°F
Example		A	D	E6010	E	E9018-M	E8016-C2	HY-80 Base Plate

Shielded Metal-Arc Welding
E6012, E6013
E6010, E6011
E9018-M, E11018-M
E8016-C2, E8018-C2

Submerged Arc Welding
Conventional 2 pass
Conventional Multilayer
Special technique

Gas Metal-Arc
CO_2, CO_2-O_2
Argon, argon-CO_2

Electroslag, Electrogas
As welded
Normalized

Note: The areas enclosed by dotted lines indicate that there is some uncertainty whether welds will meet the requirements.

Figure 15-16 GENERAL TRENDS IN THE NOTCH-TOUGHNESS OF WELD METALS DEPOSITED BY VARIOUS WELDING PROCESSES

selecting welding processes when the weld metal needs to meet a certain notch toughness level.

Weld metals which meet the Class 1 requirement can be obtained with all welding processes listed except:

    (1) When using E6012 and E6013 electrodes and

    (2) Electroslag and electrogas processes (in the as-welded condition).

Two-pass submerged-arc welding with conventional wire and flux in heavy plates may not meet the requirement unless proper procedures are used. To meet the Class 2 requirement, welding procedures should be selected as follows:

    (1) Shielded metal-arc welding: E6010 electrodes can be used.

    (2) Submerged-arc welding: Multilayer procedures with conventional wire and flux may be used. For two-pass welding, special wire and flux must be used.

    (3) Gas metal-arc welding: CO_2- and CO_2O_2-shielded processes do not appear to meet the requirement, but argon and argon-CO_2 shielded processes will be satisfactory.

    (4) Electroslag and electrogas welding: These processes may be used when welds are normalized.

Only selected welding processes deposit weld metals which meet the Class 3 requirement, as follows:

    (1) Shielded metal-arc welding: Low-hydrogen low-alloy electrodes E9018-M, E10018, E11018-M, and E12018-M will deposit weld metals which can meet the requirement in the as-welded condition.

(2) Submerged-arc welding: Special techniques must be used.

(3) Gas metal-arc welding: Argon and argon-CO_2 (with high argon content) shielded processes must be used.

Weld metals which meet the Class 4 requirement can be deposited by a limited number of processes, as follows:

(1) Shielded metal-arc welding: Low-hydrogen nickel-steel electrodes E8016-C2 and E8018-C2 will deposit weld metals which meet the Class 4 requirements in the stress-relieved condition.

(2) Gas metal-arc welding: Argon and argon-CO_2 (with high argon content) - shielded processes appear to be satisfactory.

(3) Submerged-arc welding: With the use of special techniques, submerged-arc welding may be satisfactory.

No welding process can be guaranteed to deposit weld metals as tough as HY-80 steel base metal. The most promising welding process in this respect is the inert-gas metal-arc process. The inert-gas tungsten-arc process also deposits weld metals with a high notch toughness; however, the low deposition rates involved make this process unattractive for the fabrication of commercial structures.

References

(1) B. Augland and N. Christensen, "Some Observations on the Notch Toughness of Submerged Arc Weld Deposits," British Welding Journal, Vol. 8, No. 10, 1961, pp. 473-476.

(2) Burden, C. A., Garstone, J., and Lucey, J. A., "Electro-Slag Welding of Relatively Thin Plate," British Welding Journal, Vol. 11, No. 4, 1964, pp. 148-155.

(3) Current Welding Processes, American Welding Society, New York, 1964.

(4) Delong, W. T., "The New 1964 AWS-ASTM Specifications for Mild Steel and Low-Alloy Steel Arc Welding Electrodes," Welding Journal, pp. 87-99, February 1965.

(5) "Final Report of the Board to Investigate the Design and Methods of Construction of Welded Steel Merchant Vessels," July 15, 1946. Government Printing Office, Washington, D.C. 1947, also printed in part in The Welding Journal, Vol. 26, No. 7, pp. 569-619, 1947.

(6) Jackson, C. E., "The Phenomenal Growth of Welding Technology," The Welding Journal, Vol. 42, No. 3, Page 216, 1963.

(7) Kaku, S., "Unified Requirements for Hull Structural Steels and Some Problems on Their Welding Procedures," Journal of the Japan Welding Society, Vol. 33, No. 4, 1964, pp. 11-16.

(8) King, J. R., "Electron Beam Welding: What It Can--and Can't--Do," Materials Engineering, pp. 66-68, July 1968.

(9) Kockn, B., "CO_2 Welding Is Today A Qualified Process Within Shipbuilding," Welding Journal, pp. 1005-1015, December 1965.

(10) Lewis, W. J., et al. "Submerged-Arc Welding HY-80 Steel," The Welding Journal, Vol. 39, No. 6, Research Supplement, 1960, pp. 266s-272s.

(11) Lewis, W. J., Faulkner, G. E., and Rieppel, P. J., "Flux and Filler Wire Developments for Submerged-Arc Welding of HY-80 Steel," The Welding Journal, Vol. 40, No. 8, Research Supplement, 1961, pp. 337s-345s.

(12) Martin, D. C., "Advanced Metals Joining Technology," a publication of of the Naval Research Laboratory, pp. 175-195.

(13) Masubuchi, K., "Welding Problems in Shipbuilding," Marine Technology, Vol. 6, No. 1, 1969, pp. 66-75.

(14) Masubuchi, K., Monroe, R. E., and Martin, D. C., "Interpretive Report on Weld-Metal Toughness," Welding Research Council Bulletin No. 111, Welding Research Council, January 1966, and Document No. X-514-69, Commission X of International Institute of Welding (1969).

(15) Masubuchi, K., unpublished manuscript.

(16) Meier, J. W., "Partial Vacuum...the Key to Production Electron Beam Welding," Metal Progress, pp. 63-64, July 1966.

(17) Meister, R. P., and Martin, D. C., "Narrow-Gap Welding Process," British Welding Journal, Vol. 13, No. 5, 1966, pp. 252-257.

(18) Metals Handbook, American Society for Metals, Metals Park, Ohio.

(19) "Method of Testing and Approval of Electrodes for Welding Mild and Low Alloy High Tensile Steel," Welding in the World," Vol. 1, No. 1, 1963, pp. 2-17

(20) Mills, E. G., Atterbury, T. J., Cassidy, L. M., Eiber, R. J., Duffy, A. R., Imgram, A. G., and Masubuchi, K., "Design, Performance, Fabrication, and Material Considerations for High-Pressure Vessels," RSIC-17, Redstone Scientific Information Center, Redstone Arsenal, Alabama (March, 1964).

(21) Mishler, H. M., and Randall, M. D., "Underwater Joining and Cutting," Battelle Research Outlook, Vol. 1, No. 1, pp. 17-22 (1969).

(22) Nelson, J. W., Randall, M. D., and Martin, D. C., "Development of Methods of Making Narrow Welds in Thick Steel Plates by Automatic Arc Welding Processes," Final Report to Bureau of Ships, Department of the Navy, on Contract NObs-86424, Battelle Memorial Institute, March, 1964.

(23) Pellini, W. S., "Notch Ductility of Weld Metal," The Welding Journal, Vol. 35, No. 5, Research Supplement, 1956, pp. 217s-233s.

(24) Rote, R. S., "Investigation of the Properties of Electroslag Welds," The Welding Journal, Vol. 43, No. 5, 1964, pp. 421-426.

(25) Sagan, S. S., and Campbell, H. C., "Factors Which Affect Low-Alloy Weld-Metal Notch-Toughness," Welding Research Council Bulletin, Series No. 59, April 1960.

(26) Sekiguchi, H., et al. "Impact Properties of Weld Metals Deposited With the CO_2-O_2 Gas Shielded Arc Welding Process-Part I," Journal of the Japan Welding Society, Vol. 32, No. 9, 1963, p. 732, abstract of a paper presented at the National Fall Meeting, 1963.

(27) Smith, D. C., "Development, Properties and Usability of Low Hydrogen Electrodes," The Welding Journal, Vol. 38, No. 9, Research Supplement, 1959, pp. 377s-392s.

(28) "Tentative Methods for Testing Weld Metal Deposited by Submerged-Arc Welding (Mild Steel)," Welding in the World, Vol 1, No. 1, 1963, pp. 28-37.

(29) Wah, T., Editor, "A Guide for the Analysis of Ship Structures," PB 181 168, U. S. Department of Commerce Office of Technical Services, 1960.

(30) Watkinson, F., "Notch Ductility of Mild-Steel Weld Metal," British Welding Journal, Vol. 6, No. 4, 1959, pp. 162-174.

(31) Welding Handbook, Sections 1 through 5, American Welding Society, New York.

APPENDIX

A CONDENSED REPORT
ON
UNDERWATER CUTTING AND
WELDING STATE OF THE ART

J. J. Vagi, H. W. Mishler, and M. D. Randall

Battelle Memorial Institute
Columbus Laboratories
Columbus, Ohio

This appendix was condensed at M.I.T. from a report prepared
by Battelle Memorial Institute. We are grateful to Captain
W. F. Searle of the Naval Ship Systems Command for allowing
us to prepare this condensed version.

SECTION 1. UNDERWATER CUTTING

Underwater Cutting Processes

Since there are numerous processes available for underwater cutting, consideration of the factors which influence selection of the proper process for a specific application is important. These factors were summarized by Fey[7] and Rodman[29] and are given in Table A-1. Advantages of some of these processes are reviewed in Table A-2. It is important to recognize that probably all fuel gases are dangerous deck cargos. Explosives used in explosive cutting must be handled carefully with trained and experienced personnel. Danger from electric shock exists with all of the arc welding and cutting processes. Fey[8] states that gas cutting processes are sometimes specified when there is a possibility of damage from electrolysis if electric cutting were used. Probably consideration also should be given to the possibility of danger from stray electrical currents to equipment or personnel in areas remote from the cutting operation.

Oxygen Cutting

Fundamentally, the underwater oxygen cutting processes are the same as those used above water,[9,10,11,30] and are based on the ability of high-purity oxygen to react rapidly with iron that has been heated to the kindling temperature. When cutting with these processes, the steel is preheated to the kindling temperature and then a high-velocity stream of high-purity oxygen is directed at the preheated metal to commence cutting. The cutting action depends on chemical reactions between iron (Fe) and oxygen (O, O_2). Three separate chemical reactions take place during cutting and each of these releases large amounts of heat necessary to maintain the cutting action. The chemical reactions and heat liberated during the reactions are shown by the following chemical equations:[35]

$$Fe + O \rightarrow FeO + heat \text{ (63,800 calories)} \quad \text{(first reaction)}$$

$$3 Fe + 2O_2 \rightarrow Fe_3O_4 + heat \text{ (267,800 calories)} \quad \text{(second reaction)}$$

$$2 Fe + \frac{3}{2} O_2 \rightarrow Fe_2O_3 + heat \text{ (196,800 calories)} \quad \text{(final reaction)}$$

Only the first and second reactions occur when cutting thin plate; the final reaction takes place only when cutting heavier plate. The heat generated during these reactions melts some of the iron adjacent to the cut. This molten iron is blown away by the oxygen stream which also contains some previously formed iron oxide. These new iron layers also become oxidized and liberate heat. Since these oxidation processes are not instantaneous, heat developed at the upper level of the kerf is liberated at the lower level.

TABLE A-1.   FACTORS IN THE SELECTION OF UNDERWATER
CUTTING PROCESSES

1. Nature of the work (determined by thorough and complete inspection)

2. Depth of work

3. Equipment availability

4. Cutting electrode availability (ceramic rod, because of light weight, may be preferable if supply transport is a problem)

5. Oxygen availability

6. Fuel availability

7. Cutting rate

8. Type of material to be cut

9. Material condition

10. Visibility

11. Other factors which may affect the work

12. Inherent properties of the cutting processes

13. Skill and experience of the diver and his crew

14. Susceptibility of parts to electrolysis

15. Effect of electric current on parts or personnel

TABLE A-2. ADVANTAGES OF VARIOUS CUTTING
PROCESSES[2,8,29]

Oxygen-Arc Process, Tubular Steel Cutting Electrodes

Preheating is not required
Flame adjustments are unnecessary
Applicable to all metal thicknesses
Overlapped plates can be cut
Holidays (skips) can be cut
Only one gas (oxygen) is needed
Torches are lightweight
Less training and skill are required
Higher cutting rates on thin metal
Explosive fuels are not required

Oxygen-Arc Process, Ceramic Cutting Electrodes

Low burnoff rate, long life
Short length provides easier access in confined spaces
Light weight improves transportability
Fuel gases are not required

Shielded Metal-Arc Process

Cuts ferrous and nonferrrous metals
Fuel gases are not required
Standard electrode holders can be used in an emergency if properly
  adapted
Oxygen is not required

Oxy-Hydrogen Process

Electricity is not required for cutting
Nonmetallic materials can be severed
Insulated diving equipment is unnecessary
Power generators are not required
Smaller boats can be used
There is no need to change ground connections
Higher cutting rates on thick metal

Oxyacetylene

High-flame temperature
Electricity is not required for cutting
Insulated diving equipment is unnecessary
Power generators are not required
Nonmetallic materials can be severes

TABLE A-2. (CONTINUED)

Plasma-Arc

High cutting rates in air
Fuel gases are not required
Cuts ferrous and nonferrous materials
Oxygen is not required

Explosives

Multiple cuts can be made simultaneously
Cutting rates are high
Fuel gases are not required
Oxygen is not required

Consequently, this heat is no longer available at the uppermost level where the oxidation reaction is just beginning. Thus, it is necessary to continually preheat the uppermost level of the material being cut and this is the reason all oxygen cutting torches have continuously burning preheat flames. In actual practice, not all of the iron is oxidized because it is blown away. As much as 30 percent unoxidized metal has been reported.[35] In oxygen-arc cutting, the arc takes the place of oxy-fuel gas preheat flames. Alloying elements present in small amounts in steel do not markedly interfere with cutting. When present in large amounts, alloying elements such as chromium may interfere markedly with the cutting process. With such materials, other cutting processes are found more suitable.

Oxy-Fuel Gas Cutting

The state-of-the-art survey showed that very little, if any, underwater cutting is performed in the United States, Britain, Russia, or Japan. However, there is no source for the actual extent of the use of these processes for underwater work. In offshore welding and cutting operations in the Gulf of Mexico, welding was estimated at 5 percent, oxygen arc cutting at 95 percent. Among those interviewed during the field survey, there were no reports of oxy-fuel gas cutting or knowledge of others performing these type operations. The extensive use of oxygen-arc cutting was attributed to the lower skill and less training required and faster cutting rates obtainable with the process. For this reason, very little new information has become available on oxy-fuel gas cutting processes. Nevertheless, oxy-fuel gas cutting processes for underwater use can be very important in salvage, emergency, rescue, and other special applications. However, the main reasons for the use of these processes are due to the absence of generators or other equipment for electric arc cutting. The oxy-fuel gas processes also may be required and specified when electric currents are not desirable because of the possibility of electrolysis, spark formation, or danger from electric shock.

Materials Required

Materials for cutting underwater with the oxy-fuel gas processes are the same as those used above water. The basic materials required are pressurized fuel gases, oxygen and air. Only one fuel gas, hydrogen, is generally used for underwater cutting. Acetylene has been used, but because of its instability at pressures over about 15 psi, this fuel is not used at depths greater than about 25 feet. MAPP gas (stabilized methyl-acetylene propadiene) introduced in 1964 has been used to a limited extent for underwater cutting. Other fuel gases such as propane and natural gas also have been used, but for various reasons, these gases are not generally considered effective for underwater cutting. These cutting fuels are listed in Table A-3. Comparison of some of the properties of some of these gases are given in Table A-4.

TABLE A-3.   FUELS USED FOR UNDERWATER OXY-FUEL GAS CUTTING

Fuel	Chemical Formula	Reference
Acetylene	C_2H_2	32,36
Benzol-Alcohol	–	32
Ethane	C_2H_6	32
Ethylene	C_2H_4	32
Gasoline	–	
Hydrogen	H_2	32,36
MAPP Gas	–	–
Methane	CH_4	32
Naptha	–	28
Natural Gas	–	–
Propane	C_3H_8	32,36

TABLE A-4. COMPARISON OF PROPERTIES OF FUELS FOR UNDERWATER CUTTING [5,6]

	Acetylene	Hydrogen	Stabilized Methyl-acetylene Propadiene	Natural Gas	Propane
Safety					
Shock sensitivity	Unstable	Stable	Stable	Stable	Stable
Explosive limits in oxygen, %	3.0-93	–	2.5-60	5.0-59	2.4-57
Explosive limits in air, %	2.5-80	4.0-75	3.4-10.8	5.3-14	2.3-9.5
Max. allowable regylator pressure, psi	15	–	(225 psi at 130 F)	Line	Cyl-inder
Burning velocity in oxygen, ft/sec	9.8	21.7	7.9	8.2	5.5
Tendency to backfire	Considerable	–	Slight	Slight	Slight
Toxicity	Low	–	Low	Low	Low
Reactions with common materials	Avoid alloys with more than 67% Cu	–	Avoid alloys with more than 67% Cu	Few Restrictions	
Physical Properties					
Specific gravity of liquid (60/60 F)	–	–	0.576	–	0.507
Pounds/gal liquid, 60 F	–	–	4.80	–	4.28
Cu ft/lb of gas, 60 F	14.6	–	8.85	23.6	8.66
Specific gravity of gas (air=1) at 60 F	0.906	0.1	1.48	0.554	1.52
Vapor Pressure, 70 F, psi	–	–	94	–	120
Boiling range, F, 760 mm Hg	-84	-423	-36 to -4	-161.5	-50
Flame temperature in oxygen, F	5,589	5090	5,301	4,600	4,579

Although hydrogen and acetylene are the most commonly used underwater cutting fuels, other gases are included for comparison.

When gases are subjected to sufficiently high pressures as would be the case when cutting at great depths, the gases may be converted to liquids. However, there is a temperature above which it is impossible to liquefy the gas regardless of how great a pressure is applied. This temperature is the critical temperature and the minimum pressure required to bring about liquefaction is the critical pressure. Critical temperatures and pressures of gases that are commonly encountered in oxy-fuel gas cutting are given in Table A-5.

Hydrogen

Oxy-hydrogen cutting has been used to depths of about 130 feet.[29,33] Because of its low density, large volumes are consumed in the course of underwater cutting operations. To supply a continuous demand and minimize stoppage of work for changing cylinders, manifold cylinder arrangements are recommended with regulators which provide constant flow.[11] Hydrogen can be used to the greatest depths at which oxy-fuel gas cutting has been performed to date.

Acetylene. Acetylene has been used for cutting at depths less than 25 feet. The gas is not recommended for use at depths greater than 25 feet because the acetylene pressure will approach and exceed 15 psi pressure at greater depths. Acetylene becomes unstable at pressures greater than 15 psi and may detonate.

MAPP Gas. MAPP gas is a trade name of the Dow Chemical Company for stabilized methylacetylene propadiene. The gas was introduced in 1964 and has been used for a wide variety of above water cutting applications. It has also been used for cutting underwater to shallow depths and the gas is reported to be much safer than acetylene.[39] One 70 pound cylinder does the work of 5 or more acetylene cylinders; it has been shown to be as stable to shock as water[5] and is stable at pressures up to 350 psi.[7] Physical properties are similar to those of propane.[12] Fey[5] reports that MAPP gas can take the place of acetylene for underwater cutting and can be used to depths of several hundred feet. It will cut through paint, rust, and laminated steel. Lahm[17] reports less difficulty in gas cutting overlapped plates with MAPP gas than with hydrogen. Hydrogen often causes blowback from the seond plate whereas MAPP gas cuts through both plates readily. Lahm also reports that the coupling distance with MAPP gas is not so critical as with hydrogen.

TABLE A-5.   CRITICAL CONSTANTS FOR CUTTING GASES

Name	Critical Temperature, °F °C		Critical Pressure, Atmosphere
Acetylene	+97	+36	62
Air	-122	-141	37
Ethane	+90	+32	49
Ethylene	+50	+10	51
Hydrogen	-400	-240	13
Methane	-117	-83	46
Oxygen	-182	-119	50
Propane (n-)	+207	97	42
Argon[a]	-188	-122	48
Helium[a]	-450	-268	2

[a] for comparison purposes

Other Fuels. Information on the characterisitcs of other fuels in underwater cutting is limited. Liquid fuels appear to have been used in rare instances prior to and during World War II. Galerne[8] reported that oxygen-gasoline cutting was very efficient. Rhier[28] reported that oxygen-naptha cutting was also very efficient. Both fuels were considered very dangerous and there are no known uses of these materials in underwater cutting today.

Oxygen. Oxygen for cutting operations should have a purity of 99.5 percent or higher. Unless this high purity is maintained, the efficiency of cutting will be reduced. A 25 percent reduction in cutting speed results from a one percent decrease in oxygen purity. The quality of cut will be reduced and the amount and tenacity of the adhering slag will be increased. At oxygen purities of less than 95 percent, the operation becomes one of melting and washing instead of the desired cutting.

Generally, manifold oxygen cylinders are desirable to minimize cylinder change interruptions. However, a single cylinder is adequate if the cutting operation is limited.

Air. Air in underwater cutting operations is used to form an air bubble within which cutting is initiated and performed. Consequently, compressed air facilities are required to supply air at sufficiently high pressure to overcome the increased pressure at cutting depths. With at least one commercial torch, oxygen may be substituted for air with equally good results if the air supply is inadequate.[11] The "Seafire[a]" and "Vixen[b]" torches in wide use in Great Britain are designed specifically for use with a shield of oxygen instead of air;[31] thus, this torch eliminates the need for air hoses and air supply equipment.

Materials Cut

Practically all of the materials of ship construction have been cut with the oxy-fuel gas, oxygen-arc and other cutting processes. Some processes are more capable of cutting certain types of materials than others. All of the underwater cutting processes are capable of cutting steel but the oxy-fuel-gas and oxygen-arc processes are most widely used for severing this material. Cast iron, wrought iron, aluminum, brass, bronze and other non-ferrous and oxidation resistant materials are, at best, difficult to cut by these processes.

Carbon steels can be cut readily by the oxy-hydrogen or oxygen-arc methods. Stainless steel is more difficult to cut by these methods. Steels containing Cr and Ni are not readily oxidized and retard cutting. Graphite in cast iron has the same effect. Brass is not oxidized by the oxygen jet of a conventional torch.[29] For nonferrous metals, such as brass, copper, bronze or aluminum, other processes may be preferable. Explosives have been used to cut steel (1/4-inch) and light alloys (3/4-inch) to depths up to 180 feet.[38]

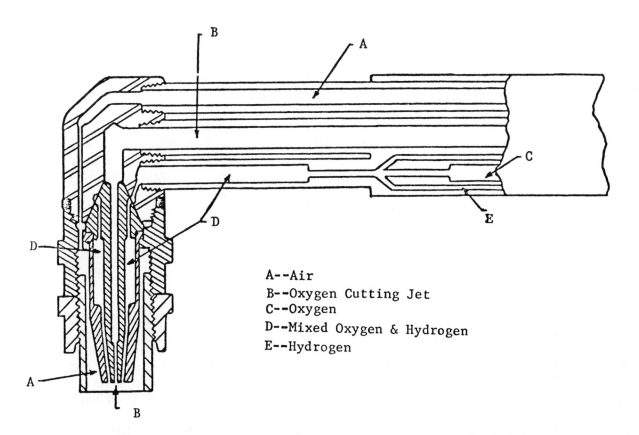

A--Air
B--Oxygen Cutting Jet
C--Oxygen
D--Mixed Oxygen & Hydrogen
E--Hydrogen

a. Gas Passages

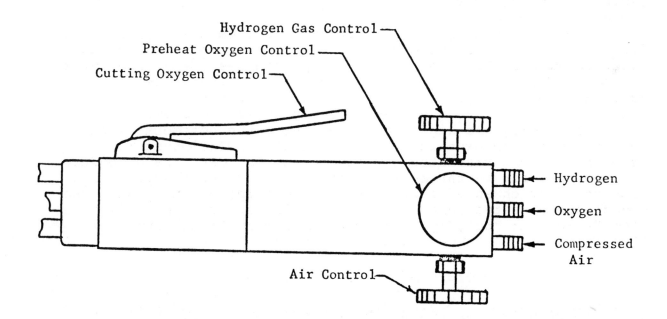

b. Gas Controls

FIGURE A-1. UNDERWATER GAS CUTTING TORCH, GAS PASSAGES
            AND GAS CONTROLS

In many instances it becomes necessary to cut away old sheet piling with marine growth and rust. Before attempting any under- water cutting on old or encrusted steel, it is necessary first to remove growth, scale, and surface rust. Light growth and scale can be removed with a hand scraper and chipping hammer; heavy growth and hard scale is best removed with pneumatic tools. Even after one surface has been cleaned, a heavy rust scale may be present on the other side of the steel plate. Such scale slows the cutting process and consumes additional oxygen and rods. This condition can sometimes be remedied by striking the area to be cut with a heavy sledge hammer to loosen the scale on the opposite side. New steel is more easily cut, as it normally requires no cleaning.

Cutting Techniques

Torch cutting can be started at the edge of a plate or in the central portion of the plate.[2] The torch operator usually keeps one hand well forward on the torch, the other hand on the high- pressure oxygen valve. When starting in either position, the torch head is directed at a local area so that the preheat flames heat a spot to the kindling temperature. Better results usually are obtained when the cutting oxygen stream is turned on gradually instead of turning it on rapidly. When starting at the central portion of a plate, the torch head is raised slightly while the cutting oxygen stream is gradually increased to avoid blowing sparks or slag back into the tip. This technique is called the "blowing through" technique.

When advancing the cut, the air shield is held against the plate being cut and the torch is slid on this shield in contact with the steel being cut as the cut progresses.[10,11] The torch is moved along the line of cut at a uniform speed, fast enough to continue the cutting action but slow enough to cut completely through the metal.[1] When the torch speed is too slow, the material will cool below the kindling temperature, the hot metal will be quenched, and the cutting action will stop. The operation then will have to be restarted. When the cut is lost, the procedure recommended for restarting is to start a new cut alongside and 1/2 to 3/4-inch ahead of the end of the original cut. When the torch speed is too rapid, the plate will not be completely cut. The "skips" or "holidays" are very difficult to cut out.

Oxygen-Arc and Arc Cutting

In principle of operation, oxygen-arc cutting is similar to oxy-fuel gas cutting. The major difference is that preheating is accomplished with an electric arc instead of a preheat oxy-fuel gas flame. In operation the arc is established between a hollow tubular electrode and the metal being cut. The cutting oxygen jet is delivered through the electrode bore to furnish additional heat (from the chemical reaction of oxidation) and to erode the molten metal away by the jetting of the high-speed oxygen stream. Cutting is performed directly in the water and no air shield is used.

Oxygen-arc cutting is classified by the American Welding Society (AWS) as an oxygen cutting process because the actual cutting is performed by the oxygen stream. The AWS defines the process as follows:

> Oxygen-arc cutting is an oxygen cutting process wherein the severing of metals is affected by means of the chemical reaction of oxygen with the base metal at elevated temperatures, the necessary temperature being maintained by means of an arc between an electrode and the base metal.

The oxygen-arc process also is referred to as the arc-oxygen process. In this report, the AWS terminology and definitions will be used.

Applications and Materials Cut

The oxygen-arc cutting process has been used extensively by the naval services in connection with salvage and other operations for many years. Since the development of offshore oil fields, the process also has found extensive use among commercial organization in repairs and maintenance of offshore oil rigs.

Underwater oxygen-arc cutting is especially satisfactory for cutting of steel in thickneeses from sheet gages to about 3 inches. Cast iron and nonferrous metals are not readily cut by this method because the materials are not readily oxidized. The main effect of the oxygen stream in cutting these materials is in the mechanical effect of blowing the molten metal out of the kerf. Ronay[22] claimed that the process was applied to cast iron, stainless steel, and nonferrous metals. When cutting oxidation resistant materials, the operator is required to constantly manipulate the oxygen stream to mechanically blow metal out of the kerf. In addition, the operator is required to constantly manipulate the electrode in and out of the kerf to push out the molten metal that was not removed by the oxygen stream. Ronay further states that air can be used in this manner instead of oxygen for underwater cutting of brass or bronze.

Typical applications for the underwater oxygen-arc cutting process include the severing of propeller shafts, cutting sheet piling to desired levels, severing of offshore oil rig supports, and other underwater cutting jobs.

Cutting Techniques

Techniques have been developed for use with each of the arc-type cutting processes. In general, the oxygen-arc cutting

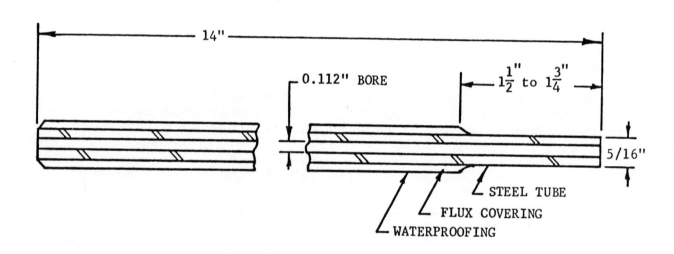

FIGURE A-2. TYPICAL STEEL TUBULAR ELECTRODE FOR UNDERWATER
OXYGEN-ARC CUTTING

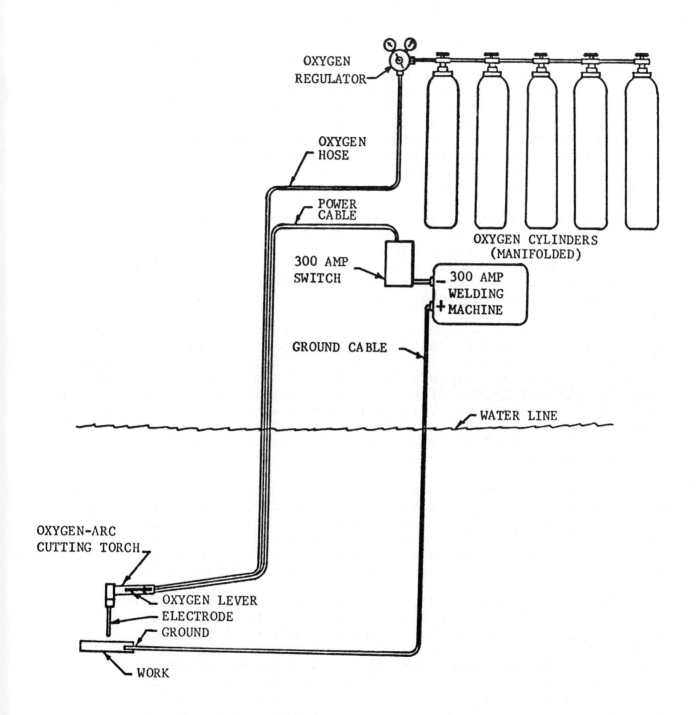

FIGURE A-3. TYPICAL ARRANGEMENT OF UNDERWATER OXYGEN-ARC

CUTTING EQUIPMENT

process is considered to require less skill and training than either the oxy-fuel gas or shielded metal-arc cutting processes.

Oxygen-arc Cutting. In oxygen-arc cutting with tubular steel electrodes, the cut is strated by placing the electrode in contact with the work, starting the oxygen flow, and switching the current on to start the arc. With the steel tubular electrode, cutting is performed with the steel core burning away slightly faster than the covering. Immediately following arc initiation, therefore, the electrode is withdrawn momentarily for a short distance to permit the core to burn back under the covering. After the cutting action progresses through the plate thickness, the torch is advanced along the line of cut.

When cutting steel plate thicker than 1/4 inch, the electrode is dragged along the line of cut with the tip of the electrode slightly in the kerf and pressed against the advancing lip of the cut. The electrode should always be positioned perpendicular to the plate surface. This technique is called the "drag" technique and is illustrated in Figure A-4.[30]

When cutting 1/4 inch or thinner plate the electrode tips should be maintained in contact with the plate surface under light pressure and not pushed into the kerf against the advancing lip of the cut. Alternately, the electrode can be inclined 45 degrees pointing in the direction of cut so the effective thickness being cut is increased. This permits the application of greater pressure to the electrode. Ceramic tubular electrodes also are used with the tip of the electrode maintained in light contact with the work surface.

For piercing holes in steel plate with the oxygen-arc process, the technique for starting the cut is the same as when starting the cut at the edge of a plate. After the steel tube core has melted back inside the covering, the electrode is slowly pushed into the hole until the hole is formed completely through the plate. The hole piercing technique is illustrated in Figure A-5.[30]

When cutting cast iron, nome nonferrous metals and other oxidation resistant materials, the only useful purpose served by the oxygen stream is the mechanical action of blowing molten metal out of the kerf. Since these materials resist oxidation, there is no chemical reaction between the oxygen stream and the metal being cut. Therefore, only the arc produces the heat required to melt the base metal. Only a very localized area in the immediate vicinity of the arc is melted. The operator, there-fore, must manipulate the tip of the electrode and the oxygen stream to melt and remove material from the kerf. A short sawing stroke is recommended.

Shielded metal-arc cutting is superior to oxygen-arc cutting of cast iron and nonferrous materials. The technique used for shield metal-arc cutting is the same as the oxygen-arc technique for cutting cast iron and nonferrous metals. Metal is removed

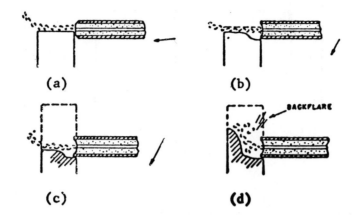

FIGURE A-4. DRAG TECHNIQUE OF UNDERWATER OXYGEN-ARC CUTTING

OF STEEL WITH STEEL TUBULAR ELECTRODE

(a)   Electrode pressed lightly in direction shown
      by arrow, oxygen valve opened and current
      switched on.

(b)   Electrode held stationary, withdrawn momentarily
      if necessary to start the arc. When cut is
      started for full thickness of plate direction
      of pressure is changed to direction shown by
      arrow and pressure is increased.

(c)   Light, firm pressure continued as cutting pro-
      ceeds. Electrode tip is maintained against
      lip of advancing cut.

(d)   Incomplete cut resulting from faulty manipula-
      tion causes backflare. Advance is halted and
      incomplete cut is completed.

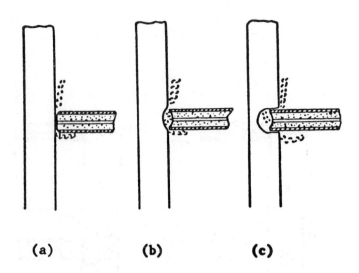

FIGURE A-5.   TECHNIQUE FOR PIERCING HOLES IN STEEL PLATE

USING THE UNDERWATER OXYGEN-ARC CUTTING

PROCESS WITH STEEL TUBULAR ELECTRODES

   (a)   Electrode tip is held in light contact with
         the plate surface, oxygen valve opened and
         current switch on.

   (b)   Electrode is held stationary, withdrawn
         momentarily if necessary to start the arc
         and melt the steel tube back inside the
         covering.

   (c)   Electrode is pushed slowly into the hole
         until the plate is pierced.

by arc melting the material in the vicinity of the arc and using a short-stroke sawing motion to push the metal out of the kerf. This technique is illustrated in Figure A-6.[30]

Special attention is given to switching off electric current when cuts are completed with the arc processes. When the electrode has been consumed to the point where a change to a new electrode is necessary, the diver-welder signals "current off" to this tender. The electrode is maintained in cutting position until the tender switches current off with the knife switch and informs the diver. In naval service, these safety precautions are mandatory in order to help protect the diver from electric shock. The diver then proceeds with terminating the work or with changing the electrode. When the electrode collet or nut is released, the stub is ejected by the oxygen stream[22] or is removed manually and a new electrode is inserted and tightened in the holder.

Other Underwater Cutting Processes

Although the oxygen and arc-cutting processes have been used conventionally for underwater cutting, advances have been made in adapting other welding and cutting techniques for underwater cutting. Processes which have been developed recently, i.e., since World War II, and have been used, at least experimentally, for underwater cutting, include the following:

1. Explosive cutting
2. Plasma-arc cutting
3. Electron beam cutting
4. Gas metal-arc cutting
5. Chemical cutting (with chemical fuels and oxidizers)
6. Burning bar cutting
7. Laser cutting*

Few have developed to a stage where they afford a convenient manual cutting method for use by the diver. Explosive cutting is a good example of a process that has been developed in recent years and is widely used for practical underwater cutting operations although less extensively than the conventional processes. Plasma-arc cutting has also been used in practical underwater cutting operations. Gas-metal-arc cutting has been used in practical cutting operations in the U.S.S.R. but not apparently in this country. The remaining processes have been used only to a limited degree in experimental underwater cutting work. With the exception of explosive cutting, extensive work is still required to develop these processes for satisfactory underwater use. Even with explosive cutting, efforts are needed to establish standard materials, procedures, techniques, and safety precautions to fully utilize the advantages of the process. At present, it appears

---

*Laser cutting is included because of interest expressed during field interviews. There are, however, no known data on underwater cutting with lasers.

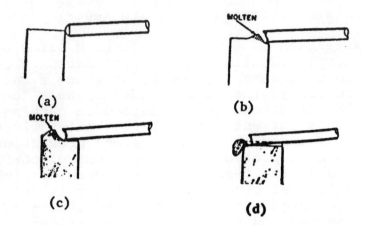

FIGURE A-6.   TECHNIQUE FOR UNDERWATER SHIELDED METAL-ARC

CUTTING OF THICK PLATE

(a)   Electrode tip is held in light contact with
       the plate edge and current is switched on.

(b)   Electrode tip is withdrawn momentarily, then
       advanced slowly across the edge of the plate.

(c)   Molten metal pushed across edge of plate.

(d)   Molten metal is pushed off edge of plate
       and electrode is returned immediately
       to the starting edge for the next stroke.

that gas-metal-arc and plasma cutting have the most to offer in possible advancing the underwater cutting art. The gas-metal-arc process in use in the U.S.S.R. appears to be further developed for underwater cutting than the plasma-arc cutting process and it has been used to much greater depths. The gas-metal-arc process also has been evaluated more extensively for underwater welding. There is, however, only meager information available regarding the use of these processes for underwater work.

SECTION 2.   WELDING IN A WET ENVIRONMENT

The term "wet welding" is used in reference to methods in which welding is performed underwater in the wet environment. The only attempt made to exclude water is by using or generating a gaseous shield in the vicinity in the electrode or torch tip. The welding torch, electrodes, materials, and the diver are completely exposed to the water. The most widely used underwater process of this type is shielded metal-arc welding. The gas-metal-arc welding process has been studied extensively and used for practical under-water wet-welding applications in the U.S.S.R., and development work is in progress in the United States. The electron-beam welding process also has been studied to a limited degree for underwater welding to establish feasibility of underwater welding with the process. Resistance welding has been used for many years for underwater welding materials like tantalum, which are sensitive to heat. However, the "wet" resistance-welding processes have always been performed on land in only a few inches of water. These processes and their past applications to underwater work are described in the following.

Shielded Metal-Arc Welding

The most widely used underwater wet-welding process is the shielded metal-arc welding process. This is the same method that is so widely used for above-water welding and commonly referred to as "covered-electrode" and "stick-electrode" welding. The equipment and techniques used for underwater welding with the shielded metal-arc process have remained virtually unchanged since developed in the 1940's. However, a program was initiated recently in the U.S. to develop welding electrodes expressly for underwater usage.

Arc-Welding Equipment

The equipment recommended for a complete underwater shielded-metal-arc welding outfit will contain the following equipment: [2,3]

1.   D-C welding power supply
2.   Cables for holder and ground clamp

3. Double pole knife switch
4. Fully insulated underwater welding electrode holder
5. Ground clamp
6. Diving outfit equipped with communications
7. Divers eyeshield
8. Divers dry rubber mittens
9. Weighted wire brush
10. Chipping hammer
11. Scraper
12. Electrodes
13. Miscellaneous hand tools.

Power supplies. Power supplies that have been used for underwater shielded metal-arc welding are d-c generators ranging in
ratings from 200 to 600 amperes. The preferred power source for
Naval service is a 300-ampere d-c generator but a 200-ampere power
supply is permissible is case of emergency when a 300-ampere unit
is not available; a 300-ampere a-c unit also is permissible providing that good safety practice with a-c power is followed. Fey[20]
recommends 300 ampere d-c generators to depths to 200 feet and 350
to 400 ampere d-c generators for greater depths. The Royal Navy
Manual specifies a d-c generator with a 70 to 75 volt open-circuit
voltage, and a maximum intermittent output of 300 to 400 amperes.
Emerson reports the case of a 600 ampere d-c generator for welding
at 600 foot depths in the Gulf of Mexico. It is generally agreed
that welding underwater requires about 25 percent more power than
welding in air. Generators in use for underwater welding may be
gasoline, diesel, or electrically driven. When greater power is
needed, two or more generators may be connected in parallel.

In studies of the use of a-c and d-c power for underwater
welding, Hipperson[13] states that straight-polarity direct current
was superior to alternating current power. To maintain arc
stability, use of a choke with sufficient inductance to promote
rapid voltage recovery after a short circuit is recommended. Owing
to poor visibility in underwater work, short-circuitry is a common
occurrence when the operator attempts to strike an arc. To protect
the welding cable, electrode holders, and connections from elec-
trolyses, welding is performed using straight polarity, direct
current. For this type of operation, the electrode is connected
to the negative terminal of the generator. These terminals are
generally marked "electrode," "negative," or "-." The ground
cable is connected to the positive terminal of the generator.
These are usually marked "ground,""positive," or "+". A convenient
way to check polarity was described earlier in the section on
power supplies for oxygen arc cutting. The Royal Navy Manual
specifies that the generator case must be grounded to the ship's
hull.

Electrode holders. Electrode holders designed specifically
for underwater welding are available commercially and from Naval
stocks for use in Naval service. Standard electrode holders with
metallic jaws and holders improvised from standard pipe fitting,
although not recommended, can be used in emergencies providing

that they are properly insulated. Most oxygen-arc cutting torches are equipped with welding electrode adaptors so that these torches also may be used for underwater welding.

Cables. The minimum recommended size of welding cable is 2/0 for use in Naval service and this size is generally regarded as minimum in commercial practice. A 10-foot length of 1/0 cable attached to the electrode holder is permitted because it has been found desirable to aid the diver in manipulating the electrode holder. The 2/0 welding cable is an extra flexible grade conforming to Navy Department Specifications. Cables are available in various lengths ranging upward from standard 50-foot lengths for Naval service and they may have connectors attached. Emerson reported the use of 600 feet of 4/0 cable and 150 feet of 2.0 cable for experimental shielded metal-arc welding at a depth of 600 feet. Voltage drops die to cable length were shown earlier in Figure A-7, and the output of the welding power supply must be adjusted to compensate for these drops. Dirty or loose connections will increase the voltage drop further. To protect working personnel from shock hazards and retard corrosion of the connection due to electrolysis, all connections should be fully insulated from the water by several wrappings of rubber or plastic tape. Mills[4] recommends that supply and ground cables should have the same current caryying capacity and these cables should be as short as possible to minimize the voltage drop. Cable connections to the ground clamp should be either bolted tightly or brazed. The ground clamp should be attached to the work as close as possible to the weld joint. As welding progresses the ground clamp should be relocated to maintain close proximity to the weld joint.[21] The same hand tools that are used above water will often be needed in underwater welding. The need for scrapers, chisels, hammers, chipping hammers, weighted wire brushes, therefore, should be considered carefully when planning the work.

Materials Used

The primary materials used for underwater welding include covered electrodes for "wet" welding and filler wires, shielding gases and cooling water for "dry" welding. Since most of the underwater welding work is performed on steel, the following section is concerned mainly with steel.

Covered Electrodes. Early electrodes for underwater welding were very simple, and were either a bare wire or a varnish-coated bare wire.[11] Covered electrodes later were used for underwater welding because of their improved performance when welding and the covered electrode is now the only one in use. No covered electrode has been designed specifically for underwater welding. The electrodes in use were selected from those that were commercially available for above-water use and adapted for underwater applications. In operation, the flux covering burns away more slowly than the core to help provide proper arc length--thereby greatly simplifying the welding operation. The remaining sheath chips away on contact with the parent metal.[1] Under these conditions, the arc operates

A-24

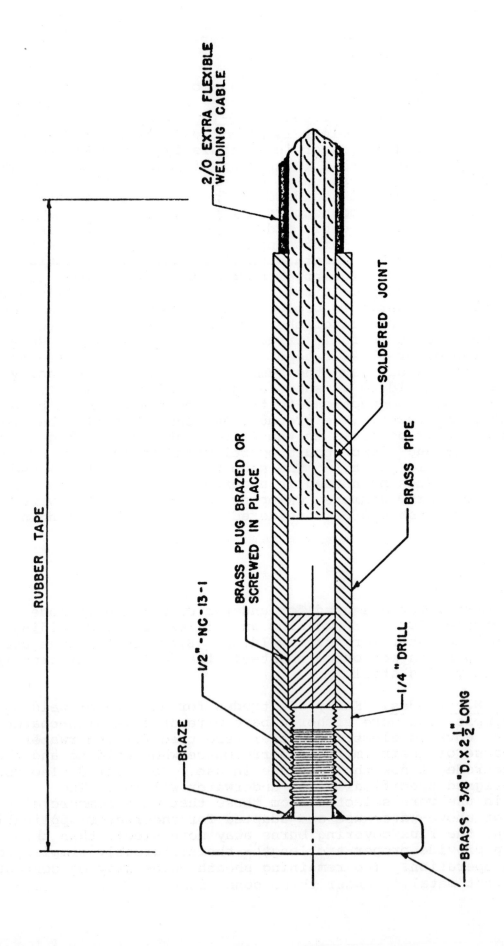

RUBBER TAPE

2/0 EXTRA FLEXIBLE WELDING CABLE

SOLDERED JOINT

BRASS PLUG BRAZED OR SCREWED IN PLACE

BRASS PIPE

1/2"-NC-13-1

1/4" DRILL

BRAZE

BRASS - 3/8"D. X 2½" LONG

FIGURE A-7. UNDERWATER WELDING ELECTRODE HOLDER BUILT FROM STANDARD SHOP ITEMS

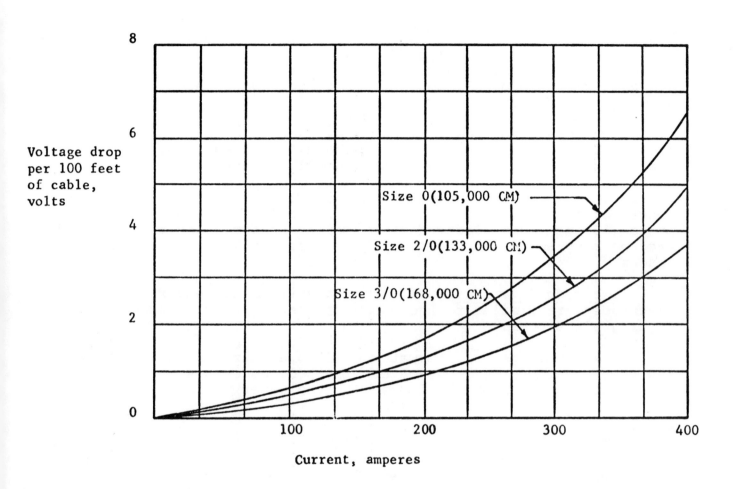

FIGURE A-8.   VOLTAGE DROP IN POWER SUPPLY CABLES

partly in a cavity protected from the surrounding water. Silva claims that the arc-core temperatures range from 5000 to 50,000 K.

In general, electrodes used for above-water welding can be used underwater.[21] AWS type E6013 electrodes have proved to be satisfactory and this type of electrode with a waterproof coating is available commercially and is the most commonly used. AWS E6010 and E6011 electrodes have been used but to a much lesser extent. Typical electrodes that are used for underwater welding are listed in Table A-6. Chemical compositions of core wire and weld deposits are given in Table A-7.

Electrode sizes that are normally used in underwater work are 1/8, 5/32, and 3/16 inch in diameter and 14 inches long. However, Madatov[19] found that longer electrodes (17.7 inches) provided improved productivity when coupled with iron powder in the flux coating. Whether the increases productivity was due to the longer length or to the iron powder was not reported. Some British firms prefer 18 inch long electrodes. Others[13] prefer a shorter electrode because they are easier to manipulate, particularly in confined spaces.

Electrode Coverings. Electrode coverings for underwater welding electrodes are similar to those used for steel tubular underwater cutting electrodes.

Flux coverings on underwater welding electrodes should perform the same functions as above-water electrode coverings in addition to satisfying requirements imposed by the water environment.

Materials Welded

The kinds of materials that have been welded underwater are far more limited than those that have been cut. Published literature and personal contact with diver-welders and ocean engineering organizations showed that only mild steel, ship hull steels, and pipeline steels have been welded underwater. There were no reports of other ferrous alloys or of nonferrous alloys being welded underwater. However, other materials of construction are bing used for the first time and the diver-welder may be faced with the need for welding these materials in the near future. For example, Lewis[18] reported that in 1965, the first aluminum pipeline for the Gulf of Mexico offshore gas system was installed. White[38] reports concern about weld repairs in materials which require special heat treatment of welds to achieve desired properties. Since there are at present no means for heat treating these materials underwater, the steels may be more subject to brittle fracture which would result in catastrophic failure before the ship can reach facilities for proper repair. New materials also are being used in submarines and undersea vehicles which could conceivably require welding. The materials used in these vehicles include:[14]

TABLE A-6. UNDERWATER WELDING ELECTRODES

Manufacturer	Manufacturer's Designation	Size, inch	Remarks	Reference
Diving Equipment & Supply Company	None	--	Waterproofed by manufacturer	4
Russian	EPO-55		--	20
Hobart	Hobart 10 (E6010)		Modified by user	
Hobart	Hobart 13 (E6013)		Modified by user	
Arcair	Cat. No. 50-034	3/16 x 14	50 lb. boxes 7.5 pcs/lb Waterproofed by manufacturer	
Arcair	Cat. No. 50-984	5/32 x 14	50 lb. boxes 10.8 pcs/lb Waterproofed by manufacturer	
Murex	--	1/8,5/32,3/16	Waterproofed by manufacturer	
Westinghouse	SW (E6013)	5/32,3/16	Waterproofed by user, formerly Flexarc SW	2
Craftsweld	--	5/32,3/16	Waterproofed by manufacturer	
Lincoln	Fleetweld 37	1/8,5/32	Waterproofed by user	2
UWR[a]	Murex Murod	5/32,3/16	Waterproofed by manufacturer	

a Underwater Welders and Repairers, Ltd.

TABLE A-7. CHEMICAL COMPOSITION OF UNDERWATER SHIELDED-METAL-ARC WELDING ELECTRODES[19, 37]

Electrode Designation	Nominal Size	Welding Conditions	Principal Elements	Chemical Composition, w/o		
				Parent Metal	Core Wire	Weld Deposit
EPO-55[a] (USSR)	--	ac	C	0.17	0.08	0.07
			Si	trace	0.03	0.17
			Mn	0.29	0.42	0.55
DPO-55[a] (USSR)	--	dc	C	0.17	0.08	0.08
			Si	trace	0.03	0.14
			Mn	0.29	0.42	0.56
EPS-5[b]	--	--	C	0.17	0.08	0.26
			Si	0.03	0.03	0.16
			Mn	0.42	0.42	0.63
EPS-52[b]	--	--	C	0.17	0.08	0.12
			Si	0.03	0.03	--
			Mn	0.34	0.34	0.51
SW (formerly Flexarc SW)	--	--	C	--	0.10-0.15	--
			Si	--	0.03 Max.	--
			Mn	--	0.40-0.60	--

[a] Depths to 10 m (32' 10")

[b] Depths to 20 m (65' 2")

TABLE A-8.   RECOMMENDED AND SUBSTITUTE UNDERWATER
WELDING ELECTRODES AND OPERATING CONDITIONS[2]

Electrode Designation	Size, Inch	Position[b]	Current[a], amperes	Time for 12-Inch Burnoff, seconds
Recommended Electrodes				
Westinghouse Flexarx SW	5/32	H	170-210	56-44
		V	170-210	56-44
		O.H.	170-190	56-50
	3/16	H	220-260	59-50
		V	220-260	59-50
		O.H.	190-210	66-61
Lincoln Fleetweld 37	3/16	H	220-260	60-49
		V	220-260	60-49
		O.H.	200-220	66-60
Substitute Electrodes				
A. O. Smith Smithweld	3/16[c]	H	220-260	55-45
		V	220-260	55-45
	1/4[c]	H	250-290	86-77
Metal and Thermit Murex Alternex	3/16[c]	H	210-250	55-45
		V	210-250	55-45
		O.H.	190-210	60-55
	1/4[c]	H	230-270	90-73
Hollup Sure-weld "C"	3/16[c]	H	200-240	83-73
		V	200-240	83-73
Reid-Avery Raco 7	1/4[c]	H	210-230	74-65
		V	210-230	74-65
Metal and Thermit Murex Type "A"	3/16[c]	H	200-240	55-49
		V	200-240	55-49
		O.H.	170-190	61-57
General Electric G.E. W-25	3/16[c]	H	180-220	61-51
		V	180-220	61-51

a   For DC only; add 10 percent for AC.

b   H = horizontal, V = vertical, O.H. = overhead

c   These electrodes shall not be tested in 5/32-inch size.   In most
cases 5/32" will be satisfactory.

TABLE A-9. UNDERWATER WELDING CONDITIONS

Base Metal	Joint Type	Electrodes Type	Size	Welding Current, amps	Open Circuit Voltage	Welding Voltage (volts)	Depth	References
1/8" mild steel to ship hull	-	E6011[a]	1/8 in.	160	-	-	-	85
	-	E6011	5/32 in.	200	-	-	-	85
N.S.	-	-	5/32 & 3/16	300 max.	80	25-35	-	89
Steel, 0.20C, 0.62Mn	Fillet Weld	Touch Welding	8 swg	210	85	-	-	72
Steel, 0.20C, 0.62Mn	Fillet Weld (slight weave)	BSS 639 Type A	6 swg	230	85	-	-	72
	(b)	-	-	-	100	-	-	72
Steel	-	TsN-P	0.16 in.	200-220	-	-	6-1/2 ft	45
	-	TsN-P	0.20 in.	250-270	-	-	6-1/2 ft	45
Steel, 0.2-0.24 inch	Flat Butt	EPO-55	0.16 in.	240-260	-	-	-	75
	Flat lap or tee	"	"	260-280	-	25-30	-	75
	Vertical butt[a]	"	"	250-270	-	-	-	75
	Vertical lap or tee[c]	"	"	280-300	-	-	-	75
Steel, 0.031-0.039 inch	Flat butt	EPO-55	0.16 in.	240-280	-	-	-	75
	Flat lap or tee and vertical butt[c]	"	"	280-300	-	25-30	-	75
	Vertical lap or tee	"	"	300-320	-	-	-	75
Steel, 0.031-0.039 inch	Flat butt	EPO-55	0.2 in.	300-320	-	-	-	75
	Flat lap or tee	"	"	320-340	-	25-30	-	75
Steel	-	Craftweld UW-electrode	3/16 in.	225-280	-	-	50 ft	71

a E6013 was preferred but not available

b Increases in open-circuit voltage produced smoother welding conditions but not better quality welds

c Vertical welds were deposited downward

TABLE A-10. PROPERTIES OF UNDERWATER WELDS

Base Plate	Electrode	Joint Type	Hardness, BHN			(continued)
			Weld	HAZ	BM	
Mild Steel	(Electrode not stated)	--	190	230	135	
"D"Steel	(Electrode not stated)	--	190	350	185	
3/8"mild steel	(0.13-0.2C electrode)	--	164	--	117	
Mild Steel		Single-V Butt	170-190	170-181	142-161	
Mild Steel		Single-V Butt	159-182	155-176	119-132	
Mild Steel		Single-V Butt	158-204	174-189	140-144	
Steel	(TsN-P electrode)	--	--	--	--	
Steel	(UONI-13/45-P electrode)	--	--	--	--	
St. 3 steel	EPO-55	--	185	--	--	
St. 4 steel	EPO-55	--	185	--	--	
St. 3 steel	EPS-5	--	--	--	--	

a Transverse-tension test results must be interpreted with great care
b Results varied with position of specimen and notch.
c Angle of bend was 120-124o for same test pieces.

TABLE A-10 (continued)

Weld Joint Tension-Test Results			Impact Properties		Bend-Test Results	
Yield Strength, psi	Ultimate Strength psi	Percent Elongation	Method	ft.lb.	Initial Cracks	Complete Failure
37,400	52,800	--	Izod	6-30[b]	15°	35°
35,700	55,000	--	Izod	6-30[b]	6°	30°
--	--	--	--	--	70°	--
46,200	68,200	9-1/2(2 in.)	--	--	--	--
40,700	51,000	23 (2 in.)	--	--	--	--
49,500	67,800	8 (2 in.)	--	--	--	--
--	46-48 Kg/mm²	--	--	--	--	60-65
--	39-41 Kg/mm²	--	--	--	--	35-40
Joint 59,900	Weld Metal 71,100	--	--	--	--	70-90[c]
Joint 66,800	Weld Metal 71,100	--	--	--	--	70-90[c]
--	54,800	--	--	--	--	--

Aluminum Alloys

    A-356-T6
    A-7079-TL

Steels

    A-36, A-212, A-285 Steels
    Algoma 44 Alloy
    HY-50, HY-80, HY-100, HY-150
    Ni-Cr-Mo Steels
    T-1

Whether all of these materials can be welded under water in times of emergency still needs to be determined.

Techniques

The principal applications for "wet" welding in Naval services are the welding of patches over holes, repairing of small cracks and making small attachments to ship hulls where fillet welds are required, in commercial work, attachments to underwater structures, and repair of damaged pipelines. Because of the difficulty of maneuvering about underwater, it is desirable to have a simple method for making the needed welds. Over the years, the Navy Department developed the "self-consuming" technique for underwater welding with electrodes.

In operation, the electrode is positioned close to the parent metal ready for welding. The welder-diver calls for "current on" and the tender closes a knife switch. The welder-diver then initiates the arc by a touch starting technique. As soon as the arc is struck, a bubble forms instantaneously and products from the electrode coating form a black, opaque cloud. The bubble expands and envelops the tip of the electrode. The arc then exists within the bubble. Madatov[20] claims that the bubble around the arc forms due to:

1. Products from electrode material
2. Products from parent material
3. Formation of steam
4. Disassociation of water
5. In seawater, from steam evaporated by heating the parts due to leakage currents immediately before arc initiation.

With the covered-electrodes used in modern underwater welding practice, the covering burns away more slowly than the core wire. The result is that the covering extends for a distance beyond the end of the core wire. This operating characteristic is desirable in that it promotes arc stability by preventing water from contacting the tip of the melting electrode and it provides an automatic control of arc length. When the electrode is manipulated correctly,

it is impossible to short-circuit the tip of the electrode to
the work. The extension of the covering allows the operator to
"drag" the tip of the electrode along the joint being welded
without the need to exercise precise control of arc length. The
operator is required only to maintain the line of welding and the
speed of welding. This technique is referred to as the "self-
consuming" technique. Less skill is required on the part of the operator to produce acceptable welds. The sense of touch and
the boiling sounds eminating from the welding arc is used by many
divers to determine whether the welding operation is proceeding
in a normal manner, [16] particularly when visibility is poor. With
this technique, no attempt needs to be made on the part of the
operator to hold the electrode off of the plate surface to
maintain an arc.

   Poor Fit-Up. The Navy Department also has developed a "feeding
in" technique, as an adjunct to the "self-consuming" technique, for
fillet welding lap joints with a large gap where it is impossible
to bring the plates into a tight overlap. This technique involves
feeding the "self-consuming" electrode into the joint at a faster
rate than the standard method. The filler metal fills the gap and
solidifies rapidly when used underwater. In air, the filler
metal tends to "run" in this type of joint. When used underwater,
the weld bead freezes rapidly and running is effectively prevented.
With this technique filler welds have been made successfully with
gaps as wide as 1/8 inch.

   Repairing Cracks. Techniques for repairing cracks underwater
are very similar to those used in air. The accepted procedure is
to drill or burn holes at each end of the crack. These holes act
as crack stoppers to prevent extension or propagation of the crack.
A patch is fitted over the plate and fillet welded to hold the
patch in place and maintain watertight integrity.

                        The requirement for over-
head welding can be minimized by positioning the patch so the
edges are inclined 45 degrees from horizontal. Experience has
shown that welds deposited directly over cracks are very likely to
open up again directly through the weld.

   Fillet Welds. Fillet weld joints are especially adaptable to
the self-consuming technique because they provide a natural groove
for guiding the electrode along the weld. String bead deposits
are preferred in order to avoid arc interruptions and slag entrapment that would result with weave-bead deposits. When making fillet
welds, the preferred orientation of the electrode is in a position
which bisects the weld with the electrode held at an angle of 30
degrees to the plate surfaces in the direction of welding; the
electrode points in a direction toward the already deposited weld
metal. Depending on the application and the skill of the diver-
welder, the angle of inclination of the electrode can be varied
from 15 - 45 degrees. Upon completion of the weld pass or when the
electrode is consumed, the diver-welder calls for "current off"
and the tender opens the knife switch. The electrodes then may be

changed. Before continuing welding, the end of the previous weld deposit must be cleaned of slag in order to restart the arc. The new deposit should slightly overlap the end of the previous deposit.

The self consuming welding technique can be applied to fillet welds in the horizontal flat, vertical and overhead positions. In order to improve visibility when welding in the horizontal flat position, it is usually better for the diver to weld in a direction toward himself. When welding in the vertical position, welds should be made from the top down. When welding is performed in the overhead position, welding should be performed so as to minimize interference from bubbles. The techniques required will depend on the individual situations.

Welding Results

When considering means for underwater welding, some measure of weld quality is desirable. The methods used for evaluating underwater welds is determined to a great extent by the application for which the welds were made. When raising sunken vessels, it is necessary only to hold the plate in place and provide air or water-tight integrity. These requirements are not so stringent as when welds must withstand lifting loads, or if the repair is subjected to wave action, or buffeting by the seas. In offshore oil structures, welded attachments are expected to have a service life of about 10 years.

Information on properties of underwater welds and service requirements is limited. In past practice, it will be recalled that for marine service, welds made underwater were considered as only a temporary measure until the vessel could reach facilities for inspection and repair under more suitable facilities. Consequently, there has been little need for extensive investigations of underwater weld properties or on the effects of welding.

Gases Evolved. All divers are trained to take precautions against the explosion of entrapped gases. Explosive gases may be contained within the structure being welded or they may be generated by the welding operations. In the naval service, it is mandatory that prior to welding in closed compartments, corners and other areas where gas may be trapped, means be provided for venting the gases generated during welding to eliminate the possibilities of explosion. Gases generated by welding can be very explosive. Possible sources of these gases are:

1. Decomposition of electrode coverings and waterproofing materials
2. Metal vapors from electrode and parent materials
3. Formation of steam
4. Decomposition of water

Welding Rates. Welding rates have been reported in a general
way. The Navy Cutting and Welding Manual reports that under
normal conditions using the "self-consuming" technique about 8
inches of 3/16 inch fillet weld can be made for each 10 inches
of 3/16 inch electrode, with a tight lap joint. This burnoff
reduces to about 6 inches of weld for each 10 inches of electrode
when using the "feeding in" technique when there are large gaps in
the lap. The burn-off rate for several electrodes and several
positions was reported earlier. There were no other useful data
found on productivity in underwater welding work.

Mechanical Properties. There is general agreement that welds
made in water can be expected to have lower mechanical properties
than similar welds made in air. A rule of thumb is that welds
made underwater in steel will have about 80 percent of the strength
and 50 percent of the ductility of similar welds made in air.[2,21,32,35]
The reduced properties are credited to various effects resulting
from the rapid quenching in water. Rapid solidification of the
molten weld metal promotes slag entrapment and grain refinement.[11]
Avilov claims that the weld metal may contain slag but in amounts
similar to welds made in air. The rapid freezing of the weld
metal surface prevents the escape of hydrogen from the weld metal.
Trapped hydrogen causes pores and cavities and may lead to either
immediate or delayed cracking. Porosity in high-carbon steel was
reported as particularly bad, being continuous and interconnected
along the length of the weld.

Gas Metal-Arc Welding

The gas metal arc welding process has been used for underwater welding in the U.S.S.R. and is under development in the
United States[24] and in Italy.

Bare Wire

Bare filler wire is used with the gas metal arc process.
These wires generally range in size from about 0.025 to 0.093
inch in diameter when welding underwater. The filler wire for
gas metal arc welding is available in long lengths, usually
several hundred feet. This wire is level wound on a spool. In
operation, the filler wire is automatically fed through a hand
torch into the joint being welded. The welding arc is established
between the end of the electrode wire and the parent metal in the
joint.

Shielding Gases

Each of the processes described above requires a gaseous
shield to protect the weld puddle and the hot electrode wire.

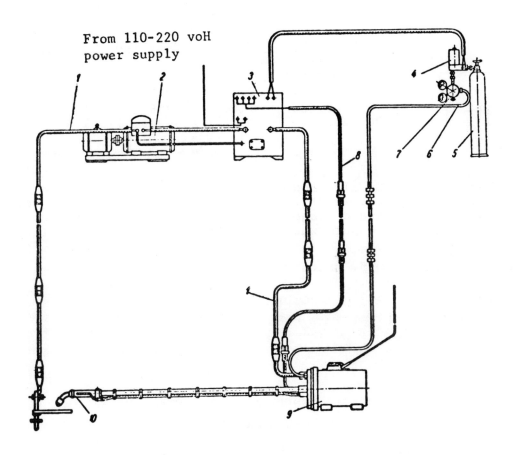

From 110-220 voH
power supply

FIGURE A-9.   GAS METAL-ARC WELDING EQUIPMENT ARRANGEMENT FOR

UNDERWATER WELDING

1. Welding cable
2. Power supply
3. Control unit
4. Heater and flow gage
5. Gas cylinder

6. Gas hose
7. Reducing valve
8. Welding cable
9. Gas reservoir and wire
    feed unit
10. Welding head

TABLE A-11. SEMI-AUTOMATIC UNDERWATER GAS METAL-ARC WELDING CONDITIONS WITH CO_2 SHIELDING GAS[a]

	Welding Position						
	Flat				Vertical		
Depth, m	10	15	30	60	10	30	60
Arc voltage, V	38-42	40-42	46-48	54-56	38-40	44-46	50-52
Current, A,d-c[b]	190-200	200-220	200-220	210-220	170-190	180-200	190-210

[a]1.2 mm diameter Sv-10GS electrode wire; Russian PS-500 welding converter having a drooping characteristic and high inductance, 300 amp welding power supply. Wire feeder speed range capability is 2-16 m/min., supply voltage= 220 volts. The CO_2 pressure was 0.5-1.0 atm above the hydrostatic pressure.

[b]Polarity not stated; U.S. practice is reverse polarity.

TABLE A-12. SEMIAUTOMATIC UNDERWATER GAS METAL-ARC WELDING CONDITIONS FOR UNSHIELDED ARC WELDING[a]

	Welding Position					Vertical	Overhead
	Flat						
Depth, m	10	15	20	40	60	15	15
Arc Voltage, V	34-36	38-40	40-42	46-48	48-52	36-38	39-42
Current, A,d-c[b]	180-200	180-220	190-220	190-220	190-210	160-180	170-180

[a]1.2 mm diameter Sv-10GS electrode wire; Russian PS-500 welding converter having a drooping V/A characteristic and high inductance, 300 amp welding power supply, wire feeder speed range capability was 2-16 m/min, supply voltage = 220 volts. The CO_2 pressure was maintained sufficient to ensure no flooding of the wire feed unit.

[b]Polarity not stated; U.S. practice is reverse polarity.

The gases generally used for this purpose are argon, helium, CO_2, or mixtures of these gases. Occasionally, additives such as oxygen, may be contained in the gas supply to promote arc stability, weld metal fluidity or other desirable features. CO_2 shielding gas has had very little application for underwater welding in the United States. Argon, helium or mixtures of argon and helium are used. The Russian literature, however, reports extensive use of CO_2 for shielding gas with the gas metal arc welding process in "wet" applications.

SECTION 3.   DRY ENVIRONMENT UNDERWATER WELDING PROCESSES

The term "dry" welding is used in reference to techniques which utilize underwater enclosures from which water is removed and in which the welding operation is performed in a gaseous atmosphere. Welding within these chambers is performed at pressures greater than standard atmosphere pressure. Caissons also have been used. These are constructed with a shaft extending from an enclosure to somewhat above the water surface. Depths at which caissons have been used, therefore, seem to be limited by the length of shaft that is needed.

The "dry environment" welding technique also has been referred to as a "hyperbaric" welding technique. Although each of these terms is generally descriptive, neither of these terms is strictly correct. The atmosphere in the chamber is far from dry because the gaseous atmosphere is in contact with the water and rapidly saturates to become very humid. Water vapor condenses at a high rate on the chamber wall and operators have reported "rain" in the chambers. When present in the welding atmosphere, water vapor can result in the lowering of weld properties. Hyperbaric welding is an applicable term but it does not necessarily distinguish underwater chamber-type welding from other processes used under-water or above water in pressurized chambers. The water and gas bubble surrounding the arc in wet welding are also at hyperbaric pressures. Welding at high, or hyperbaric, pressures also has been performed above water in pressurized, controlled-atmosphere welding chambers.

Processes that have been used for "dry" welding in under-water applications include gas tungsten arc (TIG) welding, gas metal arc (MIG) welding and shielded metal arc welding. Each of these processes are in wide use in above-water production applications, but they have had only limited application in underwater work. Major applications are in making pipeline repairs and hot taps and for repair of offshore structures. Much of the information on underwater welding with these processes is considered proprietary and little published data was found on the subject. Reports have been received, however, indicating that the welding arc in a high-ambient pressure atmosphere is considerably different from those in normal above-water atmospheres.

Dry Environment Underwater Welding Chambers

Detailed information on welding procedures used with dry-environment underwater welding chambers is limited. Published reports relevant to these techniques usually provide only very general information regarding the welding process. Organizations engaged in work with these chambers consider the details of their processes and procedures as proprietary information. In view of the high costs of development and fabrication costs, their reluctance to disclose what is considered proprietary information is understandable. Therefore, the following information concerning dry welding practices can only be general.

Chambers in use for dry underwater welding are illustrated in Figure A-10 which shows only a few of those in actual use.[20,24,34] Several types of chambers are in existance. One type is for hot tap operations. Another type employs special rams to help provide precise pipe alignment.[94] Another type can be clamped around the support legs of offshore structures. Some of these chambers are expected to be used most advantageously at depths of over 400 feet. These chambers are only a part of an entire system which is designed to support and sustain the workers engaged in the overall underwater operation. The entire system consists of a support vessel carrying needed supplies and equipment, a submersible diving chamber which carries the workmen to the proximity of the welding enclosure, and the welding enclosure.

In practice, divers descend to the work area to inspect the work, usually some part of a pipeline, to be performed and select a location for the enclosure. When necessary to uncover the pipe, provide a clear area in which to set the enclosure and to provide space for entry from underneath by the workmen, the covering material is generally removed by jetting with a high-speed stream of water. Guide wires that are attached to the pipe to assist the workmen in guiding and placing the enclosure over the pipe. Welder-divers then descend in a pressurized submersible chamber and when the proper depth is reached, they leave the chamber and seal the enclosure doors or ports around the pipe. In one system, water is displaced out the bottom of the chamber to a desired water level by means of pressurized gas supplied from the surface. The welder-divers then enter the chamber from the bottom and switch to special breathing masks supplied from the surface. One version of these enclosures carries tools in watertight compartments. These are removed to proceed with the welding operation. Future systems are under construction which will permit the workers to transfer directly from the submersible chamber directly into the enclosure.

With regard to welding, the atmosphere* within the chamber varies and information on the exact gas mixture composition is considered proprietary. One organization reports that one gas, mainly nitrogen, is used to a depth of 200 feet. Beyond this depth, helium is used.[34] Oxygen is maintained at a level low enough to prevent ignition of combustible materials but ample for

---

* Sometimes referred to as background gas for the welding operation.

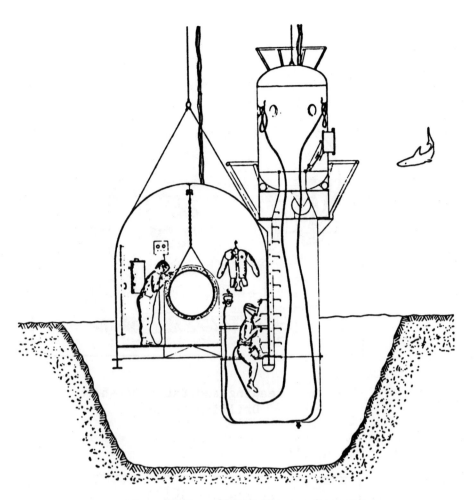

a. Taylor Diving and Salvage Company

b. Ocean Systems, Inc.

Figure A-10   DRY ENVIRONMENT UNDERWATER WELDING CHAMBERS

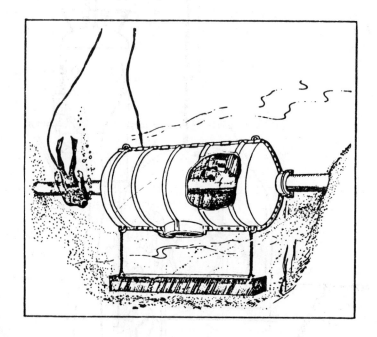

**c. Reading and Bates Offshore
Drilling Company**

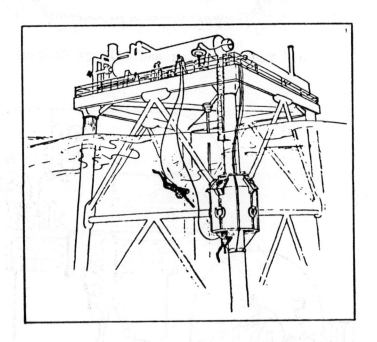

**d. Reading and Bates Offshore
Drilling Company**

FIGURE A-10. (Continued)

for the diver-welder to breath should his personal breathing
apparatus fail. The enclosure is continually ventilated to
control gases and reduce humidity. In normal operations the
enclosure is manned by two diver-welders.

Welding has been performed using the TIG (gas tungsten-arc)
welding process with argon as the shielding gas.[34] In operation,
the arc is established between the tip of a tungsten electrode
and the parent metal at the joint. The heat of the arc melts the
parent metal as the torch is manipulated along the line of welding.
At the same time, the end of a hand-held bare filler wire is
manipulated in the vicinity of the weld puddle and arc so as to
melt droplets from the end of the filler wire and deposit them into
the weld joint. The welds are inspected by radiographic techniques
to ensure that they meet the requirements of the American
Petroleum Institute Specification API 1104. Wallace and Morrisey[34]
claim that all welds made with the TIG process at depths in the
order of 300 feet were of much higher quality than equivalent welds
prepared on the surface. Their conclusion tends to confirm earlier
results of Barker, who conducted work on the effects of high ambient
pressures on quality of welds made with the gas tungsten arc and
gas metal arc welding processes. Barker showed that there was a
high-pressure argon. The reduction in porosity is even more
noticeable with high pressure helium. Barker's work was discussed
earlier in the section on "Underwater Environment."

The gas metal arc welding process also has been used for
dry-environment underwater welding using the short-circuiting
metal transfer mode of operation.[27] With this process transfer
of metal is achieved by a combination of resistance heating and
arc melting. When the wire advances and contacts the work, the
current rises to heat the end and melt the end of the wire. The
current also exerts a pinch effect to pinch the droplet off the
end. A small arc is then re-established between the weld puddle
and the wire advances to the work again.

Welding in dry-environment underwater chambers has been
accomplished with the shielded metal arc welding process. However,
no data was available on the welding procedures or results obtained.

REFERENCES

1. Avilov, T. I., "The Static Characteristics of the Welding Arc Under Water," Svarochnoe Proizvodstvo, No. 5, 16-17, 1959.

2. Bureau of Ships, Navy Department, Underwater Cutting and Welding Manual, 250-692-91, 1954.

3. Craftweld Equipment Corporation, Operating Suggestions for Underwater Welding, Long Island City, New York, New York.

4. Desco, Product Catalogue, Diving Equipment and Supply Company, Milwaukee, Wisconsin, 30, 1968.

5. Dow Chemical Company, "MAPP Industrial Gas," (Brochure), Midland, Michigan, 1966.

6. Dow Chemical Company, "Underwater Cutting (with MAPP Gas)," Midland, Michigan.

7. Fey, W., Ocean Engineering Handbook, Section 4, "Underwater Cutting and Welding," McGraw-Hill Book Company, 1969.

8. Galerne, A., Private communication, International Underwater Contractors Inc., Flushing, New York, 1968.

9. Hach, F., Private Communication, Harris Calorific Company, Cleveland, Ohio, July 1968.

10. Harris Calorific Company, "Underwater Cutting Torch," Harris, No. 28-U, Cleveland, Ohio, 1968.

11. Harris Calorific Company, "Instructions for Underwater Cutting," Cleveland, Ohio.

12. Hembree, J. D., Belfit, R. W., Reeves, H. A., et al., "A New Fuel Gas - Stabilized Methylacetylene-Propadiene," Welding Journal, 42, (5), 395-404, May 1963.

13. Hipperson, A. J., "Underwater Arc Welding," Welding Journal, 22 (8), 329s-332s, August 1943.

14. Interagency Committee on Oceanography of the Federal Council for Service and Technology, "Undersea Vehicles for Oceanography," ICO Pamphlet No. 18, U. S. Government Printing Office, Washington, D. C.

15. Iron Age, Metalworking Goes Underwater," 200 (24), 91-93, December 14, 1967.

16. Kemp, W. N., "Underwater Arc Welding," Transactions of the Institute of Welding, 8 (4), 152-156, November 1945.

17. Lahm Jack F., Private Communication, Continental Diving Service, Morgan City, Louisiana, September 10, 1968.

18. Lewis, R. D., "Offshore Pipe Line Work Booming," Pipe Line Industry, 24 (1), 26-29, January 1966.

19. Madatov, N. M., "Electrodes with Iron Powder in the Coating for Underwater Welding," Welding Production, BWRA translation, No. 8, 25, 1962.

20. Madatov, N. M., "Special Features of Underwater Torch Welding," Automatic Welding, 15 (9), pp. 52-55, 1962.

21. Morrill, J. R., "Underwater Welding," Steel, 117 (23), pp. 112-113, December 3, 1945.

22. Murex Welding Products, "Murex Electrodes, Type C-T for Underwater Cutting and for Open-Air Cutting of Cast Iron and High Alloy Steels," Department of Airco Welding Products, Division of Airco, Inc., Union, New Jersey (Brochure).

23. Oil and Gas Journal, "Caisson Speeds Underwater Pipeline-Repair Job," 60 (4), p. 37, January 22, 1962.

24. Pipeline Engineer, "Submarine Makes Quick Work of River Repair," 34 (5), pp. 310-312, May 1962.

25. Pipe Line Industry, "Caisson Makes Underwater REpairs Look Routine," 13 (4), pp. 21-22, October 1960.

26. Pipe Line Industry, "Pipeliners Use 'Submarine' to Repair and Re-Coat Pipe in 40 feet of Water, 16 (3), pp. 44-47, March 1962.

27. Reading and Bates Offshore Drilling Company, "Weld-Cone; Underwater Welding in a Dry Environment," (Brochure), Tuslsa, Oklahoma.

28. Rhier, Max, Private communication, Harvey, Louisiana, September 10, 1968.

29. Rodman, M. F., "Underwater Cutting of Metals," Journal of the American Welding Society, 7 (23), pp. 603-669, July 1944.

30. Ronay, B., and Jensen, C. D., "U.S. Navy Developments in Underwater Cutting," Journal of the American Welding Society, 25 (3), pp. 201-209, March 1946.

31. Royal Navy Diving Manual, Chapter 54, "Underwater Welding and Cutting."

32. Silva, E. A., "Welding Processes in the Deep Ocean," Naval Engineers Journal, 80 (4), pp. 561-568, August 1968.

33. Spraragen, W., and Claussen, G. E., "Underwater Cutting, Arc Cutting, The Oxygen Lance, and Oxygen Deseaming and Machine," A Review of the Literature to January 1, 1939, Journal of the American Welding Society, 19 (7), p. 81s, 1940.

34. Wallace, K. W., and Morrissey, G., "Dry-Weld Modes in an Underwater Habitat," Offshore, 28 (10), p. 67, September 1968.

35. Welding Handbook, "Welding Cutting and Related Processes," Fifth Edition, Section 3, Chapter 47, "Oxygen Cutting, Auxiliary Cutting Processes," and Chapter 48, "Arc Cutting," 1964.

36. Welding Handbook, "Underwater Cutting," Chapter 24, Third Edition, American Welding Society, New York, New York, pp. 518-549, 1950.

37. Westinghouse Electric Company, "Westinghouse SW, Data Sheet," Pittsburgh, Penssylvania.

38. White, P. A., and Common, R. P., "Underwater Ship Maintenance in the Royal Navy, Divers' Hand and Power Tools - Research and Development for the Admiralty Experimental Diving Unit," 1967 Symposium on Underwater Welding, Cutting and Hand Tools, Battelle Memorial Institute, Columbus Laboratories, Columbus, Ohio.

39. Woollard, R. F., Private communication, Dow Chemical Company, Midland, Michigan, February 5, 1968.